Networked Control Systems

Fei-Yue Wang • Derong Liu
Editors

Networked Control Systems

Theory and Applications

Fei-Yue Wang, PhD
Chinese Academy of Sciences
Beijing 100080
P. R. China

Derong Liu, PhD
Department of Electrical and Computer Engineering
University of Illinois at Chicago
Chicago, IL 60607
USA

ISBN 978-1-84996-756-3 e-ISBN 978-1-84800-215-9

DOI 10.1007/978-1-84800-215-9

British Library Cataloguing in Publication Data
Networked control systems : theory and applications
1. Automatic control 2. Computer networks
I. Wang, Fei-Yue II. Liu, Derong, 1963-
629.8'9

Cover design: eStudio Calamar S.L., Girona, Spain

Printed on acid-free paper

9 8 7 6 5 4 3 2 1

springer.com

In Memory of

George Nikolaos Saridis

(November 17, 1931 - October 29, 2006)

The Founding President of the IEEE Robotics and Automation Society
A Pioneer in Intelligent Robotic Systems and Intelligent Machines
The Visionary of Intelligent Control

Preface

The accelerated integration and convergence of communications, computing, and control over the last decade has inspired researchers and practitioners from a variety of disciplines to become interested in the emerging field of networked control systems (NCS). In general, a NCS consists of sensors, actuators, and controllers whose operations are distributed at different geographical locations and coordinated through information exchanged over communication networks. Some typical characteristics of those systems are reflected in their asynchronous operations, diversified functions, and complicated organizational structures. The widespread applications of the Internet have been one of the major driving forces for research and development of NCS. More recently, the emergence of pervasive communication and computing has significantly intensified the effort of building such systems for control and management of various network-centric complex systems that have become more and more popular in process automation, computer integrated manufacturing, business operations, as well as public administration.

Control over a communication network is not a new concept in automation. From tele-operation for space and hazardous environments to process regulation with distributed control systems, control systems with communications have already been developed and utilized in applications of real-world problems for more than 30 years. However, there are many factors that distinguish the current NCS and previous control with communications. Two of them are the most significant: (1) in the previous control with communications, the network is specialized and dedicated for the timeliness of information exchange and stability of process operation, while in the current NCS the network is general-purpose and public for various irrelevant yet concurrent applications, and thus real-time communication and stable operation are no longer ensured; (2) the functionality of the NCS from the previous to current has been diversified tremendously, from pure control to a variety of control and management or administrative functions, ranging from resource allocation, event scheduling, to task organization, etc., involving concept and methods from control and communication engineering, operations research, computer science, and management science.

Demands on diversity, complexity, and real-time performance for networked operations have brought new technological challenges to NCS. Today, many fundamental questions regarding the stability of interconnected dynamical systems, the effects of communication on the performance of control systems, etc., remain open and to be answered. Even from the perspective of control field alone, we need to think about what the new direction for research and application in this age of connected world would be. One potential approach is to extend the concept of "code on demands" with agent programming to "control on demands" with agent-based control (so called ABC). In other words, can we liberate control algorithms that are fixed to plants to be controlled to control agents that are free and mobile in a connected world? Once this is accomplished, various innovative methods based on connectivity can be employed for control and management, e.g., using "local simple, remote complex" principle to design low cost yet high performance and intelligent NCS that require less computing power, small memory space, and little upgrading. Indeed, there are many new, exciting, and challenging ideas, problems, and concepts in the emerging field of networked control systems.

This book is a follow up effort after the publication of the special issue on "Networking, Sensing, and Control for Networked Control Systems: Architectures, Algorithms, and Applications" in the *IEEE Transactions on Systems, Man, and Cybernetics–Part C*, vol. 37, no. 2, Mar. 2007. The book includes eleven chapters written by leading experts in NCS and addresses some of the questions and problems discussed above.

We start the book with two review chapters. The first chapter by Gupta and Chow provides an overview of NCS, its history, issues, architectures, components, methods, and applications. A case study of NCS with test-bed system iSpace is also described in this chapter. Chapter 2 by Wang presents the history and issues of agent-based control and management for NCS from the perspective of his own research group. He argues and calls for a paradigm shift from control algorithms to control agents so that agent-based control can be established as the new control mechanism for operation and management of networked devices and systems. The goal of his agent-based approach is to transform "code on demand" in programming into "control on demand" in control, and provides a platform for designing and building low cost but high performance networked equipment in the age of connectivity.

The remaining chapters address specific issues in design, analysis, and implementation of NCS. In Chapter 3, considering the fact that the design and implementation of many digital systems have been based on the emulation of idealized continuous-time blocks, and in analogy with sampled-data control system design, Tabbara, Nesic, and Teel explore an emulation-based approach to the analysis and design of NCS. For this purpose, they survey a selection of emulation-type NCS results in the literature and highlight the crucial role that scheduling between disparate components of the control systems plays.

They then detail several different properties that scheduling protocols need to verify together with appropriate bounds on inter-transmission times such that various notions of input-output stability of the nominal "network-free" system are preserved when deployed as an NCS. This could be an important method for designing NCS in the future. In Chapter 4, Liu addresses issues in analysis and design of NCS based on a novel control strategy, termed networked predictive control. The stability of the closed-loop networked predictive control system is analyzed. The analytical criteria are obtained for both fixed and random c ommunication time delays. The on-line and real-time simulation of networked predictive control systems is presented in detail.

In Chapter 5, Yue, Han, and Lam discuss the design of robust H_∞ controllers and H_∞ filters for uncertain NCS with the effects of both network-induced delay and data dropout taken into consideration. In this chapter, a new analysis method for H_∞ performance of NCS is provided by introducing slack matrix variables and employing the information of the lower bound of the network-induced delay. Numerical examples and simulation results are given to illustrate the effectiveness of their proposed method. In Chapter 6, Nikolakopoulos, Panousopoulou, and Tzes propose a switched output feedback control scheme for networked systems, and apply the scheme to client–server architectures where the feedback control loop is closed over a general purpose wireless communication channel between the plant (server) and the controller (client). To deal with network delay effects, a linear quadratic regulator (LQR)-output feedback control scheme is introduced, whose parameters are tuned according to the variation of the measured round trip latency times. The overall scheme resembles a gain scheduler controller with the latency times playing the role of scheduling parameter. The proposed control scheme is applied in both experimental and simulation studies to an NCS over different communication channels.

Yang and Zhang in Chapter 7 have developed a guaranteed cost networked control (GCNC) method and established the corresponding stability for Takagi–Sugeno (T–S) fuzzy systems with time delay. Both analytical studies and simulation results show the validity of their proposed control scheme. A robust H_∞ networked control method for T–S fuzzy systems with uncertainty and time delay is also presented in this chapter, along with sufficient conditions for robust stability with H_∞ performance. In Chapter 8, Sun and Wu have proposed a discrete-time jump fuzzy system for the modeling and control of a class of nonlinear NCS with random but bounded communication delays and packets dropout. In this chapter, a guaranteed cost control with state feedback is developed by constructing a sub-optimal performance controller for the discrete-time jump fuzzy systems in such a way that a piecewise quadratic Lyapunov function (PQLF) can be used to establish the global stability of the resulting closed-loop fuzzy control system. When not all states are available, an output feedback controller is designed. For the NCS based on the mixed networks, a neuro-fuzzy controller is developed. Simulation examples are carried out to show the effectiveness of their proposed approaches. Chen in Chapter 9 investigates the boundary control of damped wave equations using a boundary measurement in an NCS setting. In his approach, induced delays in this networked boundary control system are lumped as the boundary measurement delay. The Smith predictor is applied to this problem and the instability problem due to large delays is solved and the scheme is proved to be robust against a small difference between the assumed delay and the actual delay. He also analyzes the robustness of the time-fractional order wave equation with a fractional order boundary controller subject to delayed boundary measurement.

The last two chapters address two basic methods for NCS. In Chapter 10 Li and Wang discuss the coordination mechanism of multi-agent systems using an adaptive velocity strategy. In previous works, much attentions and correlative efforts for swarm intelligence have been focused on constant speed models in which all agents are assumed to move with the same constant speed. In this chapter, they have proposed an adaptive velocity model with a more reasonable assumption in which every agent not only adjusts its moving direction but also adjusts its speed based on the degree of direction consensus among its local neighbors. The adaptive velocity model provides a powerful mechanism for coordinated motion in both biological and technological multi-agent systems. In Chapter 11, Yu and Wang study the robust synthesis problem for strictly positive real (SPR) transfer functions. By using the complete discrimination system (CDS) for polynomials, complete characterization of the (weak) SPR regions for transfer functions in coefficient space is given. They have proposed an algorithm for robust design of SPR transfer functions. Their algorithm works well for both low-order and high-order polynomial families.

Finally, as the editors of the book, we would like to express our sincere appreciation to all authors for their time and effort, and to Springer's Engineering Associate Editor Oliver Jackson for his patience and great help. Although every effort has been made to include a wide spectrum of methods and applications in this emerging field, a book like this can only include a rather small number of selected chapters, and we must say that the coverage here is by no means comprehensive.

Fei-Yue Wang
Chinese Academy of Sciences, Beijing, China
The University of Arizona, Tucson, AZ, USA

Derong Liu
The University of Illinois at Chicago
Chicago, IL, USA

January, 2008

Contents

List of Contributors

YangQuan Chen
Department of Electrical and Computer Engineering
Utah State University
Logan, UT 84322, USA
yqchen@ece.usu.edu

Mo-Yuen Chow
Department of Electrical and Computer Engineering
North Carolina State University
Raleigh, NC 27695, USA
chow@ncsu.edu

Rachana A. Gupta
Department of Electrical and Computer Engineering
North Carolina State University
Raleigh, NC 27695, USA
ragupta@ncsu.edu

Qing-Long Han
Faculty of Informatics and Communication
Central Queensland University
Rockhampton, QLD 4702, Australia
q.han@cqu.edu.au

James Lam
Department of Mechanical Engineering
University of Hong Kong
Hong Kong, P. R. China
james.lam@hku.hk

Wei Li
Department of Automation
Shanghai Jiao Tong University
Shanghai 200240, P. R. China
liweil@sjtu.edu.cn

Guo-Ping Liu
Department of Engineering
University of Glamorgan
Pontypridd, CF37 1DL, UK
gpliu@glam.ac.uk

Dragan Nešić
Department of Electrical and Electronic Engineering
University of Melbourne
Parkville, Victoria 3052, Australia
d.nesic@ee.unimelb.edu.au

George Nikolakopoulos
Electrical and Computer Engineering Department
University of Patras
Patras, Achaia 26500, Greece
gnikolak@ece.upatras.gr

Athanasia Panousopoulou
Electrical and Computer Engineering Department
University of Patras
Patras, Achaia 26500, Greece
apanous@ece.upatras.gr

Fuchun Sun
State Key Laboratory of Intelligent Technology and Systems
Department of Computer Science and Technology
Tsinghua University
Beijing 100084, P. R. China
fcsun@mail.tsinghua.edu.cn

Mohammad Tabbara
Department of Electrical and Electronic Engineering
University of Melbourne
Parkville, Victoria 3052, Australia
m.tabbara@ee.unimelb.edu.au

Andrew R. Teel
Department of Electrical and Computer Engineering
University of California
Santa Barbara, CA 93106, USA
teel@ece.ucsb.edu

Anthony Tzes
Electrical and Computer Engineering Department
University of Patras
Patras, Achaia 26500, Greece
tzes@ece.upatras.gr

Fei-Yue Wang
Institute of Automation
Chinese Academy of Sciences
Beijing 100080, P. R. China
and
Department of Systems & Industrial Engineering
University of Arizona
Tucson, AZ 85721, USA
feiyue@sie.arizona.edu

Long Wang
College of Engineering
Peking University
Beijing 100871, P. R. China
longwang@pku.edu.cn

Xiaofan Wang
Department of Automation
Shanghai Jiao Tong University
Shanghai 200240, P. R. China
xfwang@sjtu.edu.cn

Fengge Wu
State Key Laboratory of Intelligent Technology and Systems
Department of Computer Science and Technology
Tsinghua University
Beijing 100084, P. R. China
wfg02@mails.tsinghua.edu.cn

Dedong Yang
School of Information Science and Engineering
Northeastern University
Shenyang 110004, P. R. China
ydd12677@163.com

Wensheng Yu
Institute of Automation
Chinese Academy of Sciences
Beijing 100080, P. R. China
wensheng.yu@mail.ia.ac.cn

Dong Yue
School of Electrical and Automation Engineering
Nanjing Normal University
Nanjing 210042, P. R. China
medongy@njnu.edu.cn

Huaguang Zhang
School of Information Science and Engineering
Northeastern University
Shenyang 110004, P. R. China
zhanghuaguang@ise.neu.edu.cn

1

Overview of Networked Control Systems

Rachana A. Gupta and Mo-Yuen Chow

North Carolina State University, Raleigh, NC 27695, USA
ragupta@ncsu.edu, chow@ncsu.edu

Abstract. Networked control systems (NCS) have been one of the main research focuses in academia as well as in industrial applications for many decades. NCS has taken the form of a multidisciplinary area. In this chapter, we introduce NCS and the different forms of NCS. The history of NCS, different advantages of having such systems are the starting points of the chapter. Furthermore, the chapter gives an insight to different challenges which come with building efficient, stable and secure NCS. The chapter talks about different fields and research arenas, which are part of NCS and which work together to deal with different NCS issues. A brief literature survey concerning each topic is also included in the chapter. iSpace is the test-bed for NCS and it attends the practical issues and implementation of NCS. At the end, iSpace at ADAC is presnted as a case study for NCS with different experimental results.

Keywords. Networked control systems, time-sensitivity, intelligent space, UGV navigation.

1.1 Introduction

A control system is a device or set of devices to manage, command, direct or regulate the behavior of other devices or systems. In engineering and mathematics, control theory deals with the behavior of dynamical systems. Although control systems of various types date back to antiquity, a more formal analysis of the field began with a dynamics analysis of the centrifugal governor, conducted by the famous physicist Maxwell in 1868 entitled "On Governors." A notable application of dynamic control was in the area of manned flight. The Wright brothers made their first successful test flights on December 17, 1903 and were distinguished by their ability to control their flights for substantial periods (more so than the ability to produce lift from an airfoil, which was known). Control of the airplane was necessary for flight safety. For many years, researchers have given us precise and optimum control strategies emerging from classical control theory, starting from open-loop control to sophisticated control strategies based on genetic algorithms.

The advent of communication networks, however, introduced the concept of remotely controlling a system, which gave birth to networked control systems (NCS). The classical definition of NCS can be as follows: When a traditional feedback control system is closed via a communication channel, which may be shared with other nodes outside the control system, then the control system is called an NCS [15]. An NCS can also be defined as a feedback control system wherein the control loops are closed through a real-time network. The defining feature of an NCS is that information (reference input, plant output, control input, etc.) is exchanged using a network among control system components (sensors, controllers, actuators, etc., see Fig. 1.1).

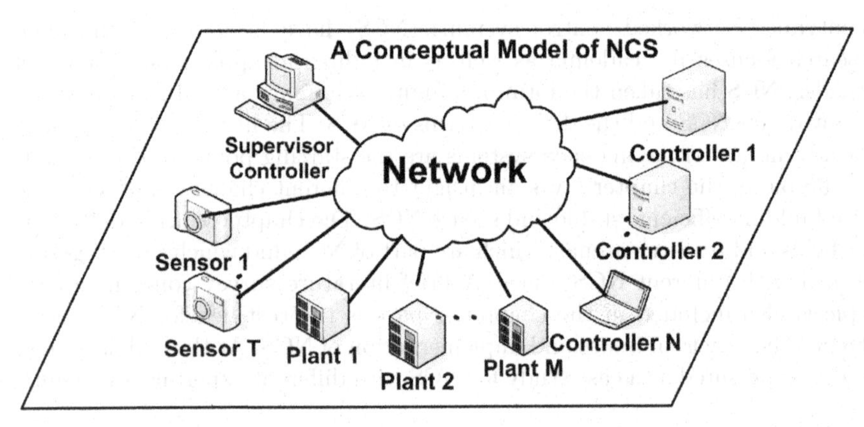

Fig. 1.1. A typical networked control system

1.1.1 Advantages and Applications of Control over Network

For many years now, data networking technologies have been widely applied in industrial and military control applications. These applications include manufacturing plants, automobiles, and aircraft. Connecting the control system components in these applications, such as sensors, controllers, and actuators, via a network can effectively reduce the complexity of systems, with nominal economical investments. Furthermore, network controllers allow data to be shared efficiently. It is easy to fuse the global information to take intelligent decisions over a large physical space. They eliminate unnecessary wiring. It is easy to add more sensors, actuators and controllers with very little cost and without heavy structural changes to the whole system. Most importantly, they connect cyber space to physical space making task execution from a distance easily accessible (a form of tele-presence). These systems are becoming more realizable today and have a lot of potential applications [16, 20], including space explorations, terrestrial exploration, factory automation (Fig. 1.2), remote diagnostics and troubleshooting, hazardous environments, experimental facilities, domestic robots, automobiles, aircraft, manufacturing plant monitoring, nursing homes or hospitals, tele-robotics (Fig. 1.3) and tele-operation, just to name a few.

Fig. 1.2. Factory automation

Fig. 1.3. Unmanned ground vehicle navigation (image courtesy of Space and Naval Warfare Systems Center, San Diego)

1.1.2 Brief History of Research Field of NCS

The advent of the Internet gave a huge base for millions of smaller domestic, academic, business, and government networks, which together carry information and services, such as electronic mail, online chat, file transfer, interlinked web pages and other documents of the World Wide Web. In the last few years, there has also been a tremendous increase in the deployment of wireless systems, which has triggered the development and research of distributed NCS. As the concept of NCS started to grow because of its potential in various applications, it also provided many challenges for researchers to achieve reliable and efficient control. Thus the NCS area has been researched for decades and has given rise to many important research topics. A wide branch in the literature focuses on different control strategies and kinematics of the actuators/vehicles suitable for NCS [2], [19], [26], [45]. Another important research area concerning NCS is the study of the network structure required to provide a reliable, secured communication channel with enough bandwidth, and the development of data communication protocols for control systems [2], [23], [35]. Collecting real-time information over a network using distributed sensors and processing the sensor data in an efficient manner are important research areas supplementing NCS. Thus NCS is not only a multidisciplinary area closely affiliated with computer networking, communication, signal processing, robotics, information technology, and control theory, but it also puts all these together beautifully to achieve a single system which can efficiently work over a network. For example, a robot which is in the eastern part of the world can be controlled by a person sitting in the USA (Fig. 1.4) [8].

Some of the well-known research institutes and research labs working in NCS are listed below.

Advanced Diagnosis, Automation and Control (ADAC) Laboratory at North Carolina State University (http://www.adac.ncsu.edu/).

Alleyne Research Group at University of Illinois at Urbana-Champaign (http://mr-roboto.me.uiuc.edu/).

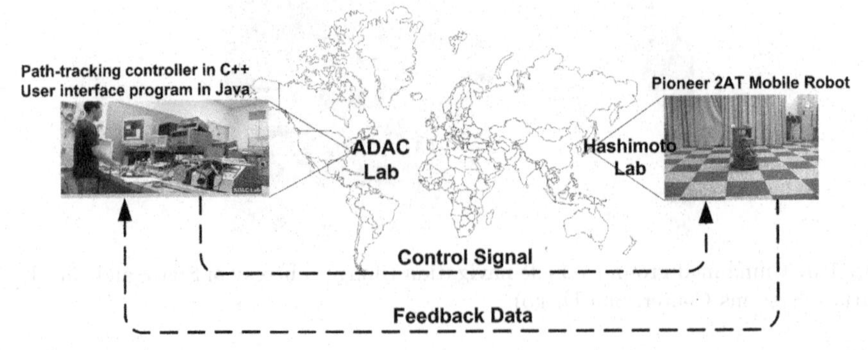

Fig. 1.4. Remote mobile robot path-tracking via IP setup between ADAC lab (USA) and Hashimoto lab (Japan)

Networked Control Systems Laboratory at University of Washington (Seattle) (http://www.ee.washington.edu/research/ncs/index.html).

Center for Networked Communicating Control Systems (CNCS) at University of Maryland at College Park (http://www.isr.umd.edu/CNCS/).

Network Control Systems Laboratory at National Taiwan University (http://cc.ee.ntu.edu.tw/~ncslab/).

Interdisciplinary Studies of Intelligent Systems at University of Notre Dame (http://www.nd.edu/~isis/).

1.2 NCS Categories and NCS Components

Generally speaking, the two major types of control systems that utilize communication networks are (1) shared-network control systems and (2) remote control systems. Using shared-network resources to transfer measurements, from sensors to controllers and control signals from controllers to actuators, can greatly reduce the complexity of connections. This method, as shown in Fig. 1.5, is systematic and structured, provides more flexibility in installation, and eases maintenance and troubleshooting. Furthermore, networks enable communication among control loops. This feature is extremely useful when a control loop exchanges information with other control loops to perform more sophisticated controls, such as fault accommodation and control. Similar structures for network-based control have been applied to automobiles and industrial plants.

On the other hand, a remote control system can be thought of as a system controlled by a controller located far away from it. This is sometimes referred to as tele-operation control. Remote data acquisition systems and remote monitoring systems can also be included in this class of systems. The place where a central controller is installed is typically called a "local site," while the place where the plant is located is called a "remote site."

There are two general approaches to design an NCS. The first approach is to have several subsystems form a hierarchical structure, in which each of the subsystems contains a sensor, an actuator, and a controller by itself,

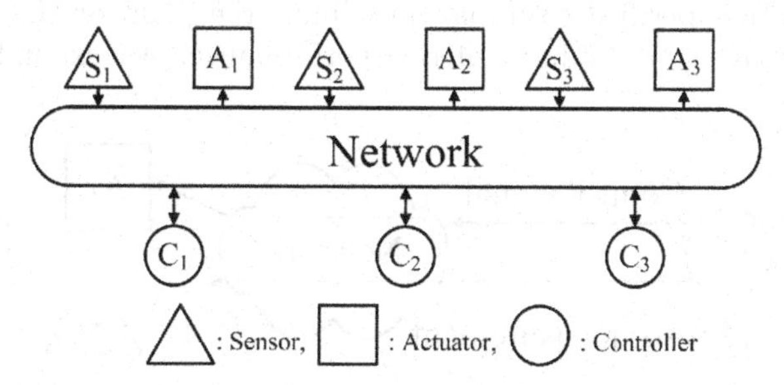

Fig. 1.5. Shared-network connections

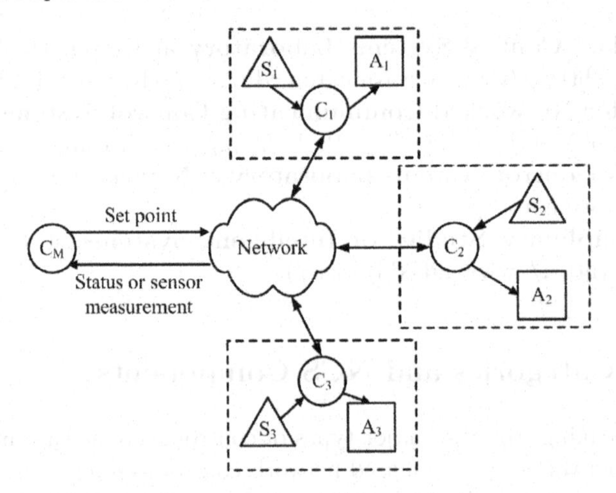

Fig. 1.6. Data transfers of hierarchical structure

as depicted in Fig. 1.6. These system components are attached to the same control plant. In this case, a subsystem controller receives a set point from the central controller CM. The subsystem then tries to satisfy this set point by itself. The sensor data or status signal is transmitted back via a network to the central controller.

The second approach of networked control is the direct structure, as shown in Fig. 1.7. This structure has a sensor and an actuator of a control loop connected directly to a network. In this case, a sensor and an actuator are attached to a plant, while a controller is separated from the plant by a network connection.

Both the hierarchical and direct structures have their own pros and cons. Many networked control systems are a hybrid of the two structures. For example, the remote teaching lab is an example that uses both structures [7], [10].

Networked control applications can be divided into two categories: (1) time-critical/time-sensitive applications and (2) time-delay-insensitive applications. In time-delay-sensitive applications, time is critical, i.e., if the delay-time exceeds the specified tolerable time limit, the plant or the device can either be damaged or produce inferior performance. An example of time-

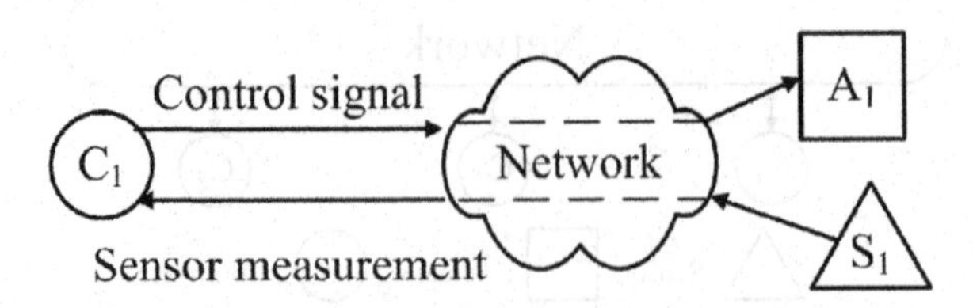

Fig. 1.7. Data transfers of direct structure

delay-sensitive applications is tele-operation via networks for fire-fighting operations, undersea operations, and automated highway driving. On the other hand, time-delay insensitive applications are tasks or programs that run in real time but whose deadlines are not critical. Examples of these applications are e-mail, ftp, DNS, and http. We will briefly mention many advantages of networked control systems, tele-operation being the most evident and tangible one. Let us categorize the NCS according to the amount of human interference in the loop.

(1) Tele-operated systems with human operator–In this case, a human operator from one location controls the actuators (robots, arms, unmanned vehicles) at different locations. The feedback to the operator is mainly visual (video or real-time image). The precision and accuracy of the system operation also depends upon operator skill including system precision, feedback delay and accuracy, signal distortion. This can also be called the *human supervisory control* [33]. Therefore for such tele-operated systems, many times, the human operators are required to be trained to operate the system. There are various applications of such systems like distributed virtual laboratories, remote surgery systems [14], field robotics, etc. Such systems therefore suffer from issues like human perception accuracy, force feedback to the operator, network delay, control prediction, ergonomics, security, system portability, etc. [4], [34]. There are also many tools developed for accurate operator feedback such as virtual reality (VR), interactive televisions, 3-D visualization environment, etc. [4]. One of the VR environments developed by Alfred E. Mann Institute at USC is used to simulate the movement of prosthetic limbs and human limbs. Its main use is to prototype the control of the prosthetic systems and fit the control to patient needs. It also allows the patients to train in VR to operate their prosthetic limbs (Fig. 1.8).
(2) Tele-operation without human intervention–In such systems the intelligence is built inside the controller modules. The sensor data and actuator feedback data is directly fed to the controller over the network. This can also be called the *autonomous networked control system*. The supervisory controller is not a human in this case. A human can act as an external user which can choose tasks or specify some manual control commands. Such systems are therefore not dependent on human perception and do not require operators to be skilled or trained. However developing intelligent and efficient data processing and controlling algorithms for supervisory control is very important. Supervisory controllers can use techniques such as machine learning, neural networks and artificial intelligence algorithms to take intelligent operation decisions. In this case, sensor data fusion and actuator bandwidth optimization and scheduling are equally important issues to be considered.
(3) Hybrid control: Main controller and actuator have distributed intelligence to increase the efficiency of network operations.

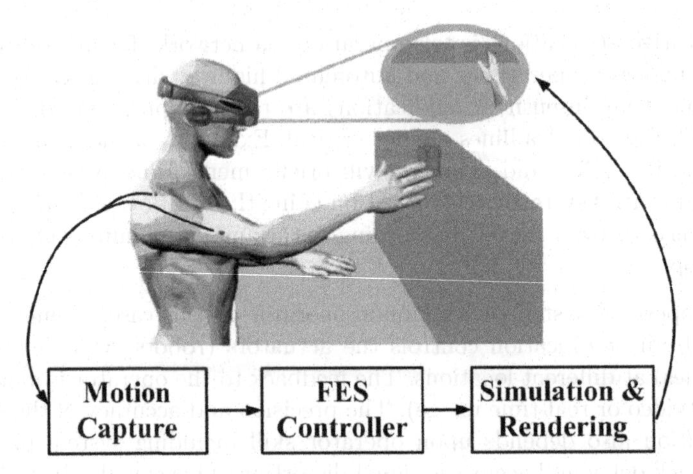

Fig. 1.8. Virtual reality (image courtesy Alfred E. Mann Institute, USC)

Here in this chapter we will focus mainly upon time-sensitive supervisory networked control systems.

1.2.1 NCS Components

Whatever the arrangement or modalities used for connecting and configuring the hardware and the software assets in order to actualize a networked control system that has certain capabilities, the components used have to enable four functions which form the basis of the function an NCS is required to project. These basis functions are information acquisition (sensors), command (controllers), and communication and control (actuators).

1.2.2 Information Acquisition in a Network

As the name suggests information acquisition requires us to study sensors, data processing, and signal processing. There is a growing excitement about the potential application of large-scale sensor networks in diverse applications such as precision agriculture, geophysical and environment monitoring, remote health care, and security [9]. Rapid progress in sensing hardware, communications and low-power computing has resulted in a profusion of commercially available sensor nodes. NCS suggests collecting the relevant data using distributed sensors in the network to study the system under control. Sensor data can be in any form starting from small numbers representing temperature, pressure, weight, etc. or in chunk form such as images, arrays, videos streams, etc. This raises important questions like:

(a) Bandwidth requirements for the data transfer in the network.
(b) Data collection strategies in the case of a number of sensors.

(c) Cheap, reliable and energy efficient sensors which can easily be added to the NCS.

Sensor fusion and sensor networks [11], [40] are very wide research fields which help improve information acquisition in a network. Developing middleware and operating systems for sensor nodes to send data efficiently in the network [29], [30], information assurance [28], energy efficient sensor nodes [44] and sensitivity of the data are the key research foci related to information acquisition in a network. Sensor networks hold the promise of facilitating large-scale, real-time data processing in complex environments.

Image data is used for applications like surveillance [9], robot navigation [27], target tracking [32] and tele-operation, etc. With the advancement in the field of computer vision and image processing, there are many sophisticated algorithms available to process images for pattern recognition and feature extraction. Many systems and algorithms have been developed using visual and other local sensing capabilities to control ground and aerial vehicles [12], [31].

1.2.3 Control of Actuators over a Network

One of the biggest advantages of a system controlled over a network is scalability. As we talk about adding many sensors connected through the network at different locations, we can also have one or more actuators connected to one or more controllers through the network. For many years now, researchers have given us precise and optimum control strategies emerging from classical control theory, starting from PID control, optimal control, adaptive control, robust control, intelligent control and many other advanced forms of these control algorithms. Applying all these control strategies over a network however becomes a challenging task. We will study different issues to be considered for successful and efficient operation of an NCS in the next section.

1.2.4 Communication

The communication channel being the backbone of the NCS, reliability, security, ease of use, and availability are the main focus when choosing the communication or data transfer type. In today's world, plenty of communication modes are available from telephone lines, cell phone networks, satellite networks and, most widely used, Internet. Sure enough, the choice of network depends upon the application to be served. Internet is the most suitable and inexpensive choice for many applications where the plant and the controller are far from each other (as shown in Fig. 1.4, where the controller is in USA and the robot to be controlled is in Japan [7]). The controller area network (CAN) is a serial, asynchronous, multi-master communication protocol for connecting electronic control modules in automotive and industrial applications. CAN was designed for applications needing high-level data integrity and data rates of up to 1 Mbps. Many manufacturing plants have a complete line of products enabling industrial designers to incorporate CAN into their applications.

For years, wireless LANs having been supporting enterprise applications, such as warehouse management and mobile users in offices. With lower prices and stable standards, homeowners are now installing wireless LANs at a rapid pace. LANs for the support of personal computers and workstations have become nearly universal in organizations of all sizes. Even those sites that still depend heavily on the mainframe have transferred much of the processing load to networks of personal computers. Perhaps the prime example of the way in which personal computers are being used is to implement client/server applications. Back-end networks are used to interconnect large systems such as mainframes, supercomputers, and mass storage devices. The key requirement here is for bulk data transfer among a limited number of devices in a small area. High reliability is generally also a requirement.

GPS systems can be used to localize vehicles all over the planet. Military applications, surgical and other emergency medical applications, however, can use dedicated optical networks to ensure fast speed and reliable data communication.

1.3 NCS Challenges and Solutions

After having an overview of different categories, components and applications of NCS, we now describe the different challenges and issues to be considered for a reliable NCS.

We can broadly categorize NCS applications into two categories as (1) time-sensitive applications or time-critical control such as military, space and navigation operations; (2) time-insensitive or non-real-time control such as data storage, sensor data collection, e-mail, etc. However, network reliability is an important factor for both types of systems. The network can introduce unreliable and time-dependent levels of service in terms of, for example, delays, jitter, or losses. Quality-of-service (QoS) can ameliorate the real-time network behavior, but the network behavior is still subject to interference (especially in wireless media), to routing transients, and to aggressive flows. In turn, network vagaries can jeopardize the stability, safety, and performance of the units in a physical environment [21], [36]. A challenging problem in the control of network-based systems is the network delay effects. The time to read a sensor measurement and to send a control signal to an actuator through the network depends on network characteristics such as topology and routing schemes. Therefore, the overall performance of an NCS can be affected significantly by network delays. The severity of the delay problem is aggravated when data loss occurs during a transmission. Moreover, the delays do not only degrade the performance of a network-based control system, but they also can destabilize the system.

(1) Stability in Control and Delay Compensation

For many years now, researchers have given us precise and optimum control strategies emerging from classical control theory, starting from PID control, optimal control, adaptive control, robust control, intelligent control and many other advanced forms of control algorithms. But these control strategies need to be modified according to the application requirements as well as for them to reliably work over a network to compensate for delays and unpredictability. Fig. 1.9 displays the typical NCS model with the time delay taken into consideration. Fig. 1.10 shows the adverse effect of the network delay on a remotely controlled system. It displays the scenario where a mobile agent was asked to track a path with varying curvatures, first with local controller and later with remote controller. As we can observe, without any modifications to the controller, the mobile agent is not able to track the path, especially at the high curvature because of the network delay [7]. Instability of the system due to the network delay is therefore a very important factor to be considered in NCS. Different mathematical, heuristic and statistical-based approaches are taken for delay compensation in NCS. A gain scheduler middleware (GSM) has been developed by Tipsuwan and Chow to alleviate the network time delay effect on network-based control systems. GSM methodology estimates the network traffic and controls the gain of the whole system using a feedback processor as shown in Fig. 1.11. Yu and Yang [46] suggested a predictive control model of NCS to overcome the adverse influences of stochastic time delay, which could improve the performance through model matching and multi-step predictive output compensation. Wang and Wang [43] suggested a delay compensation controller solution with an iterative procedure of a linear matrix inequality (LMI) minimization problem, which is derived from the cone complementarity linearization algorithm.

(2) Bandwidth Allocation and Scheduling

As we talk about having multi-sensor and controlling multi-actuator systems in a network, important consideration should be given to the available bandwidth in the network. With the finite amount of bandwidth available, we want

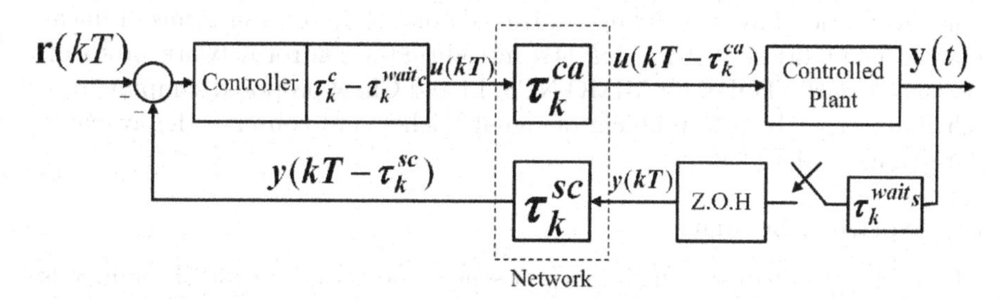

Fig. 1.9. NCS plant structure showing network delays

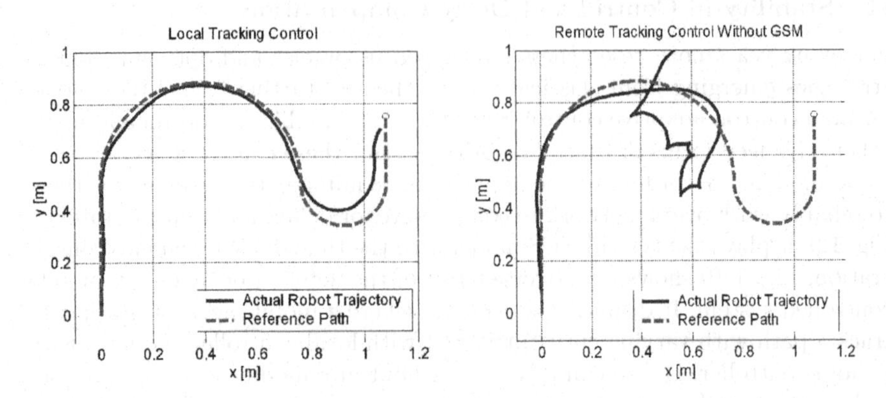

Fig. 1.10. Mobile agent trajectory (1) local control (2) remote control without delay compensation

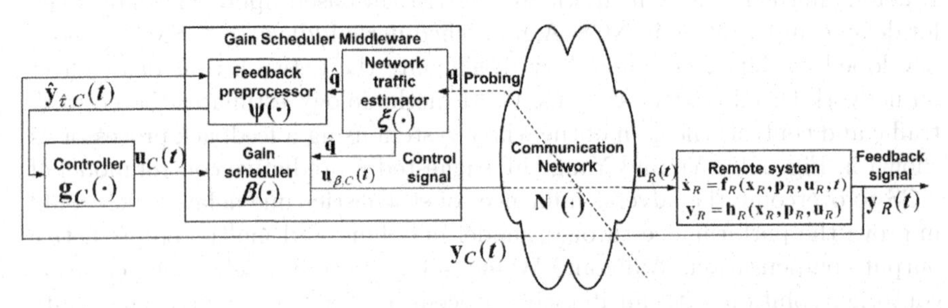

Fig. 1.11. GSM module for network delay compensation

to utilize it optimally and efficiently. This further raises the need for priority decisions and scheduling issues for controlling a series of actuators for a series of tasks [41]. Different scheduling methods and bandwidth allocation strategies have been developed for NCS over the past decade [1], [39]. There are also many tools like Petri-net modeling, integer, nonlinear, dynamic programming, AI tools, genetic algorithms developed for scheduling of networked control systems. Kim *et al.* [18] formulated a method to obtain a maximum allowable delay bound for scheduling networked control systems in terms of linear matrix inequalities. Walsh *et al.* [41] introduced a control network protocol, try-one-discard (TOD), for MIMO NCS. Li and Chow proposed sampling rate scheduling to solve the problem of signal fidelity and conserve the available data transmission [24], [25].

(3) Network Security

All this discussion of sending important sensor and actuator control commands in the network brings us to an important point of security over the network.

Any network medium especially wireless medium is susceptible to easy intercepting; it is extremely critical to protect transmitted data from unauthorized access and modifications in wireless systems. Malicious users can intercept and eavesdrop the data in transit via shared and broadcast medium. Network security includes essential elements in Internet security devices that provide traffic filtering, integrity, confidentiality, and authentication. Therefore data sharing, data classification and data/network security is of utmost concern in distributed networked control systems considering the time and data sensitive applications. In wireless systems, several security protocols such as wired equivalent privacy (WEP), 802.1x port access control with extensible authentication protocol (EAP) support are proposed to address security issues [5], [17]. Moreover, due to strong security provided by IP security protocol (IPsec) in wired networks, it is considered as a good option for wireless systems as well. However, information security and data sensitivity have not been sufficiently addressed to be applied in a real-time NCS. Very few researchers have addressed the trade-off between security addition and real-time operation of NCS. Gupta *et al.* [13] characterize the wireless NCS application on the basis of security effect on NCS performance to show the trade-off between security addition and real-time operation of NCS.

1.3.1 Integration of Components and Distribution of Intelligence

After discussing individual modules involved in NCS and possible issues related to control system, network structure and information acquisition, we come to a point of integrating the components to achieve the final goal. Fusing the global information to make intelligent decisions or to perform a particular task requires integration of different modules like data acquisition, data processing, information extraction, and actuator control. All these different modules perform tasks independently yet together making it one system. Therefore, a few of the issues faced by a network-based navigation system include data sharing, data transfer and interfacing between different modules. Thus it is evident that to improve the efficiency of an integrated networked control system, we not only have to improve each integrated module but also provide an efficient data interface between different modules.

There is a wealth of techniques available for actualizing each one of the basic function modules. A well-designed software architectural framework and middleware are critical for the widespread deployment and proliferation of networked control systems. There are a few system architectures or middleware developed to put such heterogeneous systems together. Component architecture allows individual components to be developed separately and integrated easily later, which is very important for the development of large systems. Further, such architecture promotes software reuse, since a well-designed component such as a control algorithm, tested for one system, can easily be transplanted into another similar system. At the same time suitability of the environment representation for use with the communication and command modules should also be taken into consideration, which is the key point in any practical application of NCS. Baliga and Kumar [3] developed a list of key requirements for such middleware and presented Etherware, a message oriented component middleware for networked control. Tisdale *et al.* [38] from University of California Berkeley also developed a software architecture for autonomous vision-based navigation, obstacle avoidance and convoy tracking. This software architecture has been developed to allow collaborative control concepts to be examined. These architectures represent the system at an abstract level and focus on modularization of the system to achieve flexibility and scalability in design. However, while studying all these modules separately, it is highly unlikely to find a realistic command module that jointly takes into consideration the realization of an admissible control signal when converting a task and constraints on behavior into a group of reference signals. Designing the NCS at the system level by choosing the most suitable and appropriate modules for each component of the NCS is a challenging task. To elaborate more on this point let us look at an example of NCS.

1.4 A Case Study for NCS–iSpace

Intelligent space (iSpace) is a relatively new concept to effectively use various engineering disciplines such as automation and control, hardware and software design, image processing, distributed sensors, actuators, robots, computing processors and information technology over communication networks over a space of interest to make intelligent operation decisions. It can also be considered as a large-scale mechatronic system over networks. This space can be as small as a room or a corridor or can be as big as an office, city or even a planet. ADAC lab at NCSU in Raleigh has developed a multi-sensor network-controlled integrated navigation system for multi-robots demonstrating the concept of iSpace [20]. The modularized structure of iSpace at ADAC is as shown in Fig. 1.12. The information acquisition about the space is through network cameras. The task for the robots is to reach the destination point chosen by the user through the GUI (accessible through internet). All the intelligence to generate navigation commands for the robots resides in the network controller (path generation avoiding the obstacles in the space and path tracking to reach the destination as soon as possible without hitting any of the obstacle in the space).

The system, being an NCS, observes network delay for image acquisition and command transfer from controller to the robot on wireless channels. The image processing, feature extraction and real-time path tracking algorithms are also computationally intensive. The application led to the following choice of different modules to be implemented in the network controller.

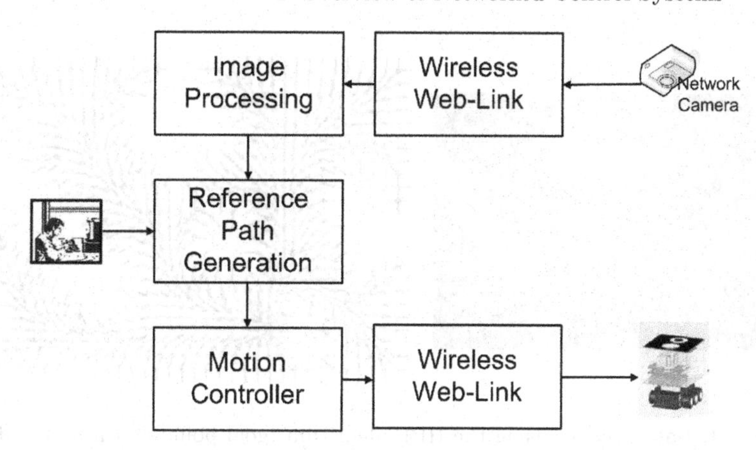

Fig. 1.12. Modularized structure of iSpace at ADAC

(1) HPF for Motion Planning

The use of potential field in motion planning was introduced by Khatib in 1985, where the obstacles were represented by the repelling force and the point of destination was represented by the attractive force. Harmonic potential fields (HPF) were introduced by Connolly to avoid the local minima in navigation. Therefore, tracking the negative gradient from the source in the harmonic potential field map will lead the robot towards the destination as shown in Fig. 1.13 created synthetically to represent obstacle boundaries by white edges and the navigation path for the robot as grey. The HPF equations are given by

$$\begin{aligned}
&\nabla^2\phi(x,y) \equiv 0,\ \ (x,y) \in \Omega \\
&\text{subject to} \\
&\phi(x,y) = 1,\ \ (x,y) \in \Gamma \\
&\text{and } \phi(x,y) = -1,\ \ (x,y) = (x_T, y_T) \\
&\text{and } \phi(x,y) = 0,\ \ (x,y) \notin \Gamma \text{ and } (x,y) \neq (x_T, y_T)
\end{aligned} \tag{1.1}$$

where ∇^2 is the Laplace operator, Ω is the workspace of the UGV ($\Omega \subset \Re^2$), Γ is the boundary of the obstacles (output of the edge detection stage), and (x_Γ, y_Γ) is the target point. The obstacle free path to the target is generated by traversing the negative gradient of (ϕ), i.e., ($\nabla\phi$).

HPF is a suitable algorithm for path planning on the network controller once the image of the actual space is acquired from the network camera as HPF is computationally fast ($O(n)$ algorithm) and it drives the mobile robot away from the boundaries of the obstacles because of the Dirichlet's settings. Fig. 1.13 shows the path planner created using HPF. All the arrows show the negative gradient direction confirming that UGV is directed away from obstacle boundaries and driven towards the goal point.

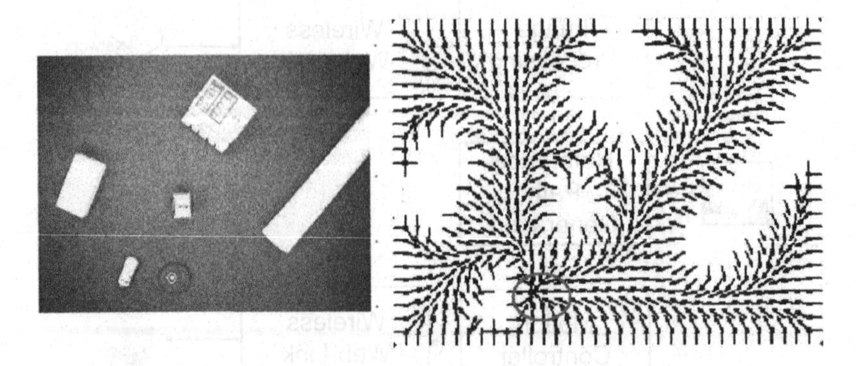

Fig. 1.13. Path planner using the HPF algorithm (goal point shown by the dot in the circle)

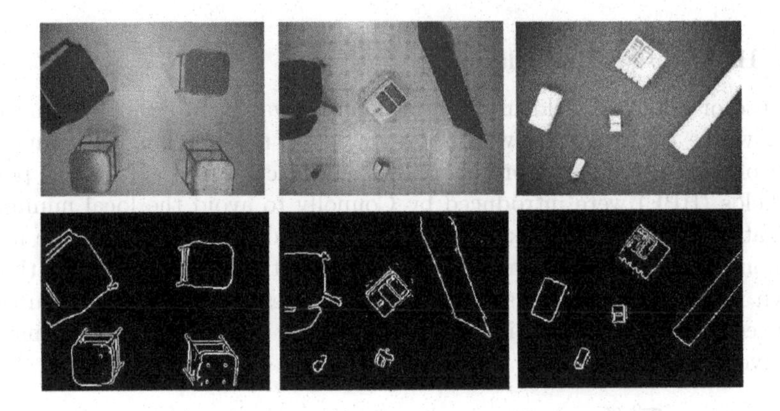

Fig. 1.14. Edge detection results

(2) Edge Detection for Boundary Detection

Converting the actual iSpace image into the raised boundaries of the obstacles is the task of information acquisition as well as data/image processing. Edge detection was used for obstacle boundary recognition. Edge detection is a classic early vision tool to extract discontinuities from the image as features. Thus all the discontinuities, which are more or less obstacle boundaries, will be represented by binary edges in the edge detected image of the UGV workspace. Results are shown in Fig. 1.14. This network-based robot navigation is developed as the research platform for NCS and it is designed for indoor environments. Therefore assuming that the system has enough control over the ambient or artificial light inside the room, cameras are calibrated and fixed, edge detection was a suitable vision feature extractor module to fit in the whole integrated navigation system. The edge maps can be mathematically described by:

$$E(x_i, y_j) = \begin{cases} 1, & \text{if } (x_i, y_j) \in \Gamma \\ 0, & \text{if } (x_i, y_j) \notin \Gamma \text{ for all } (i, j) \end{cases} \tag{1.2}$$

where $E(x, y)$ is the image representing the edge map and Γ is the set of boundary points for all obstacles in workspace.

Comparing (1.1) and (1.2), we achieved perfect data interfacing between information processing (edge detection) and motion planning (HPF) module as the output of the edge detection module is directly fed to the HPF planner without any preprocessing.

(3) Path Tracking Using Quadratic Curve Fitting Controller

A quadratic curve fitting path tracking controller is implemented as the motion controller for the UGV to traverse the path from source to the destination point after path planning using HPF is done. The basic principle of this control algorithm, as explained in [45] by Yoshizawa *et al.*, is that a reference point is moved along a desired path so that the length between the reference point and the UGV is kept at some distance (d_0). Control (velocity) commands–speed (v, in cm/s) and turn rate (ω, in rad/s)–are generated for the UGV to reach that reference position from the current position. This algorithm runs in a feedback loop until the UGV reaches its destination point tracking the reference path generated (Fig. 1.15).

The reference point needed for the path tracking algorithm to reach the destination is chosen by looking at the next negative gradient point of the HPF map. As we know, negative gradient following will lead the mobile robot towards the destination avoiding the obstacles according to the property of the HPF map.

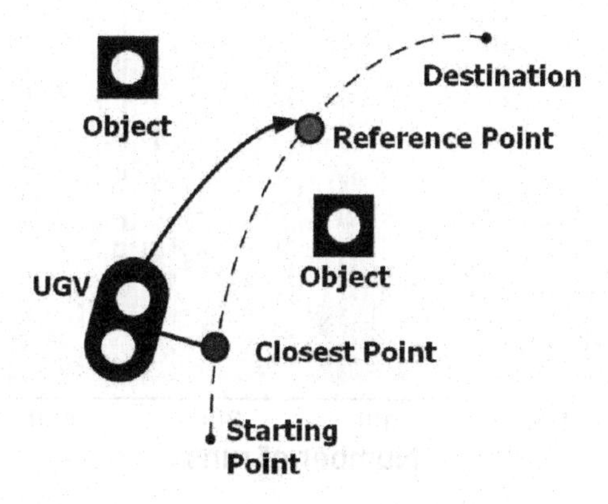

Fig. 1.15. Path tracking control using quadratic curve

(4) Network Delay

As we have been talking about network delay in the NCS, Fig. 1.16 shows the typical network delay graph observed in iSpace operation.

For network delay compensation, we used the gain scheduler middleware technique introduced by Chow and Tipsuwan [36], [37]. GSM methodology uses middleware to modify the output of an existing controller based on a gain scheduling algorithm with respect to the current network traffic conditions. The overall GSM operations for networked control and tele-operation can be summarized as follows [36], [37]. The structure is as shown in Fig. 1.11.

(i) The feedback preprocessor waits for the feedback data from the remote system. Once the feedback data arrives, the preprocessor processes it using the current values of network variables and passes the preprocessed data to the controller.
(ii) The controller computes the control signals and sends them to the gain scheduler.
(iii) The gain scheduler modifies the controller output based on the current values of network variables and sends the updated control signals to the remote system.

Thus, GSM takes care of network delay compensation satisfactorily.

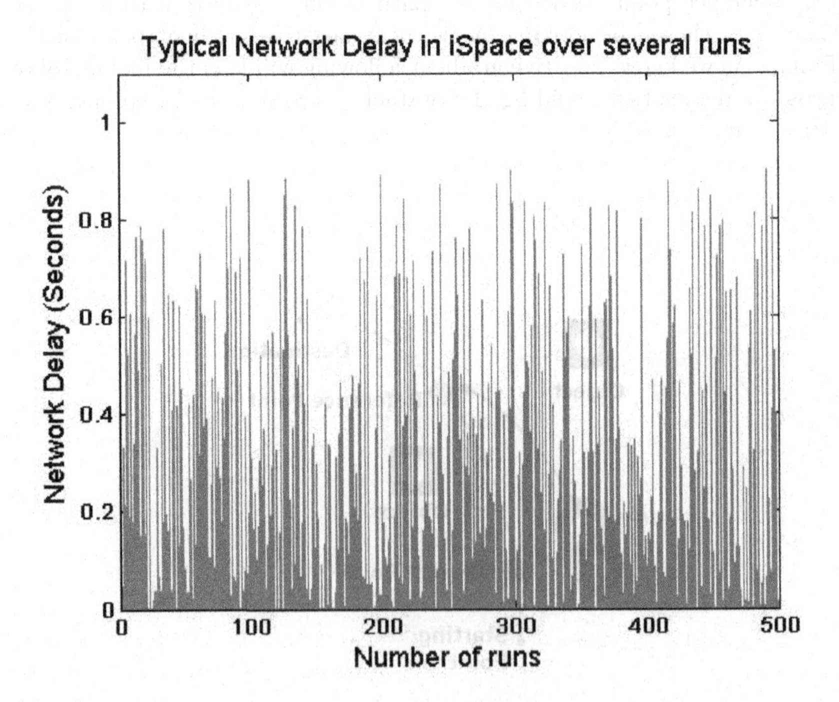

Fig. 1.16. Actual network delay

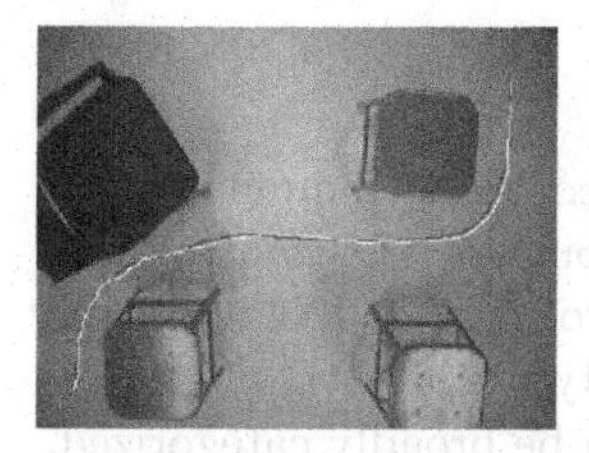

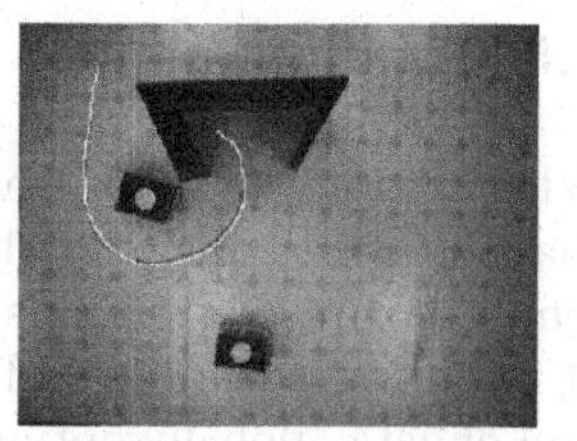

Fig. 1.17. Navigation results of the networked control navigation system for different environmental patterns. White line is the ideal path and dark line is the actual path traversed by the robot.

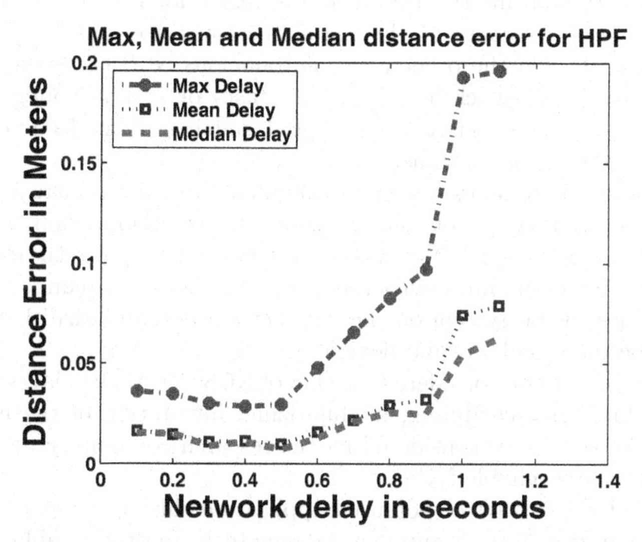

Fig. 1.18. Mean median and max distance error as a function of network delay for iSpace

The choice of individual modules to build the NCS required for indoor robot navigation was done carefully by looking at the data compatibility, environment details, and application requirements. Fig. 1.17 shows some of the experimental results using the NCS structure with edge detection and HPF for UGV navigation in ADAC lab. Fig. 1.18 shows the distance error graph as a function of average network delay for the same experiments. The distance error is the error between the robot's actual navigation path and the ideal path it should have taken with no delays.

Thus, we observe that iSpace, being one form of NCS, the choice of different components, integration of components, and network delay alleviation are important aspects of the system building. These parameters and properties decide the efficiency and operability of the NCS.

1.5 Conclusions

When a traditional feedback control system is closed via a communication channel, which may be shared with other nodes outside the control system, the control system is called a networked control system (NCS). Some of the many advantages of NCS is remote operability, scalability, global fusion of data, globally optimal solutions, etc. NCS can be broadly categorized, depending upon the multi-actuator and multi-sensor structure, as a shared network system and remote control system. It can also be categorized as a time-sensitive/real-time control system and a non-real-time/time-insensitive control system. Human intervention in the feedback loop of the NCS makes it a human supervisory controller having applications like remote operation, remote surgery, etc. On the other hand, autonomous NCS takes the human operator out of the feedback loop and only task- or system configuration-related inputs from human users are accepted, putting all the feedback data directly into the network controller.

NCS is a multidisciplinary research field affiliated with sensor fusion, data processing, control theory, computer networking, communication, security, etc. This leads to research into all fields separately and also poses the challenge of integrating all the modules efficiently. Systems software architectures are developed to design the system on the abstract level, modularize the system such that it becomes scalable and flexible.

There are issues to be considered for QoS of NCS. Network delay, stability, bandwidth allocation, scheduling, modularizing, integration of the modules are some of the key issues considered by the research community to develop an efficient, fast and reliable NCS.

However, NCS has a lot of potential applications like space explorations, terrestrial exploration, factory automation, remote diagnostic/troubleshooting, hazardous environments, experimental facilities, domestic robots navigation, automobiles, etc.

Acknowledgements

We thank Hoa Nguyen from Space and Naval Warfare Systems Center, San Diego for providing the picture of "MDARS-E carrying the 29" iSTAR UAV" (Fig. 1.3). We thank Rahman Davoodi from Alfred E. Mann Institute, USC for providing us with the description and picture of their work on virtual reality (Fig. 1.8). We would also like to acknowledge the partial support of this work by the National Science Foundation (NSF) through Grant No. IIS-0426852 and ECCS-0524519. The opinions expressed are those of the authors and do not necessarily reflect the views of the NSF.

References

1. Al-Hammouri AT, Branicky MS, Liberatore V, Phillips SM (2006) Decentralized and dynamic bandwidth allocation in networked control systems. In: Proceedings of 20th International Parallel and Distributed Processing Symposium, Long Beach, CA, 141–148
2. Arena P, Fortuna L, Frasca M, Lo Turco G, Patane L, Russo R (2005) Perception-based navigation through weak chaos control. In: Proceedings of the 44th IEEE Conference on Decision and Control, Seville, Spain, 221–226
3. Baliga G, Kumar PR (2005) A middleware for control over networks. In: Proceedings of the 44th IEEE Conference on Decision and Control, Seville, Spain, 482–487
4. Belousov IR, Tan J, Clapworthy GJ (1999) Teleoperation and Java3D visualization of a robot manipulator over the World Wide Web. In: Proceedings of IEEE International Conference on Information Visualization, London, England, 543–548
5. Borisov N, Goldberg I, Wagner D (2001) Intercepting mobile communications: the insecurity of 802.11. In: Proceedings of the Seventh Annual International Conference on Mobile Computing and Networking, Rome, Italy, 180–189
6. Chan H, Ozguner U (1994) Closed-loop control of systems over a communication network with queues. In: Proceedings of American Control Conference, Baltimore, MD, 1:811–815
7. Chow M-Y, Tipsuwan Y (2005) Time sensitive network-based control systems and applications. IEEE IES Network Based Control Newsletter 5(2):13–18
8. Chow M-Y, Tipsuwan Y (2001) Network-based control systems: a tutorial. In: Proceedings of the 27th Annual Conference of the IEEE Industrial Electronics Society, Denver, CO, 3:1593–1602
9. Chang C-K, Overhage JM, Huang J (2005) An application of sensor networks for syndromic surveillance. In: Proceedings of IEEE Networking, Sensing and Control, Tuscon, AZ, 191–196
10. Corkey PI, Ridley P (2001) Steering kinematics for a center-articulated mobile robot. IEEE transactions on Robotics and Automations 17(2):215–218
11. Durrant-Whyte H (2006) Data fusion in sensor networks. In: Proceedings of the IEEE International Conference on Video and Signal Based Surveillance, Sydney, NSW, Australia, 39–39
12. Folio D, Cadenat V (2005) A controller to avoid both occlusions and obstacles during a vision-based navigation task in a cluttered environment. In: Proceedings of the 44th IEEE Conference on Decision and Control, and 2005 European Control Conference, Seville, Spain, 3898–3903
13. Gupta RA, Chow M-Y, Agarwal A, Wang W (2007) Characterization of data-sensitive wireless distributed networked-control-systems. In: Proceedings of IEEE/ASME International Conference on Advanced Intelligent Mechatronics, ETH, Zrich, Switzerland
14. Guthart GS, Salisbury JK (2000) The IntuitiveTM telesurgery system: Overview and application. In: Proceedings of IEEE International Conference on Robotics and Automation, San Francisco, CA, 1:618–621
15. Hong Y (1998) Networked Control Systems. [Online document] available http://www.enme.umd.edu/ice_lab/ncs/ncs.html

16. Hsieh TT (2004) Using sensor networks for highway and traffic applications. IEEE Potentials 23(2):13–16
17. Karygiannis T, Owens L (2002) Wireless network security 802.11, bluetooth and handheld devices. National Institute of Technology Special Publication 800–848
18. Kim YH, Park HS, Kwon WH (1988) A scheduling method for network-based control systems. In: Proceedings of the American Control Conference, Atlanta, GA, 2:718–722
19. Léchevin N, Rabbath CA, Tsourdos A, White BA (2005) A causal discrete-time estimator-predictor for unicycle trajectory tracking. In: Proceedings of the 44th IEEE Conference on Decision and Control, and European Control Conference, Seville, Spain, 2658–2663
20. Leung WLD, Vanijjirattikhan R, Li Z, Xu L, Richards T, Ayhan B, Chow M-Y (2005) Intelligent space with time-sensitive application. In: Proceedings IEEE/ASME International Conference on Advanced Intelligent Mechatronics, Monterey, CA, 1413–1418
21. Lian F-L, Moyne J, Tilbury D (2002) Network design consideration for distributed control systems. IEEE Transactions on Control Systems Technology 10(2):297–307
22. Liberatore V (2006) Networked control systems and internet robotics. [Online Document] available http://vorlon.case.edu/~vxl11/NetBots/
23. Liberatore V (2006) Integrated play-back, sensing, and networked control. In: Proceedings of 25th IEEE Conference on Computer Communication, Barcelona, Catalunya, Spain, 1–12
24. Li Z, Chow M-Y (2005) Adaptive multiple sampling rate scheduling of real-time networked supervisory control system-Part I. In: Proceedings of the IEEE Industrial Electronics Conference, Raleigh, NC, 4605–4609
25. Li Z, Chow M-Y (2005) Adaptive multiple sampling rate scheduling of real-time networked supervisory control system-Part II. In: Proceedings of the IEEE Industrial Electronics Conference, Raleigh, NC, Seville, Spain, 4615–4620
26. Litz L, Gabel O, Solihin I (2005) NCS-controllers for ambient intelligence networks-control performance versus control effort. In: Proceedings of the 44th IEEE Conference on Decision and Control, and European Control Conference, Seville, Spain, 1571–1576
27. Mariottini GL, Pappas G, Prattichizzo D, Daniilidis K (2005) Vision-based localization of leader-follower formations. In: Proceedings of the 44th IEEE Conference on Decision and Control, and European Control Conference, Seville, Spain, 635–640
28. Olariu S, Xu Q (2005) Information assurance in wireless sensor networks. In: Proceedings of the 19th IEEE International Parallel and Distributed Processing Symposium, Denver, CO
29. Park S, Kim JW, Lee K, Shin K-Y, Kim D (2006) Embedded sensor networked operating system. In: Proceedings of Ninth IEEE International Symposium on Object and Component-Oriented Real-Time Distributed Computing, Gyeongju, Korea, 117–124
30. Park S, Kim JW, Shin K-Y, Kim D (2006) A nano operating system for wireless sensor networks. In: Proceedings of the 8th International Conference on Advanced Communication Technology, Phoenix Park, Korea, 1:345–348
31. Rathinam S, Zu K, Soghikian A, Sengupta R (2005) Vision based following of locally linear structures using an unmanned aerial vehicle. In: Proceedings

of the 44th IEEE Conference on Decision and Control, and European Control Conference, Seville, Spain, 6085–6090
32. Saeed IAK, Afzulpurkar NV (2005) Real time, dynamic target tracking using image motion. In: Proceedings IEEE International Conference on Mechatronics, Taipei, Taiwan, 241–246
33. Sheridan TB (1992) Telerobotics, automation, and human supervisory control. The MIT Press, Cambridge, MA
34. Tanner NA, Niemeyer G (2005) Improving perception in time delayed teleoperation. In: Proceedings of the IEEE International Conference on Robotics and Automation, Barcelona, Spain, 354–359
35. Tavassoli B, Maralani PJ (2005) Robust design of networked control systems with randomly varying delays and packet losses. In: Proceedings of the 44th IEEE Conference on Decision and Control, and European Control Conference, Seville, Spain, 1601–1606
36. Tipsuwan Y, Chow M-Y (2004) Gain scheduler middleware: a methodology to enable existing controllers for networked control and tele-operation–Part I: networked control. IEEE Transactions on Industrial Electronics 51(6):1218–1227
37. Tipsuwan Y, Chow M-Y (2004) Gain scheduler middleware: A methodology to enable existing controllers for networked control and teleoperation–Part II: teleoperations. IEEE Transactions on Industrial Electronics 51(6):1228–1237
38. Tisdale J, Ryan A, Zennaro M, Xiao X, Caveney D, Rathinam S, Hedrick JK, Sengupta R (2006) The software architecture of the Berkeley UAV platform. In: Proceedings IEEE International Conference on Control Applications, Munich, Germany, 1420–1425
39. Velasco M, Fuertes JM, Lin C, Marti P, Brandt S (2004) A control approach to bandwidth management in networked control systems. In: Proceedings of the 30th Annual Conference of IEEE Industrial Electronics Society, Busan, South Korea, 3:2343–2348
40. Vieira MAM, Coelho CN Jr, da Silva DC Jr, da Mata JM (2003) Survey on wireless sensor network devices. In: Proceedings of IEEE Conference on Emerging Technologies and Factory Automation, Lisbon, Portugal, 1:537–544
41. Walsh GC, Hong Y (2001) Scheduling of networked control systems. IEEE Control Systems Magazine 21(1):57–65
42. Walsh GC, Hong Y, Bushnell LG (2002) Stability analysis of networked control systems. IEEE Transactions on Control Systems Technology 10(3):438–446
43. Wang C, Wang Y (2004) Design networked control systems via time-varying delay compensation approach. In: Proceedings of the Fifth World Congress on Intelligent Control and Automation, Hangzhou, P. R. China, 2:1371–1375
44. Yamasaki K, Ohtsuki T (2005) Design of energy-efficient wireless sensor networks with censoring, on-off, and censoring and on-off sensors based on mutual information. In: Proceedings of the IEEE 61st Vehicular Technology Conference, Stockholm, Sweden, 2:1312–1316
45. Yoshizawa K, Hashimoto H, Wada M, Mori SM (1996) Path tracking control of mobile robots using a quadratic curve. In: Proceedings of IEEE Intelligent Vehicles Symposium, Tokyo, Japan, 58–63
46. Zhuzheng Y, Maying Y (2004) MFC-based control methodology in network control system. In: Proceedings of Fifth World Congress on Intelligent Control and Automation, Hangzhou, P. R. China, 2:1361–1365

2

Overview of Agent-based Control and Management for NCS

Fei-Yue Wang

Institute of Automation, Chinese Academy of Sciences, Beijing, P. R. China
The University of Arizona, Tucson, AZ, USA
feiyue@sie.arizona.edu

Abstract. Every breakthrough in technology has brought a milestone change or paradigm shift in automatic control. As electricity evolves into connectivity and we are at the edge of a connected world, what would be the corresponding changes in automation? This is the focus of discussion in this chapter. We argue and call for a paradigm shift from control algorithms to control agents so that agent-based control can be established as the new control mechanism for operation and management of networked devices and systems. The motivation is to transform "code on demand" into "control on demand" and provide a platform for designing and building low cost but high performance networked equipment in the age of connectivity. Issues related to this vision and real-world applications are addressed in the chapter based on our previous work in this direction.

Keywords. Agent-based control, control on demand, networked systems, fuzzy logic based control systems, neuro-fuzzy networks.

2.1 Introduction

The idea of control can be traced back to the origin of human civilization, but control as an independent scientific field did not begin until late 1940s when N. Wiener published his classic book on control and coined the term cybernetics for this field [1]. Before Wiener's work, the practice of control was geared toward specific problems and applications, its studies was *ad hoc* and normally considered as particular problem solving in mechanics or applied mathematics.

World War II and the space exploration afterwards provide the thrust for a full-fledged development of automatic control, and T. S. Tsien's Engineering Cybernetics [2] marked the true beginning of modern control for real-world applications. Today, both as a field in science and an area in technology, automatic control or automation has become an indispensable part of the modern society.

According to Friedland [3], we may call the period from 1868 to the early 1900s the primitive period of automatic control. It is standard to call the period from then until 1960 the classical period, and the period from 1960 through present times the modern period [4]. During those periods, our society went from agricultural, mechanical or steam engine, electrical, electronic, to computer ages, while our control theory and applications experienced the changes from *ad hoc* applied mathematical problems, frequency-domain design, state space approach, and discrete event and hybrid dynamic systems, to intelligent control algorithms.

With no exception, every breakthrough in technology has caused a milestone change or paradigm shift in automatic control. Now we have entered the age of a networked society, people are expecting and experiencing a new connected lifestyle in a connected world where you can "compute anywhere, connect anything." What would be the corresponding milestone change or paradigm shift for automatic control? To be specific, what will we design and use to control networked devices and systems in the new age?

In this chapter, we call for a paradigm shift toward agent-based control (ABC) for networked systems [5]–[14]. Our basic idea is to go from "code on demand" in programming to "control on demand" in automation, and transform and "liberate" dedicated control algorithms within controlled devices to mobile control agents over interconnected networks. We believe agent-based computing and control will be the mechanism for future automation and could provide a foundation for developing next-generation control theory and application in the age of a connected world. Note that our discussion in the sequel is based exclusively on our own previous work.

2.2 From Electricity to Connectivity: Why Agent-based Control and Management for Networked Systems

Over the last century, electricity has changed our world, and automatic control has played a critical and significant role in this process of change. Now comes connectivity, and we know it has started and will continue to change our world. What would automatic control do in response this time?

In daily life, the availability of various low cost and reliable electrical appliances turned electricity from limited industrial use to infinite household applications, and consequently changed our lifestyle. We depend on electricity everywhere and all the time now. Today, the development of various network-enabled devices, or "net appliances," will turn the connectivity provided by the Internet from its currently limited use in the workplace to infinite applications in homes and everywhere in the near future, thus leading us to the true Connected Lifestyle in a Connected World.

Many network-enabled devices are already available on the market and in daily life now. The central issue to be addressed here is how to develop intelligent networked systems with low cost but reliable and with high performance. Are we going to use the traditional control methods through incremental but small modification and improvement? Or do we need some "revolutionary" changes in our control and management for networked systems, or at least a paradigm shift in the mechanism and implementation of automatic control?

In the age of electrical and microchip industries, various control algorithms have been developed to run industrial and household devices, many modeling and synthesis methods, from linear to nonlinear, deterministic to stochastic, have been established for the design of such algorithms. A control algorithm for real execution is always designed as an integral part of a device, it resides and functions within the device, and once out of a device, it becomes "lifeless" or "meaningless." Normally, as the performance requirement increases, the corresponding complexity and cost of a control algorithm will increase too. For example, to make a network-enabled refrigerator behave intelligently with enhanced functionalities, a sophisticated and complicated control algorithm demanding more memory space and more processing power, and thus high cost, must be developed and implemented.

In the era of Internet and connected world, the connectivity and mobility provided by the Internet offer us an opportunity to control network-enabled devices with a new technology, i.e., mobile agent-based control method. In this method, a control algorithm is decomposed into many simple task-orientated control agents distributed over a network, normally a wide area one. Control agents that can be deployed and replaced over the network as operating conditions vary will run network-enabled devices. In this way, a network-enabled device can operate on a "control-on-demand" basis, i.e., it will need to host only the operating agents, not all possible agents that are needed for its overall operation. This is similar to the idea of "code-on-demand" in the mobile agent technology, which has been successfully used in many areas [5]–[14]. Therefore, the memory space and processing power requirement for agent-controlled network-enabled devices will be less than those operated by traditional control algorithms. This is significant because network-enabled devices are normally embedded systems for which memory space and processing power are key factors in determining their cost. Since manufacturers or service providers can develop and maintain control agents over the network efficiently, mobile agent-based control can lead to low cost but reliable and intelligent network-enabled devices for end-users or consumers. This is the foundation for the agent-based control theory.

One may naturally ask what will happen to a network-enabled device when its network is disconnected or not working. When the electricity is not available, one has to use candles instead of lights, i.e., back to the pre-electrical time. Similarly, when the connectivity is not available, one will have a network-enabled device operated by its current or "default" control agents with limited functionality and performance, back to the electrical age from the network era. In this case, instead of an intelligent device, you would have a dumb but working device. In other words, whenever the connectivity is available, network-enabled devices can demonstrate intelligent and high performance. Without the connectivity, they would behave like the conventional isolated devices in the electrical age. Therefore, the connectivity to network-enabled devices is like the electricity to conventional devices.

We believe the step from control algorithms to control agents is a natural development of control engineering in the Internet era and the connected world. It would make control become an independent entity instead of an affiliated function in system design. The development of a theoretical framework for agent-based control systems would be significant in advancing knowledge in control engineering from the network and information technology aspect. It also has a much broader and significant impact on many real-world industrial and household applications, especially for home automation systems, traffic control systems, vehicle electronic systems or telematics, etc., where real-time requirements are not extremely demanding, systems have a long "resting" or idle period, and network connectivity is available or emerging.

Last but not least, an additional advantage of utilizing agent-based control is the reduced demand for upgrading than using conventional control algorithms. As will become obvious from the hosting mechanism for agents, there is no need for upgrading as long as the largest agents (in terms of memory space and computing power requirements) can be hosted by the device.

In view of those advantages, as electricity evolves into connectivity, we should go from control algorithms to control agents.

2.3 Hosting Mechanism and System Architecture for ABC

The operating mechanism of agent-based control is illustrated in Fig. 2.1 in comparison with that of traditional control systems. In a traditional control system, a control algorithm is constructed as an integral part of each isolated device to be controlled. In an agent-based control system, the operation of a network-enabled device is carried out by control agents distributed and moving over networks.

Fig. 2.2 details the hosting mechanism for control agents. First, a number of "default control agents" reside in a device to ensure its basic operation and performance in the case that connectivity is not available. Second, the device hosts a number of "executing control agents" which are "optimal" for its operation under the current situation. In general, the control process performed by the executing agents can be divided into three steps as follows,

(1) situation assessment,
(2) arbitration, and
(3) control fusion.

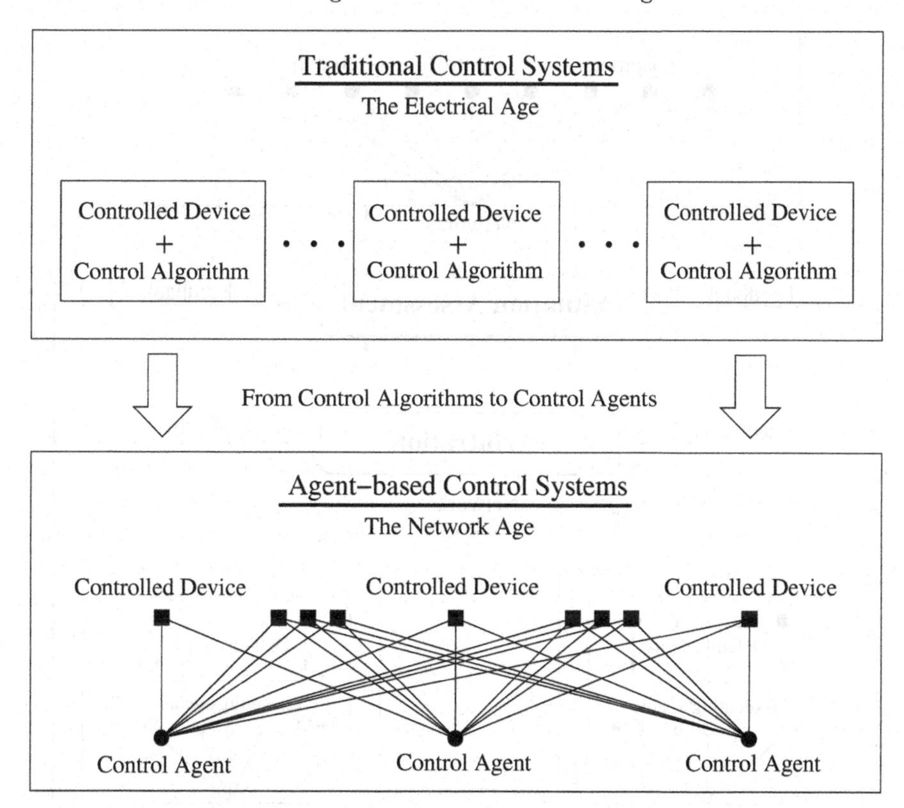

Fig. 2.1. Control agents vs control algorithms: the operating mechanism

The function of situation assessment is to classify the current external and internal system status into particular cases upon which control agents can make decision on their actions. Executing control agents will compete for the right to make the decision for a particular case. An arbitrator will determine which executing control agents are appropriate to participate in the decision process. If only one control agent is selected by the arbitrator, that agent will control the device on behalf of all executing control agents. If multiple agents are selected, they will make their individual decisions and a control fusion algorithm will combine those decisions into a single one for the control of the network-enabled device. The arbitrator also decides whether some executing agents should retire and be replaced by new control agents. New control agents are located outside the device at a remote global control agent center and are deployed by a regional control agent dispatcher upon request by some arbitrator. Both executing and default control agents can also be recalled directly through the network by the remote global control agent center. Simple and effective algorithms for situation assessment, arbitration, and control fusion are key issues for the success of agent-based control systems. In our previous work, we have used a fuzzy logic approach to address those

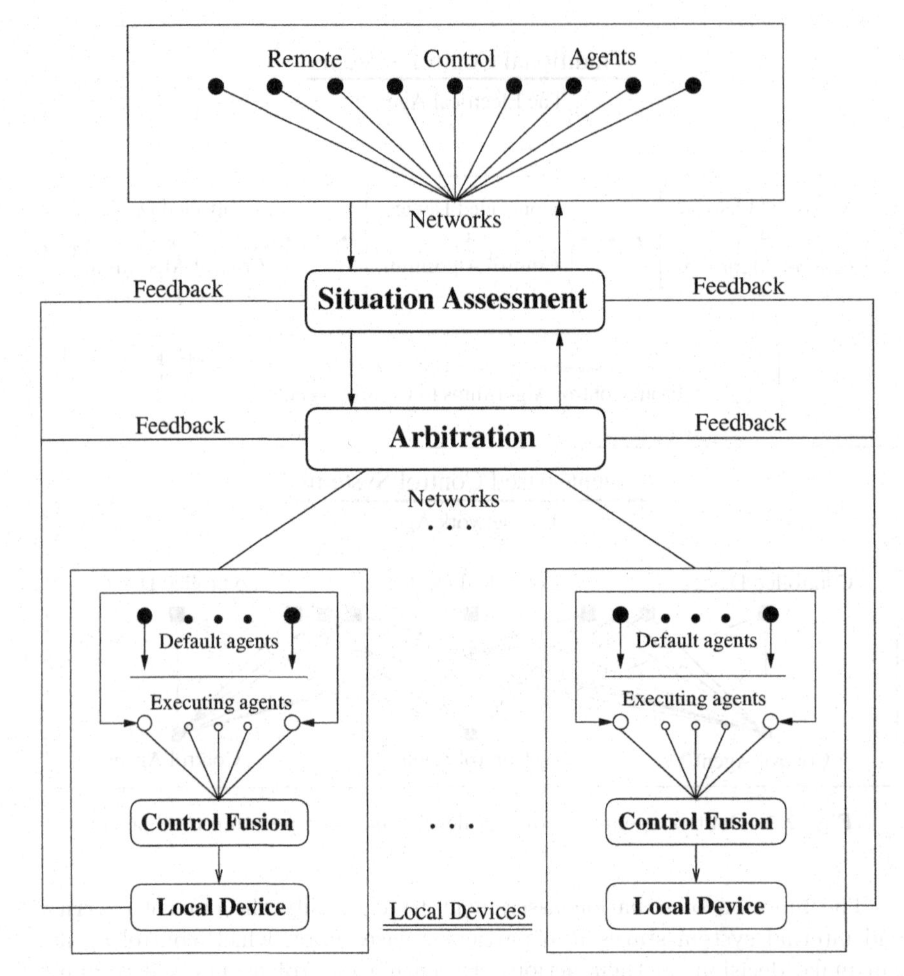

Fig. 2.2. The hosting mechanism for agent-based control

problems [15]–[26], where control behavior programs can be considered as earlier version of control agents.

To ensure a coherent control and communication mechanism among control agents, we must integrate and coordinate their activities for required function and performance. To this end, the hierarchical architecture developed for intelligent control systems [5] is utilized to divide an agent-based control system into three levels of organization, coordination, and execution for its control agents, as indicated in Fig. 2.3. In general, the function of each of the three levels can be specified briefly as follows.

Agent Organization Level: This level mainly performs reasoning and planning for task sequences and organizes control agents to achieve specified goals. It also develops, maintains, and trains control agents, and provides proto-

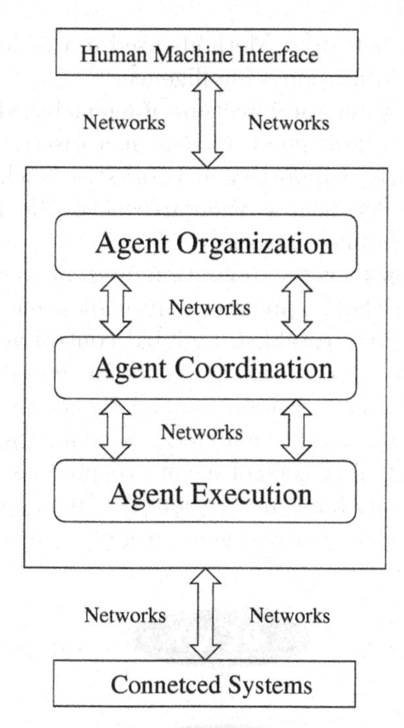

Fig. 2.3. Three levels of hierarchical agent-based control systems – organization, coordination, and execution

cols, algorithms, knowledge bases and databases for agent communication, decision-making, and learning. Methods developed in artificial intelligence, organization theory, and intelligent systems can be utilized here.

Agent Coordination Level: This level is the interface between the organization and execution levels. Generally, it consists of a dispatcher and a number of coordinators. The dispatcher receives control agents from the organization level and deploys control agents to appropriate coordinators through wide area networks in a "control-on-demand" basis. A coordinator is connected to several network-enabled devices through local area networks and will be responsible for downloading control agents to devices and enabling the possible collaboration of those agents. Normally, the dispatcher and coordinators are located in different geographical places and connected through wide area networks, the dispatcher is close to the remote center, and the coordinators are close to the network-enabled devices. Methods developed in operations research, such as dynamic programming, task allocation, event scheduling, and resource sharing should be applied at this level.

Agent Execution Level: This level consists of embedded hardware and software units for deploying, replacing, hosting, and running control agents. Generally, this level is distributed among many LAN linked workplaces that are connected by wide area networks. Methods used in this level are mainly from control systems and computational intelligence.

The corresponding system architecture of agent-based control for operating and managing network-enabled devices in a distributed environment is given in Fig. 2.4. Clearly, supporting networks for such operations involve multiple types of networks, such as wide area networks, local area networks, and sometimes *ad hoc* mobile networks.

As an example, consider the operation and management of network-enabled intelligent household appliances through agent-based control. The appliance manufacturer will establish a global control agent center (GCAC) to develop and maintain various control agents for intelligent appliances. The GCAC is connected to several regional control agent dispatchers (RCAD) via wide area networks. Each RCAD maintains a control agent repository and is responsible for dispatching control agents to millions of intelligent appliances at thousands of households in a region. At each household, a gateway is responsible for downloading control agents from a wide area network into in-

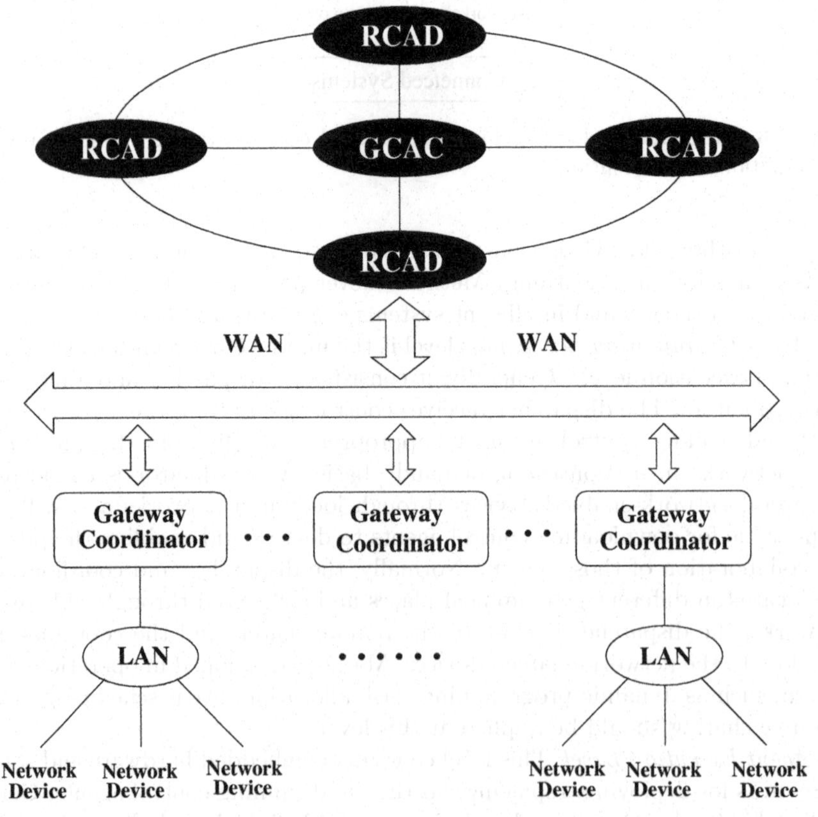

Fig. 2.4. Architecture of agent-based control for networked devices and systems

telligent appliances through a local area network, and uploading information or requests from intelligent appliances to the RCAD. Each household also has a control agent coordinator for conducting cooperative controls among intelligent appliances or even within a single appliance [58].

This mechanism will enable us to develop region-specific, household-specific, weather-specific, or application-specific intelligent appliances from initially mass-produced uniform appliances, for the purpose of energy efficiency or other high performance considerations. For example, the same appliances at neighboring houses may behave quite differently later since local environments and user habits are different and therefore very different control agents might be utilized by the neighboring houses eventually.

2.4 Design Principle for Networked Control Systems: Local Simple, Remote Complex (LSRC)

The smartness, efficiency, and performance of control agents can be greatly enhanced by the design principle for networked control systems: local simple and remote complex (LSRC). The idea is straightforward: find a way to represent a control agent in two different but somehow equivalent forms, one is simple and can be implemented by local devices at low cost with limited memory and computing power, and another is complex, of high cost and can be used at remote sites with complex machines and sophisticated algorithms. In this way, through the connectivity provided by networks, a control agent of simple form working in a local device can travel to some remote site with related data or experience after a period of task execution, and then transform itself into its equivalent complex form and start to learn or improve its task skills by using remote complex algorithms and the powerful computing facility there. After learning and performance enhancement, the control agent transforms itself back into its simple form suitable for low cost implementation and then travels back to the local device as a renewed agent.

This is very similar to the process of training human subjects, such as local factory process or machine operators, at some off site or remote schools for improved ability, enhanced skill, and better performance.

How to achieve this goal and implement the design principle of "local simply, remote complex" for networked control systems? A simple but effective way of doing this is to use the concept and algorithms of neuro-fuzzy networks developed by our group in the early 1990s [15]–[17]. Fig. 2.5 presents the structure of neuro-fuzzy networks consisting of three types of subnets: pattern recognition networks, fuzzy reasoning networks, and control synthesis networks.

A neuro-fuzzy network is constructed directly from the set of decision rules used in fuzzy control systems. The uniqueness of these kinds of networks is that they are equivalent to their original fuzzy control systems from which

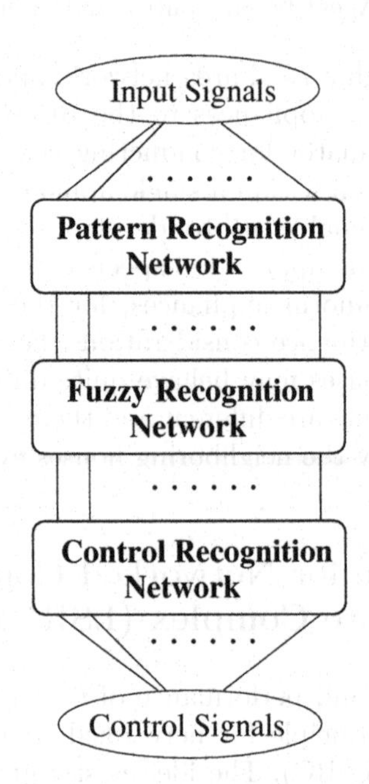

Fig. 2.5. Structure of neuro-fuzzy networks

they are built. In other words, one is able to recover the rule structure, membership functions, logic operators for fuzzy reasoning used in the original fuzzy systems. However, the complexity of the constructed neuro-fuzzy networks is much higher than that of the source fuzzy control systems. Since it is much easier and more cost effective to implement a set of fuzzy rules than a set of sub neural networks, we can use simple hardware to realize fuzzy control systems at local devices or systems, and complex machines to achieve corresponding neuro-fuzzy networks. This provides a mechanism for the realization of "local simple, remote complex" design principle for agent-based control of networked devices and systems.

Many methods of combining neural networks and fuzzy logic have been proposed in the last two decades, but our neuro-fuzzy networks are different from others in their preservation of original structures, procedures, and parameters of fuzzy logic during the combination and transformation. In our approach, we can go from fuzzy logic to neural networks and vice versa, instead of one way to neural networks with no return to fuzzy logic as in many other methods. The price for this equivalence is paid in terms of the extra layers in the network structure, the extra number of nodes in network processing, and the additional complexity of corresponding learning algorithms. However, those extras are justified by the availability of powerful computing capacity at the remote sites. Therefore, we can achieve high performance and intelligent networked devices and systems with no additional cost to users and reasonable extra burden to manufacturers or service providers. Note that extra cost might be easily recovered by manufactures or service providers through some value-added services. Actually, this could become a potential avenue for new profits.

The original motivation for our neuro-fuzzy networks was to construct neural networks using linguistic knowledge instead of digital information, and build knowledge structure into artificial neural networks so those networks are no longer like black boxes but more like biological neural networks where different regions have different processing capacities and distinctive functions [15]–[17].

In a sense, the function of rule-based fuzzy control systems is similar to that of providing lectures, manual, or verbal instructions to train new operators for assembly lines or machining processes. This will ensure that operators know how to conduct their job safely and with limited performance. To improve their skills and achieve better performance, operators must further learn with their brains, i.e., their biological neural networks, based on their job experience and knowledge bases. Clearly, the function of this human learning process is similar to that of our neuro-fuzzy networks.

This comparison offers us the confidence to apply neuro-fuzzy networks to adaptive and intelligent controls. For traditional adaptive control, a problem or concern always exists regarding the beginning and transient period of the execution. In other words, how can we jump start the adaptive control mechanism without the risk of breaking the system or deteriorating its performance significantly in the beginning or during the transition? For a neuro-fuzzy network that can easily be constructed from *ad hoc* and heuristic knowledge in linguistic forms without invoking analytic models (which are normally unknown and need to be learnt during the process of adaptation), its initial performance is safe but not optimal since its behaviors are almost identical to that of the original rule-based fuzzy control system. Therefore, we do not need to worry about the safety of using neuro-fuzzy networks for adaptive control or other kinds of intelligent control. Clearly, this feature of neuro-fuzzy networks can be extended to control agents in their implementation so that their initial behaviors are not optimal but safe and reasonable, thus providing a time period so that learning and performance improvement can be conducted as data and experiences are gained during the process of their task execution.

2.5 Modular Construction and Learning Algorithms of Neuro-fuzzy Networks for LSRC Implementation

This section describes briefly network construction and learning algorithms of neuro-fuzzy networks for "local simple, remote complex" agent-based control. Most of the materials are from our previous work presented in [17].

Procedure of Fuzzy Logic Control Systems

We will use a fuzzy logic based control system (FLCS) to represent the decision-making mechanism for a control agent. As in Fig. 2.6, an FLCS consists of five major parts: a set of fuzzy linguistic variables that describe signal patterns and control actions, a mechanism that associates crisp values of signals to linguistic variables (fuzzification), a knowledge base that specifies decision rules in terms of linguistic variables (IF-THEN rules), an inference engine for fuzzy reasoning that determines fuzzy control actions based on the knowledge base, and an algorithm that converts a fuzzy control action into a control action of crisp value (defuzzification). The detailed procedure of conducting fuzzy logic based controls is described as follows.

Consider a process monitored through a signal vector s with m readings,

$$s = (s_1, s_2, \ldots, s_m),$$

and driven by a control vector u with n components,

$$u = (u_1, u_2, \ldots, u_n).$$

Each of the sensor readings and control components is described by a set of linguistic terms, namely,

$$A_i = \{S_i^1, S_i^2, \ldots, S_i^{p_i}\} \text{ and } B_j = \{U_j^1, U_j^2, \ldots, U_j^{q_j}\}$$

for s_j and u_j, $i = 1, \ldots, m$, $j = 1, \ldots, n$, respectively. Association of a particular value x of s_i or u_j with a linguistic term Z in A_i or B_j is characterized by the concept of membership function in fuzzy set theory, denoted by $\mu_z(x)$, where $\mu_z(x)$: $X \to [0, 1]$ and X is the universe of discourse of s_i or u_j. The

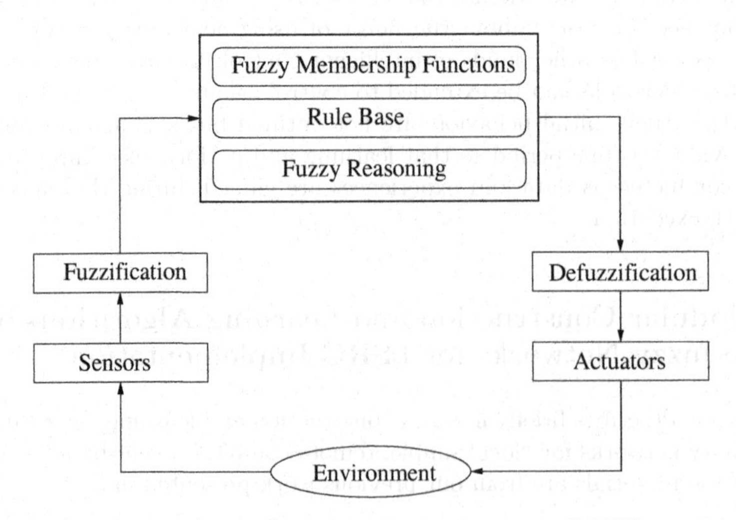

Fig. 2.6. Fuzzy logic based control systems (FLCS)

knowledge base of the FLCS consists of a number of linguistic IF-THEN decision rules. Assume that the knowledge base has R control rules,

$$\begin{aligned}
\text{Rule } 1:\ & \text{IF } s_1 \text{ is } S_1^{k_{11}} \text{ and} \dots \text{and } s_m \text{ is } S_m^{k_{m1}}, \\
& \text{THEN } u_1 \text{ is } U_1^{c_{11}} \text{ and} \dots \text{and } u_n \text{ is } U_n^{c_{n1}}, \\
& \vdots \\
\text{Rule } R:\ & \text{IF } s_1 \text{ is } S_1^{k_{1R}} \text{ and} \dots \text{and } s_m \text{ is } S_m^{k_{mR}}, \\
& \text{THEN } u_1 \text{ is } U_1^{c_{1R}} \text{ and} \dots \text{and } u_n \text{ is } U_n^{c_{nR}},
\end{aligned}$$

where $S_i^{k_{ir}} \in A_i$ and $U_j^{c_{jr}} \in B_j$ for $r = 1, \dots, R$, $i = 1, \dots, m$ and $j = 1, \dots, n$.

The preconditions of rule r form a cross product of fuzzy sets $S_1^{k_{1r}} \times \dots \times S_m^{k_{mr}}$ and the corresponding consequence is the union of n independent fuzzy sets $U_1^{c_{1r}} + \dots + U_n^{c_{nr}}$. Thus the rule can be represented as a fuzzy implication:

$$\text{Rule } r:\ S_1^{k_{1r}} \times \dots \times S_m^{k_{mr}} \rightarrow U_1^{c_{1r}} + \dots + U_n^{c_{nr}}.$$

Given a specific sensor reading $s = (s_1, s_2, \dots, s_m)$, the function of the inference engine of the FLCS is to match the preconditions of R control rules with the sensor reading and conduct fuzzy implication. Many approximate reasoning methods have been developed for fuzzy inference (Mizumoto, 1988; Lee, 1990). One of the most popular strategies is the maxmin compositional rule of inference. The inference procedure of this method can be summarized as follows.

- *Firing Strength Calculation:* The strength of firing rule r is calculated as

$$\alpha_r = \mu_{S_1^{k_{1r}}}(S_1) \wedge \mu_{S_2^{k_{2r}}}(S_2) \wedge \dots \wedge \mu_{S_m^{k_{mr}}}(S_m), \quad r = 1, \dots, R, \tag{2.1}$$

  where $\wedge$ is the conjunction (and) operator.
- *Rule Consequence Deduction:* The consequence of rule r for the jth control component is determined by

$$\mu_{C_j^r}(u_j) = a_r \wedge \mu_{U_j^{c_{jr}}}(u_j), \quad j = 1, \dots, n, \quad r = 1, \dots, R, \tag{2.2}$$

  where C_j^r is the fuzzy set for the jth control component according to rule r.
- *Resultant Control Generation:* The resultant fuzzy set C_j for the jth control component is deduced from

$$\mu_{C_j}(u_j) = \mu_{c_j^1}(u_j) \vee \mu_{c_j^2}(u_j) \vee \dots \vee \mu_{c_j^R}(u_j), \quad j = 1, \dots, n, \tag{2.3}$$

  where $\vee$ is the disjunction (or) operator.

There are several different ways to perform the fuzzy implication by using different operators for conjunction and disjunction operations.

The inference process produces a fuzzy set for each of the control components. Because a physical system must be driven by a non-fuzzy control input, a method of defuzzificaton is needed to generate a crisp control value that best represents the membership function of a fuzzy control action. Several procedures have been proposed for tackling this problem. Among them, center-of-area (COA) and weighted combination (WC) procedures are two of the most commonly used. According to these two methods, the fuzzy control C_j is defuzzified to a crisp value using the following methods.

- *Center-of-Area Defuzzification:* The strength of firing rule r is calculated as

$$u_j^* = \frac{\sum_{i=1}^{n_j} u_{ji}\mu_{C_j}(u_{ji})}{\sum_{i=1}^{n_j} \mu_{C_j}(u_{ji})}, \quad j = 1, \ldots, n, \tag{2.4}$$

where $U_j = \{u_{j1}, \ldots, u_{jn_j}\}$ is the discrete universe of discourse of u_j.

- *Weighted Combination Defuzzification:*

$$u_j^* = \sum_{r=1}^{B} \omega_{rj} u_{rj}^*, \quad j = 1, \ldots, n, \tag{2.5}$$

where

$$\omega_{rj} = \frac{\sum_{i=1}^{n_j} \mu_{C_j^r}(u_{ji})}{\sum_{r=1}^{R} \sum_{i=1}^{n_j} \mu_{C_j^r}(u_{ji})}, \quad u_{rj}^* = \frac{\sum_{i=1}^{n_j} u_{ji}\mu_{C_j^r}(u_{ji})}{\sum_{i=1}^{n_j} \mu_{C_j^r}(u_{ji})}, \tag{2.6}$$

respectively. Note that in the WC defuzzification, there is no need to calculate the fuzzy resultant control, because the WC procedure has been carried out for each individual rule and the final crisp control is a combination of individual defuzzifications weighted by the areas covered by the corresponding membership functions.

To be specific, consider the popular cart-pole problem of controlling the motion of the cart along a horizontal line so that the pole does not fall down and eventually stands vertically; see Fig. 2.7. An Internet-based experimental setup controlling this system had been developed in the middle of 1990s [68], and now similar systems are widely used for lab experiments of networked control systems over the world. For the purpose of control, a 9-rule fuzzy control system is used. Letting $s_1 = \theta$, $s_2 = \dot{\theta}$ and u be the control, these control rules are,

Rule 1: If s_1 is PO and s_2 is PO then u is PL;
Rule 2: If s_1 is PO and s_2 is ZE then u is PM;
Rule 3: If s_1 is PO and s_2 is NE then u is ZE;
Rule 4: If s_1 is ZE and s_2 is PO then u is PS;
Rule 5: If s_1 is ZE and s_2 is ZE then u is ZE;
Rule 6: If s_1 is ZE and s_2 is NE then u is NS;
Rule 7: If s_1 is NE and s_2 is PO then u is ZE;

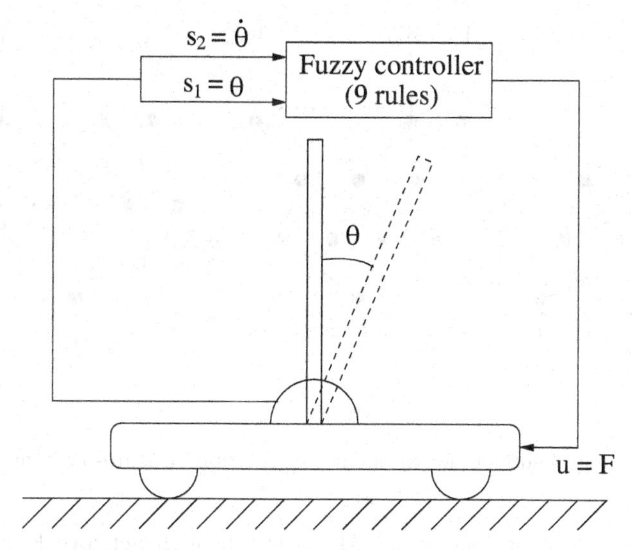

Fig. 2.7. Fuzzy logic based control for cart-pole problem

Rule 8: If s_1 is NE and s_2 is ZE then u is NM;
Rule 9: If s_1 is NE and s_2 is NE then u is NL.

Here we have $A_1 = A_2 = \{PO, ZE, NE\}$ and $B_1 = \{PL, PM, PS, ZE, NL, NM, NS\}$.

Modular Construction of Neuro-fuzzy Networks

As indicated in Fig. 2.5, the structure of the neural network implementation of FLCS consists of three modular subnets of distinctive functions.

The first network identifies patterns of input variables in terms of membership functions of linguistic terms; the second one conducts fuzzy reasoning (conjunction) by calculating the strength of firing each of the decision rules; and the third carries out the task of control synthesis by generating fuzzy control action and then defuzzifying it. Although the three neural networks are connected sequentially, it is important to point out that the construction and training of these networks can be performed independently and simultaneously, and the decision-making procedure in an FLCS is fully preserved in its network implementation. This will be clear from the following description.

Neural Subnets for Pattern Recognition. For each signal reading s_i, a neural network SN_i is constructed to match its values with the linguistic terms in the set of signal patterns A_i. In other words, the function of SN_i is to calculate membership functions $\mu^k_{s_i}(x)$ for $k = 1, \ldots, p_i$, $i = 1, \ldots, m$. Fig. 2.8(a) shows a three-layer SN_i for this purpose. Initially, this network is trained with the specified membership functions for terms in A_i. At this stage, network SN_i is not required to learn the memberships very accurately because these specified membership functions are usually very subjective. Note that if two sensor readings have an identical set of linguistic terms, they can use

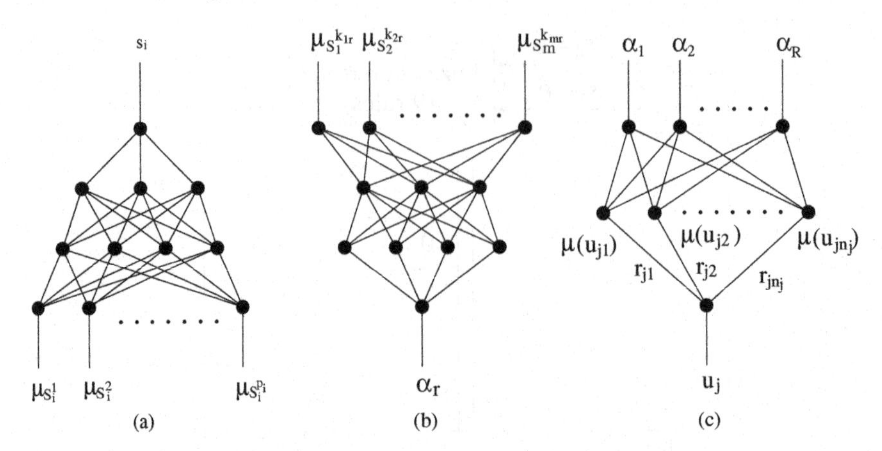

Fig. 2.8. Neural subnets for modular construction of neuro-fuzzy networks

the same subnet at the beginning. However, through network learning, the membership functions of linguistic terms can be changed adaptively later for better performance.

Neural Subnets for Fuzzy Reasoning. For each decision rule r in the knowledge base of an FLCS, a subnet RN_r, $r = 1, \ldots, R$, is used to calculate the firing strength of the rule. Thus, RN_r is actually a network implementation of the conjunction operator. Fig. 2.8(b) presents a three-layer RN_r. By changing its weights, this network could implement $S1$, $S2$, $S3$, or other triangular norms. Therefore the initial training of RN_r can be carried out by using any of these norms, or even their combinations, and the network can easily be modified for new fuzzy reasoning by the use of learning algorithms. Clearly, as long as every rule has the same number of linguistic terms in its precondition, we can choose the same fuzzy reasoning network for all the control rules at the initial stage. Note no input or output scaling is needed for network RN_r because both its input and output range from 0 to 1.

Neural Subnets for Control Synthesis. Control synthesis is the process of determining the final crisp control according to the firing strengths of rules and membership functions of linguistic terms defined for control actions. It involves steps of deducing consequences for individual rules, generating resultant fuzzy control, and then converting it into a crisp value. There are two ways to use neural networks to conduct control synthesis. The first one is to develop an individual network for each control component, and the second is to build a single network for all components. However, the actual construction of control synthesis networks is not trivial. There are several different implementation schemes. The key issue in the network construction is how to recover completely the membership functions of fuzzy control actions from a network implementation. Here we present only one of the design paradigms.

Fig. 2.8(c) illustrates a two-layer neural network CN_j for the synthesis of control component u_j, $j = 1, \ldots, n$. The first layer is introduced for calculating fuzzy controls. In this layer, a neuron is created for each of the elements in the universe of discourse U_j. Given the firing strengths of control rules, neuron k produces the value of the membership function of the resultant control at a specific u_{jk}, $1 \leq k \leq n_j$. Note that we have used the $S2$ operator for conjunction in rule consequence deduction and $T3$ for disjunction in the resultant fuzzy controls [17]. Because the logic operations have been fixed, there is no need for initial network training. The initial weights of network CN_j at this layer can be calculated from membership functions as

$$w_{jkr} = \mu_{U_j^{c_{jr}}}(u_{jk}), \quad k = 1, \ldots, n_j, \tag{2.7}$$

while disjunction operator $T3$ is implemented by a linear activation function $f(x) = x$, if $0 \leq x \leq 1$; $f(x) = 0$, if $x < 0$; and $f(x) = 1$, if $x > 1$ [17].

The second layer carries out the task of defuzzification. Initial values of weights in this layer depend on the defuzzification algorithm selected. However, if we assume $\sum_{r=1}^{R} \mu_{C_j^r}(u_{jk}) \leq 1$ then COA and WC defuzzifications will give us the same result. Therefore, we will consider only COA defuzzification. In this case, weights of the second layer are given by

$$\gamma_{jk} = u_{jk}/\sigma_j, \quad \sigma_j = \sum_{k=1}^{n_j} \mu_{jk}, \quad u_j = \sum_{k=1}^{n_j} \gamma_{jk}\mu_{jk}, \quad j = 1, \ldots, n, \tag{2.8}$$

where μ_{jk} is the value of neuron k at the first layer. These initial weights of network CN_j, $j = 1, \ldots, n$, can be changed later by learning algorithms to improve control performance. However, learning will only change the membership functions of control actions and the defuzzification algorithm, not the logic operations involved in control synthesis.

Integration of Neural Subnets: Neuro-fuzzy Networks. Once modular subnets SN_i, RN_r and CN_j have been created, the final step toward a structured neuro-fuzzy network is to connect those networks appropriately according to the original FLCS. Fig. 2.9 presents the neuro-fuzzy network for the 9-rule fuzzy controller described in the previous section. From its construction, it is very clear that the whole computation process of the integrated neural network can be divided into three stages: pattern recognition, fuzzy reasoning, and control synthesis (Fig. 2.5), resulting in a functional interpretation for a subnet and a structure of flow of knowledge thereof. Although the structured neural network, designated as FCN, can be viewed as an ordinary multilayer network with sensor readings as its input and control actions as its output, the distinctive knowledge structure embedded within this network makes it different from other neural network implementations of fuzzy logic controls. Note that as a neural network, FCN is not a fully connected one. For example, its subnet SN_i is linked only to input node s_i and has no connection with other input nodes (i.e., connection weights are zero), and subnet RN_r is linked only to one output node of subnet SN_i.

As long as the original FLCS works reasonably well, no additional training is required to put the structured neural network FCN to work. Because its

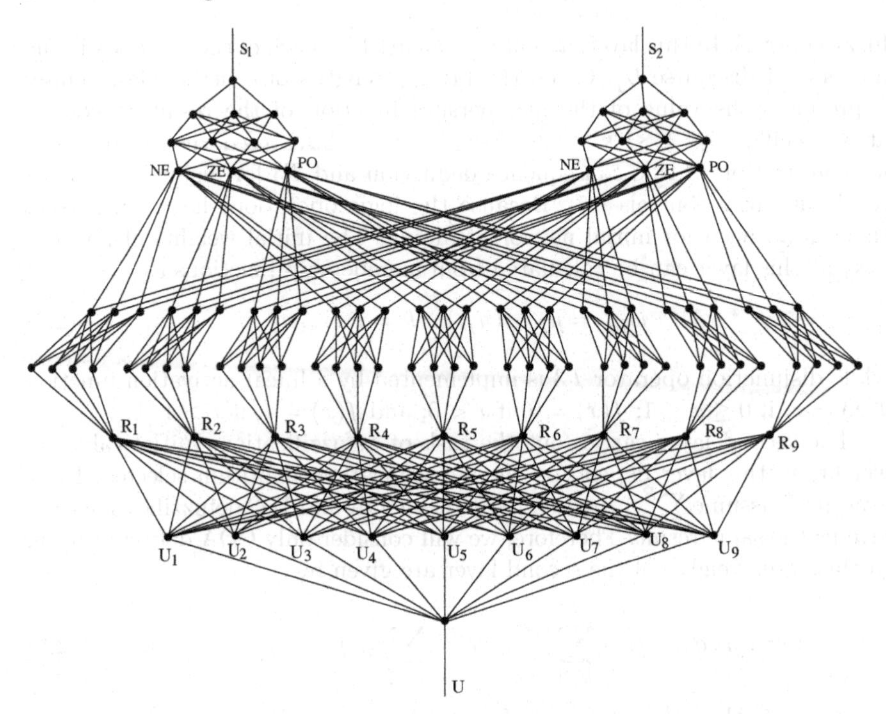

Fig. 2.9. Neural subnets for modular construction of neuro-fuzzy networks

performance should be at least as good as the original FLCS from which the network is constructed. Additional trainings and on-line learning by the network, however, can improve its performance and thus lead to refinement of the existing control rules and even generation of new rules for the original FLCS. This is discussed in detail in the following section.

Rule Refinement by Back Propagation Training

Because the whole decision-making procedure of the FLCS is preserved in network FCN, by breaking it up into subnets of pattern recognition, fuzzy reasoning and control synthesis, the actual modification in membership functions of sensor readings, operators of fuzzy reasoning, membership functions of control actions, and methods of defuzzification can be recovered separately from FCN. Therefore, the neural network implementation provides a mechanism for the refinement of fuzzy logic based control systems, a problem that has not been addressed effectively within the original context of fuzzy logic controls. Furthermore, by augmenting the network with neurons of zero initial weights, new control rules can be produced through network training or learning.

The back propagation learning algorithm developed for standard multilayer feedforward neural networks can easily be generalized to neuro-fuzzy network FCN. As for multilayer neural networks, this will enable FCN to tune its weights to match a given set of optimal input/output pairs. To find the training algorithms for FCN, we define the error function as

$$e = \frac{1}{2} \sum_{j=1}^{n} \left(u_j^d - u_j\right)^2 \tag{2.9}$$

where u_j^d is the jth component of a desired control. If we use (2.8) only for the calculation of initial values of weights γ_{jk} the rules for updating the weights of FCN are exactly the same as those for the standard multilayer neural networks, except that weights between any two neurons with no direct connection are always treated as zero. This will change the defuzzification algorithm through training. On the other hand, if (2.8) is considered as the definition of γ_{jk}, which implies no change in the defuzzification algorithm, then the rule for updating weight w_{jkr} in the first layer of CN_j has to be modified as

$$w_{jkr}(t+1) = w_{jkr}(t) + \eta\alpha_r(u_j^d - u_j)\left(\gamma_{jk} - \sum_{l=1}^{n_j} \mu_{jl}u_{jl}/\sigma_j^2\right) f'(\mu_{jk}) \tag{2.10}$$

where $f'(x) = 1$ when $0 < x < 1$, $f'(x) = 0$ otherwise; $0 < \eta < 1$ is the learning rate; and t represents the number of iteration steps in training. This result can easily be obtained from gradient calculation. In this case, error term δ_r backpropagated to the output neuron of RN_r is found to be

$$\begin{aligned} \delta_r &= \sum_{j=1}^{n} \left(u_j^d - u_j\right) \frac{\partial u_j}{\partial \alpha_r}, \\ \frac{\partial u_j}{\partial \alpha_r} &= \sum_{k=1}^{n_j} w_{jkr}\left(\gamma_{jk} - \sum_{l=1}^{n_j} \mu_{jl}u_{jl}/\sigma_j^2\right) f'(\mu_{jk}). \end{aligned} \tag{2.11}$$

After training has been completed, we can recover membership functions and fuzzy conjunction operators by breaking up FCN into subnets of pattern recognition, fuzzy reasoning, and control synthesis. Specifically, from SN_i we get the refined membership functions of signal patterns for s_i, and from RN_r the modified conjunction operator for rule r. Note that, after training, different control rules would have different conjunction operators. To obtain the updated membership functions of control actions from u_j, we need to set only one input neuron, say, α_j, of CN_j to 1, and all others to zero. In this way, output values of neurons in the first layer of CN_j present the new membership function for control term $U_j^{C_{jr}}$. Like conjunction operators, a fuzzy control action employed by two or more control rules could have different membership functions in different rules after training.

Rule Generation by Network Augmentation

In many cases, the performance of an FLCS cannot be improved further unless new control rules are added. Using network FCN, we can generate new rules automatically by augmenting FCN with neurons of zero initial weights. As illustrated in Fig. 2.10, for a single new rule, this can be achieved by adding an output neuron S_i^{new} to each subnet SN_i, and a new fuzzy reasoning subnet RN_{new} that takes S_i^{new} as its input neurons and a α_{new} as its output neuron. These new neurons represent a new control rule in the form of

$$\begin{aligned}&\text{IF } s_1 \text{ is } S_1^{\text{new}} \text{ and} \dots \text{and } s_m \text{ is } S_m^{\text{new}}\\ &\text{THEN } u_1 \text{ is } U_1^{\text{new}} \text{ and} \dots \text{and } u_n \text{ is } U_n^{\text{new}}\end{aligned}$$

where S_i^{new} and U_j^{new} are new signal and control terms with unknown membership functions. Initially, all weights associated with these neurons are zero or very small random numbers. Therefore, the new control rule has no or very little effect in the computing process of the augmented network FCN. However, after training the augmented FCN with the algorithm described above, weights associated with these neurons would take certain non-zero values. Membership functions of the new signal patterns now can be obtained from the output of neurons S_i^{new}, $i = 1, \dots, m$, while the conjunction operator of the new rule from the output of neuron α_{new}. The procedure to get membership functions of the new fuzzy control actions U_j^{new}, $j = 1, \dots, n$, is the same as the one used for the existing control terms. Clearly, multiple new rules can be generated in the same way.

As proposed in [17], the generation of new rules can be carried out either globally or locally. The global scheme is to update all weights of the augmented network FCN simultaneously according to the training algorithm, while in the local scheme, only the weights associated with the new neurons are modified and all other weights remain unchanged. Because new weights can be viewed as additional optimization variables, it is expected that the augmented network will have better control performance. Although the global scheme would have better results than the local one eventually, it may cause serious performance

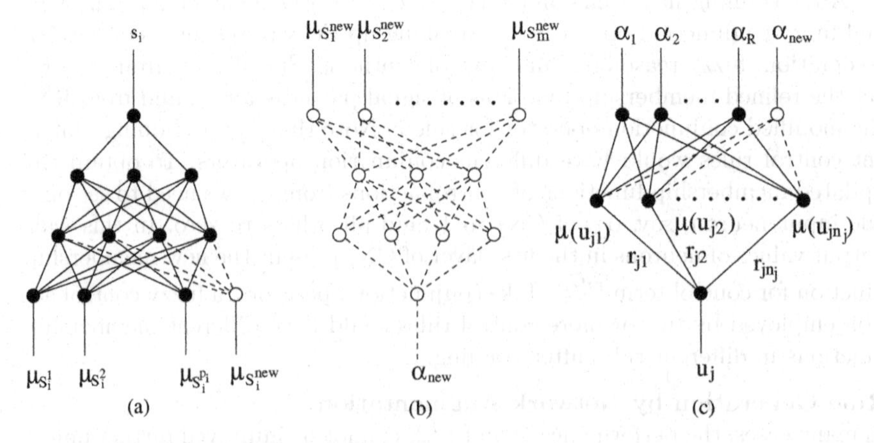

Fig. 2.10. Neural subnets for rule generation by network augmentation

degradation during the initial training stage because it starts the training process by driving the network away from a locally optimal stage. (Assume that the original network has obtained at least locally optimal weights.) On the other hand, the local learning method can at best achieve a locally optimal performance for the augmented network. However, this method can result in a smooth improvement during the training process without causing dramatic changes in control performance. A preferred strategy is to combine the global and local schemes, i.e., starting with the local one in the initial stage and then switching to the global one gradually. A simple implementation of this strategy can be accomplished by

$$w(t+1) = w(t) + \alpha(t)\eta\Delta w(t) \tag{2.12}$$

where w represents any weight that is not associated with the new neurons S_i^{new} and α is the learning rate, and

$$\alpha(t) = \begin{cases} 0, & t \leq T_L, \\ 1 - e^{-\beta(t-T_L)}, & t > T_L, \end{cases}$$

where T_L is the local learning period and β is the rate of introducing the global scheme.

Real-time Learning from On-line Task Executions

A more realistic way to achieve real-time on-line learning for neuro-fuzzy network FCN is through learning after each task execution. After all, learning is only meaningful when a task is repetitive. This is especially appropriate for FCN because it is expected that the network has a reasonable performance without any training other than learning membership functions and conjunction operators by small subnets at the very beginning.

We propose a learning scheme based on the idea of self-learning control systems developed in automatic control [17]. To this end, we assume that the process to be controlled has dynamics in the form

$$s_{i+1} = P(s(i), u(i)), \quad i = 0, 1, \ldots, K, \tag{2.13}$$

during each task execution, where P represents an unknown dynamic equation and control is determined by the neuro-fuzzy network $u(i) = \text{FCN}(s(i))$. The objective of process control is to minimize the final error,

$$e = \frac{1}{2}\|s_d - s(K)\|^2,$$

where s_d is the desired state of the process after K steps of execution.

Because the plant dynamics is unknown, a separate neural network NE, called the plant emulator, is trained to behave like the plant before we conduct learning after task execution for FCN. Training the emulator NE is similar to plant identification in control theory, except that the plant identification here is done automatically by the neural network capable of modeling nonlinear plants. Once the emulator NE is trained to match the plant dynamics P closely, we use it for the purpose of learning from on-line task executions for network FCN. Specifically we have

$$w(t+1) = w(t) - \eta \frac{\partial e}{\partial w} = w(t) + \eta (s_d - s(K)) \frac{\partial s(K)}{\partial w} \tag{2.14}$$

where w is any weight parameter of FCN, and gradient vectors are calculated recursively by the following equations:

$$\frac{\partial s(i)}{\partial w} = \frac{\partial \mathrm{NE}}{\partial s} \frac{\partial s(i-1)}{\partial w} + \frac{\partial \mathrm{NE}}{\partial u} \frac{\partial \mathrm{FCN}}{\partial w} \bigg|_{s=s(i-1), u=u(i-1)}, \quad \frac{\partial s(0)}{\partial w} = 0.$$

Therefore, gradients to be calculated from neural networks are derivatives of output with respect to input ($\partial \mathrm{NE}/\partial s$ and $\partial \mathrm{NE}/\partial u$) and weight ($\partial \mathrm{FCN}/\partial w$), respectively. The first gradient can be calculated very easily for any specific network, while the second one can be found based on (2.10) and (2.11).

As discussed in the previous subsection, automatic rule generation can also be implemented here through network augmentation by learning from on-line task execution.

Note that both modular subnets and learning algorithms are some primitive version of more efficient construction and algorithms for real applications. Based on different task requirements and performance demands, we can construct different neuro-fuzzy networks for fuzzy logic based control systems at different levels of equivalence. Corresponding learning algorithms can be developed accordingly [25].

2.6 Issues in Software, Middleware, and Hardware Platforms

Software Platforms

The system architecture for agent-based control consists of three layers that are Global Control Agent Center (GCAC), Regional Control Agent Dispatcher (RCAD), and local Gateway and Coordinator (GC). The GCAC coordinates RCADs to achieve global control agent management and to optimize the entire system performance. To address the control agent management tasks, GCAC features the following functions.

Control Agent Naming Service keeps track of control agent location to facilitate inter-agent communication.

Control Agent Factory designs and creates control agents to implement dynamic control intelligence.

Control Agent School trains and improves the performance of control agents based on their specific task experiences, task requirements, and task environments.

Control Agent Repository stores and maintains control agents.

Performance Management monitors control agents and system runtime situations and gauges their performances.

Task Dependency analyzes and decomposes the dependency relationships existing among various control and service tasks as well provides basis for failure analysis and recovery.

Resource Allocation manages system resource reservation and allocation mechanisms and implementation.

Interoperability and Legacy manages, facilitates and integrates existing out-of-date or heterogeneous control systems.

Quality of Service builds up negotiation mechanisms to guarantee soft real-time requirements, and looks up failure causes and corresponding recovery schemes based on service dependency analysis, failure recovery managements.

Security Service provides mobile agent mutual-trust mechanisms which are the base of secured agent-based control systems.

System Management provides the general system management services.

Regional Control Agent Dispatcher coordinates the regional network communications and implements control agents caching as well provide regional control agents coordination functionality.

Local Gateway and Coordinator (LGC) functions as the single integration-point for a range of agent-based services in local network environment. It works not only as a network communication gateway but also the coordinator of local control agents.

Due to the limitations of memory space and computing ability imposed on practical embedded devices, it is infeasible to make them powerful enough to host all possible control algorithms or control agents for that matter. Doing this will increase both complexity and cost of embedded devices. The emergence of mobile agents provides a novel way to address these problems, since the mobile agent is a mobile object which can migrate from one network node to another one with its code, data, and thread.

In our software platform for Agent-based Distributed Control Systems (aDCS) [8], we combine mobile agents and CORBA to form a flexible system software platform to achieve the goal of "control-on-demand" by dispatching related agents to the controller according to particular demands of various control scenarios. In this software platform, required control algorithms or control parameters have been encapsulated into mobile agents that can migrate from remote control center to the field controller or from one field controller to another. This software platform benefits from the complementary properties of mobile agents and CORBA communication infrastructure. On one hand, the mobile agent expands the static object concept of CORBA, that is, the object can move from one node to another one until its task is completed. The "dynamical adaptability" of mobile agents enables them to sense their execution environments and react autonomously to changes. The asynchronous and autonomous execution of mobile agents makes them suitable to work in agile network environments. On the other hand, CORBA provides mobile agents with necessary services: Naming Service keeps track of the location of mobile agents; Lifecycle Service defines services and conventions for creating, deleting, copying and moving CORBA objects; Externalization Service provides a standardized mechanism for mobile agents migration during which the state of mobile agents can be recorded into or re-constructed from a network data stream; and Security Service to guarantee the execution of mobile agents. Furthermore, the adoption of CORBA as the underlining communication infrastructure facilitates the integration of heterogeneous control system and legacy control systems that could be non-mobile-agent-based or even non-object-oriented.

Middleware Platforms and OSGi

Middleware support and the corresponding operating and service protocols are critical for applications of agent-based control. Since the objective of agent-based control systems is to deploy and maintain control agents that will run network-enabled devices efficiently over the network, issues related to protocols of manufacturing, delivery, and maintenance of control agents become essential and must be addressed. Open Services Gateway Initiative (OSGi), an open specification for the delivery of multiple services over wide area networks to local networks and devices, provides a reasonable foundation for further investigation. In the architecture of the OSGi model, the concept of Services Gateways is proposed, which connect smart devices together and enabling them to seamlessly communicate with each other or to receive a variety of services from providers. A service gateway acts as a bridge between an external network, typically the Internet, and a local network consisting for example of Ethernet. It provides flexible interfaces and an environment for service providers to deploy and maintain services to users at the end of local networks. Therefore, the scenario considered by OSGi is almost identical to that of agent-based control systems, and OSGi architecture and modules can be directly applied to operate network-enabled devices with agent-based control systems.

Another important issue is where to locate embedded servers for network-enabled devices when dealing with more complicated tasks. To reduce additional cost, embedded servers generally reside in network-enabled devices with limited computing ability and resources. Based on such hardware conditions, some special requirements for embedded servers arise: small memory footprint, high performance, platform independence, scalability, dynamic code-loading capability etc. Now the approach to develop and deliver the software of embedded servers becomes a problem to be solved. Based on the requirement, Java or C++ will be the choices to construct embedded servers. To some extent, an embedded server can be considered as an agent to control "control agents." That means embedded servers will also be easily delivered like other control agents. Network-enabled devices can download and install control agents (including embedded servers), and remove them when they are no longer required. Like a control agent, embedded servers might migrate among network-enabled devices in the local network and get more flexibility and scalability.

There are still some issues related to embedded environment formed by network-enabled devices. First, embedded servers are intended to run on many kinds of network-enabled devices whose different hardware features will affect many aspects of control agent implementation. How can embedded servers repel such difference to provide a common view to manufacturer or service providers? A concise and consistent operation model provided by embedded servers will be essential. It will simplify the deployment of control agents and decouple the service's specification from its implementations. A consistent operation model helps control agent developers cope with scalability issue that is significant. Consistent interfaces insure that the software components can be mixed up and still result in a stable status. On the other hand, agent-based control systems will have multiple real-world applications. The model also guarantees application independency via defining common implementation APIs, making it suitable for embedded servers of a variety of applications in different markets.

Hardware Platforms

Development of Remote Configurable and Programmable Devices (RCPD) and real-time Application Specific Operating Systems (ASOS) should be the key issue for supporting hardware platforms of agent-based control. Programmable devices are a class of general-purpose chips that can be configured for a wide variety of applications. The Field Programmable Gate Array or FPGA, In-System Programmable or iSP, as they are more widely called, are types of programmable device. FPGA, iSP, or similar devices can be programmed an unlimited number of times. Their main features include the following.

Ease of Design: Hardware Description Language (HDL) is used in the design process, such as VHDL, Verilog, and ABEL.

Lower Development Cost: Since FPGAs and iSP are re-programmable, designers can very easily and inexpensively modify their designs and implementation with no penalty.

Reduced Board Area: FPGAs or iSP offers a high level of integration.

Reconfigurable: When FPGA or iSP is used in a Reconfigurable Computing Platform, it can be configured from within an application program and triggered from outside through a network.

All those features make this kind of hardware technique ideal for implementing agent-based control systems at the execution level. Over the last few years we have developed a hardware platform for constructing mobile agent-based Distributed Control Systems (aDCS), and developed applications in process control and home automation based on this platform.

ASOS is the heart of RCPDs and agent-based control in general [27]–[29]. A systematic approach to constructing ASOS for different applications is still under development. At this stage, our approach is first to build ASOS for a specific class of applications, such as vASOS for vehicular applications, hASOS for home automation, rASOS for robotic systems, tASOS for traffic control, iASOS for intelligent spaces, etc, and then to develop a common framework and methodology for general ASOS.

2.7 Real-world Applications

Over the last one and half decades, several industrial applications have been conducted based on the idea of ABC systems. However, at this early development stage of agent-based control systems and due to the current network speed constraint, ABC's use is limited to problems where real-time demands are important but easy to meet, and system operations allow frequent and long communication over networks. The followings are some real-world examples.

A. Web-based Physical Control Experimental Systems [68]
Students can design control agents based on different control algorithms for different operating situations, download them from a local web site to control remote physical control experimental systems, such as an inverted pendulum, and receive almost real-time data feedback and delayed visual feedback. An aDCS was constructed for this system. This has been conducted in our previous work with a US NSF funded project WAVES (Web-based Audio/Video Educational Systems) for combined research and curriculum development in the area of systems dynamics and automatic control.

B. Internet-based Home Automation Systems [57]–[67]
Household appliances such as refrigerators, air conditioners, washing machines, and cameras are controllable via the Internet for security, convenience, energy efficiency and cost management purposes. An integrated home server (iHS) and a home application-specific operating system (hASOS) were constructed in this work. This has been developed based on previous work on Internet refrigerators and air conditioners in a project funded by the Chinese appliance manufacturer Kelon Electrical Group.

C. Network-based Traffic Control Systems [51]–[56]
Intersection traffic light controllers are connected through networks to a traffic operation center (TOC) and different control agents are deployed to different local traffic light controllers from the traffic operation center according to the current traffic conditions and/or weather conditions. In this system, a traffic control application-specific operating system (tASOS) and an agent-based distributed and adaptive platform for transportation systems (aDAPTS) were developed. This had been conducted in previous work on intelligent intersection signal and ramp control systems for integrated urban freeways and surface streets, a US DOT funded research project over the last decade and a recent funded project by the Knowledge Innovation Program of the Chinese Academy of Sciences.

D. Wireless Connected Vehicle Electronic Systems [45]–[50]
Vehicles are connected through wireless communications to a vehicle operation center (VOC) and different control agents are sent to different vehicles by the vehicle operation center for safety warning, vehicle health monitoring, repairing and maintenance services, emergence help, entertainment purposes, and others. A vehicular application-specific operating system (vASOS) and several other agent-based control and management systems were built in this project. A prototype system was developed based on previous work on the Lunar/Martian Robotic Vehicle Prototype Project funded by US NASA, the AutoDig Systems for Wheel Loaders funded by Caterpillar, the autonomous and intelligent vehicle project VISTA funded by the Arizona DOT, the intelligent vehicle/highway system project funded by US DOT, and a key 863 project on software development funded by the MOST and the Shandong Province of China.

On the basis and experience from those projects, a recent project, called intelligent agent spaces or iaSpace, for developing intelligent space application-specific operating systems (iASOS) and related agent-based systems has been launched. The purpose of this project is to construct a common software/hardware platform for embedded applications encountered in intelligent space research and development, such as in fixed spaces for labs, homes, offices, and business settings, mobile spaces with vehicles, aircrafts, ships, trains, and space objects, and mixed spaces on intersections, airports, harbors, stations, and border crossings. Clearly, agent-based computing and control will be the key techniques for future intelligent spaces, especially intelligent transportation spaces.

There are also several other applications in process control and automation where the idea and concept of agent-based control were originated and formulated [8], [9], and its potentials and usefulness will be continuously tested and justified.

2.8 Concluding Remarks and Future Work

Our future work will be focused on developing a comprehensive framework for agent-based control systems based on the new and emerging gateway infrastructure that connects network-enabled devices and systems together and enables them to seamlessly communicate with each other or to receive a wide variety of services from providers via the Internet or other networks in a cost effective manner. This approach to control also has great potential in operations and management of general complex systems. To this end, one must address various important issues related to the theoretical foundation, software and hardware requirements and platforms for agent-based control systems. Specifically, efforts must be made in the following aspects.

A Theoretical Framework for Agent-based Control Systems. To address issues related to system architecture, operation flow, agent construction, communication protocol, cooperation strategy, agent assignment, task scheduling, learning mechanism, etc.

A Software Platform for Agent-based Distributed Control Systems. To address issues related to programming, mobility, optimization, and software environment of control agents in distributed environments over the Internet.

A Hardware Platform for Remote Programmable and Configurable Devices. To address issues related to the hosting, manipulation, and hardware environment of control agents in network-enabled devices and systems.

The theoretical framework can be established based on mobile agent theories developed in distributed systems, especially distributed AI methods such as behavior control or programming proposed for mobile robots, and intelligent control systems established for hierarchical systems.

The software and hardware platforms can be developed based on the service gateway infrastructure specified in OSGi, a first protocol designed specifically for delivering services over the Internet. They are the keys to the technical development and application of agent-based control systems. Since OSGi is a Java-based protocol and has specific requirements for accessing network-enabled devices, it can be used to implement control agents in distributed environments effectively and easily. In our previous work, the software and hardware platforms consist of two prototypes for an integrated gateway server that is OSGi-compatible and a gateway operating system based on real-time Linux. This will be the foundation for future development of software and hardware platforms for constructing agent-based control systems.

In conclusion, agent-based control provides a mechanism for a paradigm shift in intelligent automation of networked devices and systems. Furthermore, agent-based computing and control will be the foundation for developing next-generation control theory and applications in the age of a connected world.

Acknowledgements

I would like to express my deep appreciation to my former and current students and colleagues for their time and effort in working with me in this area for the past two decades, and my special thanks to the funding supports since 1988 from the NASA, NSF, Caterpillar, BHP Copper, ADOT, and DOT in USA, and Kelon, CASIC, Shandong Province, Chinese Academy of Sciences, the National Natural Science Foundation, and MOST of China.

Finally, this chapter is written based on several of my previous papers and completed on many journeys during my travels around the world. Although no new structures and algorithms have been proposed here and many recent related works by others are omitted, I still hope this chapter has made my vision of future agent-based control clear and attractive to other researchers.

References

1. Wiener N (1948) Cybernetics: or the control and communication in the animal and the machine. MIT Press, Cambridge, MA
2. Tsien HS (1954) Engineering cybernetics. McGraw-Hill, New York
3. Friedland B (1986) Control system design: an introduction to state-apace methods. McGraw-Hill, New York
4. Lewis F (1992) Applied optimal control and estimation. Prentice Hall, New Jersey
5. Wang FY, Saridis GN (1990) A coordination theory for intelligent machines. Automatica 26(5):833–844
6. Wang FY (1998) ABCS: agent-based control systems. SIE Working Paper, University of Arizona, Tucson, AZ
7. Wang FY (1998) Application of agent-based methods in intelligent control for complex systems. SIE Working Paper, University of Arizona, Tucson, AZ
8. Wang FY (1999) aDCS: agent-based distributed control systems. PARCS Technical Report, University of Arizona, Tucson, AZ
9. Wang FY (1999) Agent programming and technology and their applications in automation. PARCS Technical Report, University of Arizona, Tucson, AZ
10. Mo P (2003) An analytic model for agent systems with Petri nets. SIE Dept, University of Arizona, Tucson, AZ
11. Wang FY, Wang CH (2002) Suggestions and related analysis for some fundamental problems in networked control systems (in Chinese). Chinese J Automation 28:171–176
12. Wang FY, Wang CH (2003) Agent-based control systems for operation and management of intelligent network-enabled devices. In: Proc 2003 IEEE International Conference on Systems Man Cybernetics, Washington, DC
13. Wang FY (2004) Intelligent control and management of networked systems in a connected world (in Chinese). J Pattern Recognition and Artificial Intelligence 17(1):1–6
14. Wang FY (2008) Petri nets for modeling agent-based control systems. Westing Publishing Co (to be published)
15. Wang FY (1992) Building knowledge structure in neural nets using fuzzy logic. In: Jamshidi M (ed) Robotics and manufacturing: recent trends in research, education and applications. ASME, New York
16. Wang FY (1993) Adaptive design of fuzzy control systems using neural networks. In: Proc First Chinese World Congress on Intelligent Control and Intelligent Automation, Beijing, P. R. China
17. Wang FY, Kim HM (1995) Implementing adaptive fuzzy logic controllers with neural networks: a design paradigm. J Intelligent and Fuzzy Systems 3(2):165–180
18. Wang FY, Kim HM, Zhou MC (1994) Dynamic back propagation for neuro-fuzzy networks. In: Proc IEEE International Conference on Electronics and Information Technology, Beijing, P. R. China
19. Kim HM, Wang FY (1994) Design of adaptive neuro-fuzzy controllers. In: Proc IEEE Int Conf on Systems Man Cybernetics, San Antonio, TX
20. Wang FY, Chen D (1994) Learning laws for neural network implementation of fuzzy control systems. In: Proc IEEE Int Conf Systems Man Cybernetics, San Antonio, TX

21. Wang FY, Huang ZY, Chen DD, Lever P (1998) Refinement and generation of decision rules through training and augmentation of neural networks. International J Intelligent Control and Systems 2(3):329–360
22. Chen L, Wang FY (2003) A new neuro-fuzzy system and corresponding learning algorithm (in Chinese). J Pattern Recognition and Artificial Intelligence 16(2):178–184
23. Shan ZF, Kim HM, Wang FY (1996) Plant identification and performance optimization for neuro-fuzzy networks. In: Proc IEEE International Conference on Systems Man Cybernetics, Beijing, P. R. China
24. Ge D, Tsutsumi M, Wang FY (1997) On dynamic error back propagation for neuro-fuzzy networks. In: Fifth Int Conf Fuzzy Theory and Technology, Research Triangle Park, NC
25. Wang FY (2001) Equivalence at different levels: on transformation between linguistic representations and neural computing models and corresponding implementation and application in networked environments. PARCS Technical Report 06-01-2001, SIE Dept, University of Arizona, Tucson, AZ
26. Lin YT, Wang FY (2005) Modular structure of fuzzy system modeling using wavelet networks. In: Proc 2005 IEEE Networking Sensing and Control, Tucson, Arizona, AZ
27. Wang FY, Wu ZH (2000) ASOS: a developing trend of embedded operating systems. Computer World 11(45):B6–B14
28. Wang FY, Zhu L (2004) Component-based construction of embedded ASOS (in Chinese). J Computer Engineering 30(3):42–43
29. Sun Y, Wang FY, Wang ZX, Qiao X, Wang KF (2006) A scheduling algorithm for vehicular application specific embedded operating systems. In: Proc 2006 IEEE Systems Man Cybernetics, Taipei, Taiwan, 2535–2540
30. Wang FY, Lever P (1994) An intelligent robotic vehicle for lunar and martian resource assessment. In: Zheng YF (ed) Recent trends in mobile robots. World Scientific, River Edge, NJ
31. Lever P, Wang FY, Chen DQ (1994) A fuzzy control system for an automated mining excavator. In: Proc IEEE International Conference on Robotics and Automation, San Diego, CA
32. Lever P, Wang FY (1994) Using fuzzy behaviors for fuzzy-goal-directed excavation tasks. In: Jamshidi M, *et al.* (eds) Intelligent automation and soft computing: trends in research, development, and applications. TSI Press, Albuquerque, NM
33. Lever P, Wang FY, Shi X (1994) An intelligent task control system for dynamic mining environments. AIME Trans Mining, Metallurgy and Exploration 174(1):165–174
34. Lever P, Wang FY (1994) Intelligent excavation control for a lunar mining system. In: Proc 4th Int Conf Robotics for Challenging Environments, Albuquerque, NM
35. Wang FY, Marefat M, Schooley L (1994) An intelligent robotic vehicle for lunar/martian applications. In: Proc 4th Int Conf Engineering, Construction, and Operations in Space, Albuquerque, NM
36. Shi XB, Wang FY, Lever P (1995) Task and behavior formulations for robotic rock excavation. In: Proc IEEE Int Symp Intelligent Control, Monterrey, CA
37. Lever P, Wang FY (1995) An intelligent excavator control system for a lunar mining system. ASCE J Aerospace Engineering 8(1)16–24

38. Lever P, Wang FY, Chen D, Shi X (1995) Autonomous robotic mining excavation using fuzzy logic and neural networks. J Intelligent and Fuzzy Systems 3(1)31–42
39. Shi XB, Wang FY, Lever PJA (1996) Experimental results of robotic excavation using fuzzy behavior control. Control Engineering Practice 4(2):145–152
40. Shi XB, Lever PJA, Wang FY (1996) Experimental robotic excavation with fuzzy logic and neural networks. In: Proc IEEE Int Conf Robotics and Automation, Minneapolis, MN
41. Shi XB, Lever PJA, Wang FY (1996) Fuzzy behavior integration and action fusion for robotic excavation. IEEE Trans Industrial Electronics 43(3):395–402
42. Shi XB, Lever P, Wang FY (1998) Autonomous rock excavation: intelligent control techniques and experimentation. World Scientific, River Edge, NJ
43. Wang FY, Lever PJ (1998) Advances in robotics and automation for hazardous environment. World Scientific, Singapore
44. Wang FY (2004) Agent-based control for fuzzy behavior programming in robotic excavation. IEEE Trans Fuzzy Systems 12(4):540–548
45. Wu L, Xu Y, Wang FY, *et al.* (1999) Supervised learning of longitudinal driving behavior for intelligent vehicles using neuro-fuzzy networks: initial experimental results. Int J Intelligent Control and Systems 3(4):443–464
46. Lin YT, Wang FY, Mirchandani PB, *et al.* (2001) Implementing adaptive driving systems for intelligent vehicles using neuro-fuzzy networks. Transportation Research Record 1774:98–105
47. Li YT, Wang FY, He F, Li ZJ (2005) OSGi-based service gateway architecture for intelligent automobiles. In: Proc 2005 IEEE Intelligent Vehicles Symposium, Las Vegas, NV
48. Sun Y, Wang FY (2005) A design architecture for OSEK/VDX-based vehicular application specific embedded operating systems. In: Proc IEEE Intelligent Vehicles Symposium, Las Vegas, NV
49. Wang FY (2005) Agent-based control strategies for smart and safe vehicles. In: Proc 2005 IEEE International Conference on Vehicular Electronics and Safety, Xi'an, P. R. China
50. He F, Wang FY, Tang SM (2006) An agent-based controller for vehicular automation. In: Proc 2006 IEEE International Conference on Intelligent Transportation Systems, 771–776
51. Liu XM, Wang FY (2003) Agent-based coordination and control of regional traffic flows (in Chinese). J Computer Engineering 30(9):45–47
52. Liu XM, Wang FY (2004) Agent-based control of traffic flows at individual intersections (in Chinese). J Computer Engineering 31(4):53–55
53. Wang FY (2005) Agent-based control for networked traffic management systems. IEEE Intelligent Systems 20(5):92–96
54. Liu XM, Wang FY (2002) An agent-based study of coordination control for metropolitan area traffic flow. In: Proc IEEE Int Conf Intelligent Transportation Systems, Singapore
55. Li ZJ, Wang FY, Miao QH, He F (2006) An urban traffic control system based on mobile multi-agents. In: Proc 2006 IEEE International conference on Vehicular Electronics and Safety, 103–108
56. Wang FY (2006) Driving into the future with ITS. IEEE Intelligent Systems 21(3):94–95

57. Wang FY (1999) Network-based neuro-fuzzy control systems for smart consumer electronics. In: Proc 5th US-Sino Conf Science and Technology, Beijing, P. R. China
58. Wang FY, Lin YT, Wu QL, Fu M, Yeo C (2000) Architecture and implementation of intelligent control systems for smart consumer appliances via Internet. In: Proc IEEE Int Conf Systems Man Cybernetics, Nashville, TN
59. Wang FY (2000) Smart appliances in amart houses: networked homes and beyond. In: Proc First Chinese Conf Science and Technology of Appliances, Shanghai, P. R. China
60. Wang FY (2001) OSGi and agent-based technology for control and management of networked appliances. In: Proc Second Chinese Conf Science and Technology of Appliances, Hangzhou, P. R. China
61. Liu MK, Wang FY (2001) Trends and the state of the art of intelligent appliances. China Appliances
62. Wang FY, Lin YT, Huang XC, Wang ZX, Jian S, Wu QL (2001) Smart control for smart consumer appliances: a neuro-fuzzy-based approach. In: Proc Int Appliance Technical Conference, Columbus, OH
63. Wu QL, Lin YT, Wang FY (2001) A mobile-agent based distributed intelligent control system architecture for home automation. In: Proc IEEE Int Conf Systems Man Cybernetics, Tucson, AZ
64. Wang FY, Lin YT, Wu QL, Fu PM, Yeo C (2000) Architecture and implementation of intelligent control systems for smart consumer appliances via Internet. In: Proc IEEE International Conference on Systems Man Cybernetics, Nashville, TN
65. Wang FY, Lin YT, Huang XC, Wang ZX, Jian S, Wu QL (2001) Smart control for smart consumer appliances: a neuro-fuzzy-based approach. In: Proc Int Appliance Technical Conference, Columbus, OH
66. Wu QL, Wang FY (2001) A mobile-agent based distributed intelligent control system architecture for home automation. In: Proc IEEE International Conf Systems Man Cybernetics, Tucson, AZ
67. Zhang HT, Wang FY, Ai YF (2005) An OSGi and agent based control system architecture for smart home. In: Proc 2005 IEEE Networking, Sensing and Control, Tucson, AZ
68. Fu M, Yeo C, Lin YT, Wang FY (2000) WAVES: web-based audio/video educational systems for real-time laboratory experiments. In: Proc IEEE Symp Advance in Control Education, Gold Coast, Australia

3

Networked Control Systems: Emulation-based Design

Mohammad Tabbara[1], Dragan Nešić[1], and Andrew R. Teel[2]

[1] University of Melbourne, Parkville, Victoria 3052, Australia
m.tabbara@ee.unimelb.edu.au, d.nesic@ee.unimelb.edu.au
[2] University of California, Santa Barbara, CA 93106, USA
teel@ece.ucsb.edu

Abstract. A common approach to the implementation of digital systems is through the *emulation* of idealized continuous-time blocks in order to be able to leverage the rich expanse of results and design tools available in the continuous-time domain. The so-called sampled-data systems are now commonplace in practice and rely upon results that ensure that many properties of the nominal continuous-time system, including notions of stability, are preserved under sampling when certain conditions are verified. In analogy with (fast) sampled-data design, this chapter explores an emulation-based approach to the analysis and design of networked control systems (NCS). To that end, we survey a selection of emulation-type NCS results in the literature and highlight the crucial role that *scheduling* between disparate components of the control systems plays, above and beyond sampling. We detail several different properties that *scheduling protocols* need to verify together with appropriate bounds on inter-transmission times such that various notions of input–output stability of the nominal "network-free" system are preserved when deployed as an NCS.

Keywords. Nonlinear systems robust stability, scheduling, emulation-based design.

3.1 Introduction

Control of a system is to influence its behavior to achieve a desired goal, often, through the use of feedback. Diagrammatically, we are often concerned with the setup depicted in Fig. 3.1: analysis of plant P with (vector) output y and design of a controller C with a (vector) control u to achieve a desired closed-loop behavior, typically, a notion of stability.

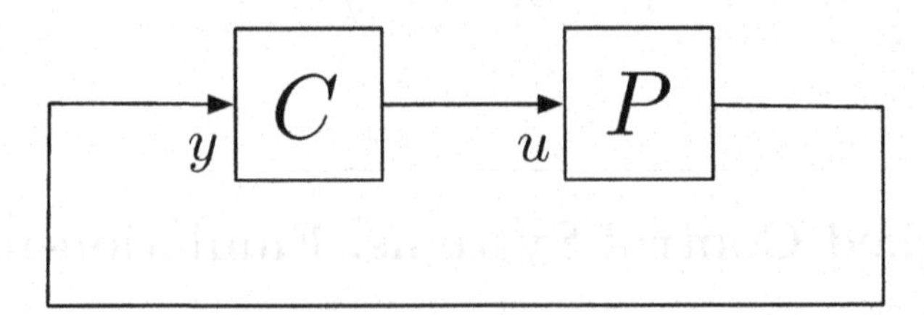

Fig. 3.1. Conceptual block diagram of feedback control

The interconnection of physical signals between controller and plant is seldom as elementary as that depicted in Fig. 3.1. Many properties of the plant including its physical size, complexity and mobile nature require the distribution of the control and observation task across multiple spatially separated *nodes*, including actuators, sensors and devices that compute the control law, connected via a *network*. For example, the system in Fig. 3.1 may potentially be implemented as in Fig. 3.2, using two output-feedback controllers C_1, C_2 and two sensors that transmit output values y_1, y_2 across a network to both controllers. Note that this implementation is suggested without specific reference to how and when and under which constraints this exchange of information takes place.

Abstractly, any set of communication channels together with a connection topology and constraints on the exchange of information across the channels that prescribe how and when information can be exchanged between nodes can be referred to as a network.

In this chapter, we restrict our attention to systems with nodes connected via a single shared communication channel or *bus* as in Fig. 3.3. The control law, plant, nodes, the bus itself and the *protocol* that describes how and when

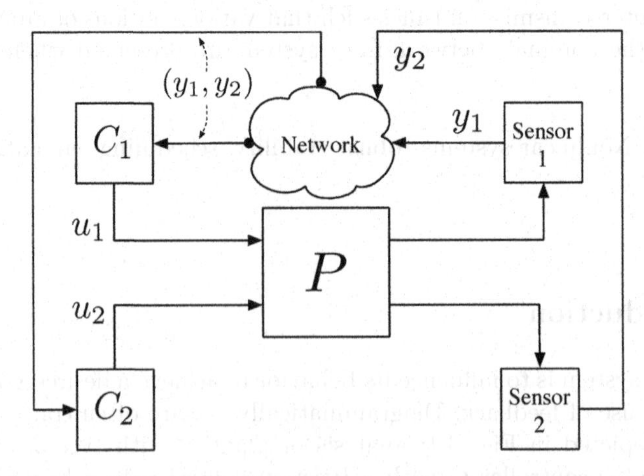

Fig. 3.2. A potential 4 node implementation of the system in Fig. 3.1 as an NCS

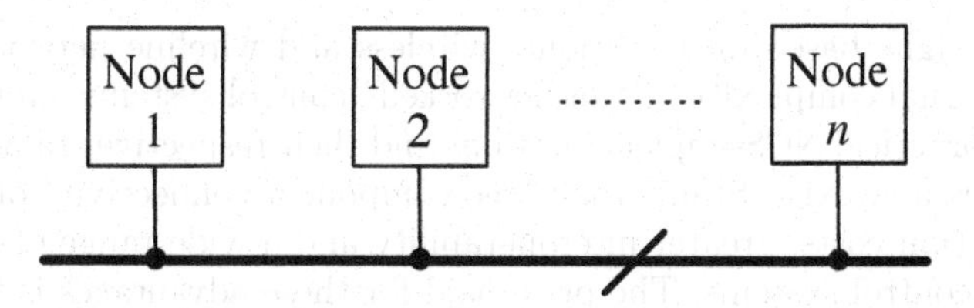

Fig. 3.3. Nodes connected with a bus network topology

information can be exchanged amongst nodes are collectively referred to as the *networked control system* (NCS).

Central to the study of NCS is the analysis and design of scheduling protocols. NCS depart from the use of dedicated point-to-point links for connectivity amongst nodes replacing some or all links with a shared network channel.

As in traditional data networks, the problem of arbitrating multiple access on the network becomes an issue, motivating the discussion of the *scheduling* of nodes and the design and analysis of scheduling protocols suitable for NCS applications. By scheduling, we mean the transmission of information across a link in the form of a discrete packet or frame.

Canonical NCS examples include so-called by-wire systems: drive-by-wire and fly-by-wire with analogies in industrial applications. Here, the *network* in NCS is thought of as in the sense of a traditional data (computer) network but the "network" may exist at a lower level of abstraction as in, for example, embedded digital control systems.

Example 3.1 (Embedded digital control systems). Transmission of controller and sensor values to and from the device executing the control law is governed by protocols of an electrical bus, e.g., a PCI bus, and typically, the scheduler of an operating system. Even if the underlying control system employs point-to-point connections from nodes to the controller, communication within the controller and its constituent components are subject to the communication constraints of various electrical buses and the operating system. □

Example 3.1 is one of the strongest motivations for studying NCS as presented in this chapter. It is perhaps taken for granted that the digital control systems designed and deployed in industry will continue to behave like their idealized continuous-time (resp., discrete-time) counterparts, besides the effects of sampling and quantization. As control systems increase in size and complexity and the levels of component integration increase, the flow of data between elements of the system is subject to constraints similar to that of a "real" network. Indeed, components of systems based on the PCI Express® architecture communicate via a switched serial network. Regardless of how controllers and sensors are connected, at least internally, every non-trivial digital control system can be thought of as an NCS.

From designs based on traditional wireless and wireline networks to the growing internal complexity of "un-networked" control systems, an increasing number of practical NCS implementations and their respective traffic scheduling protocols now exist. Standards-based component connectivity offers lower implementation costs, greater interoperability and a wide range of choices in developing control systems. The price paid for these advantages is the added complexity in the initial design and analysis of NCS. As alluded to earlier, part of this complexity comes in the form of issues of arbitration of network access amongst links, or *scheduling*, which is of fundamental importance. But above and beyond scheduling, NCS also presents the designer with the limitations of

(a) finite bandwidth of communication channels;
(b) finite precision of encoding and decoding schemes for transmitted information;
(c) pure (propagation) delays of channels;
(d) and data dropouts from unreliable channels.

These limitations are not mutually exclusive, however. As transmission rates increase, and with frame and packet sizes well in excess of machine (CPU) precision, effects of quantization and pure delay play an increasingly diminishing role in the analysis of most NCS and we forgo their consideration in this chapter. We will, however, examine models of data dropouts and unreliable channels with the Ethernet and the so-called p-persistent collision-sense multiple access (CSMA) as prime examples of such channels.

3.2 Overview of Emulation-based NCS Design

3.2.1 Principles of Emulation-based NCS Design

As stated in the introduction, scheduling and scheduling protocols are an integral part of NCS design. A survey of scheduling and various scheduling protocols is provided in [17] and stability and performance results of NCS have been examined in [8, 9, 15, 16, 17, 19]. An elementary example of a scheduling protocol, round-robin (RR), grants network access to NCS components in sequential, round-robin fashion and is used almost exclusively in practice. The aforementioned works present various alternative protocols that demonstrate a performance gain over RR in simulations and, in special cases, demonstrate the superiority of the alternative protocols analytically. The NCS design approach adopted in [8, 9, 12, 15, 16, 18], and in this chapter consists of the following steps:

(1) design a stabilizing controller ignoring the network;
(2) choose an appropriate scheduling protocol;

(3) and analyze the robustness of stability with respect to effects that scheduling within a network introduce.

The principal advantage of this approach is its simplicity – the designer of the NCS can exploit familiar tools for controller design and select an appropriate scheduling protocol and transmission rate such that the desired properties of the network-free system are preserved.

This chapter will introduce and characterize the various classes of admissible protocols for which stability results are developed but it is important to note that when the network-free system verifies a nominal stability property and an admissible protocol is chosen, stability of the resultant NCS can be achieved through sufficiently high transmission rates (or equivalently, sufficiently low inter-transmission times). Moreover, stability (robustness) properties of the NCS are actually parameterized by the transmission rate and hence, step (3) in the design process can be reinterpreted as:

(3′) choose a transmission rate (above requisite minimum) to achieve a desired degree of robust stability.

Results will be presented where this design approach is adopted with various notions of transmission rate (minimum or expected) and robust stability (uniform global exponential or asymptotic stability, L_p or L_p in-expectation or input-to-state stability).

3.2.2 Results in Perspective

Consider the following LTI control system:

$$\dot{x}_P = A_P x_P + B_P u, \qquad \dot{x}_C = A_C x_C + B_C y, \tag{3.1}$$

$$y = C_P x_P, \qquad u = C_C x_C, \tag{3.2}$$

where $x_P \in \mathbb{R}^{n_P}$, $x_C \in \mathbb{R}^{n_C}$, y, and u denote, respectively, plant state, controller state, plant output, and control, and where u has been designed ignoring the network as outlined in the previous section. In the presence of a network and an associated scheduling protocol, y and u cannot be continuously transmitted between the plant and controller. The network introduces the following limitations:

(a) transmissions occur only at specific transmission instants $\{t_i\}_{i=0}^{\infty}$; and
(b) only one logical component of the NCS is allowed to transmit (broadcast) data onto the network at a given transmission instant t_i, e.g., for a 3-output 2-input system, one component of $y = (y_1, y_2, y_3)$, $u = (u_1, u_2)$ can be transmitted.

Let $\hat{y}$ denote the "stand-in" for y available to and maintained by the device(s) that compute the control law and $\hat{u}$ denote the "stand-in" for u available to and maintained by the device(s) that actuate the plant. In effect, the NCS for the network-free system is described by

$$\dot{x}_P = A_P x_P + B_P \hat{u}, \qquad \dot{x}_C = A_C x_C + B_C \hat{y}, \tag{3.3}$$
$$y = C_P x_P, \qquad u = C_C x_C. \tag{3.4}$$

In analogy with zero-order hold sampling, $\hat{y}$ and $\hat{u}$ can be held constant between transmission instants and "reset" or updated with components of u and y as those become available and transmitted. Fig. 3.4 illustrates the situation for an NCS where two outputs are alternately transmitted in RR fashion across the network to the device(s)[3] that compute the control law and actuate the plant *at transmission instants.* RR is only one example of a scheduling protocol amongst several that we consider and one of the primary aims of the chapter will be to characterize protocol properties that capture the effects of the protocol on NCS stability.

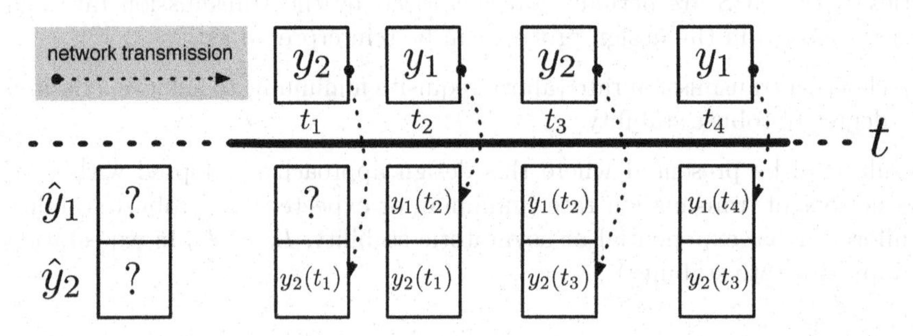

Fig. 3.4. RR scheduling of two-outputs with a zero-order hold $\hat{y}$ update policy

Although a scheduling protocol determines how the transmission of plant measurements and control values are arbitrated at transmission instants, it is useful to think of the scheduling protocol in terms of the effects on the *error* that a network induces compared to the network-free system. Indeed, if we define

$$e = \begin{pmatrix} e_y \\ e_u \end{pmatrix} = \begin{pmatrix} \hat{y} - y \\ \hat{u} - u \end{pmatrix}, \tag{3.5}$$

we can rewrite (3.3) and (3.4) between transmissions as

$$\begin{aligned} \dot{x}_P &= A_P x_P + B_P u + B_P e_u, & \dot{x}_C &= A_C x_C + B_C y + B_C e_y, \\ y &= C_P x_P, & u &= C_C x_C, \\ \dot{\hat{u}} &= 0, & \dot{\hat{y}} &= 0, \end{aligned}$$

and hence,

$$\begin{bmatrix} \dot{x} \\ \dot{e} \end{bmatrix} = \begin{bmatrix} A_{11} & A_{12} \\ A_{21} & A_{22} \end{bmatrix} \begin{bmatrix} x \\ e \end{bmatrix}, \tag{3.6}$$

[3] Since data is presumed to be broadcast across the network, the number of controller-actuator devices that actuate the plant is immaterial so long as they adopt identical policies for updating their copies of $\hat{y}$.

where

$$A_{11} = \begin{bmatrix} A_P & B_P C_C \\ B_C C_P & A_C \end{bmatrix}, \quad A_{12} = \begin{bmatrix} B_P \; B_C \end{bmatrix}, \tag{3.7}$$

$$A_{21} = -\begin{bmatrix} C_P \; 0 \end{bmatrix} A_{11}, \quad A_{22} = -\begin{bmatrix} C_P \; 0 \end{bmatrix} A_{12}. \tag{3.8}$$

These equations describe how the state and NCS error evolves between transmissions and it is clear that components of e are reset or experience "jumps" *at transmissions instants.* For example, let $e_{y,j} = \hat{y}_j - y_j$. Ignoring the effects of quantization and delay, if the jth component of y is transmitted at the ith transmission instant we have

$$\hat{y}_j(t_i) \leftarrow y_j(t_i) \iff e_{y,j}(t_i) \leftarrow 0. \tag{3.9}$$

Hence, *the effect of the scheduling protocol is to reset components of the NCS error*[4] *at transmission instants.* An NCS model in this fashion is thus completely prescribed by:

(a) NCS continuous-time dynamics as in (3.6) and depicted conceptually in Fig. 3.5;
(b) a sequence of increasing transmission instants $\{t_i\}_{i=0}^{\infty}$; and
(c) a scheduling protocol, or *error reset map* that is described via its effect on the error, e, at transmission instants.

Regarding the NCS continuous-time dynamics as fixed, we would like to characterize the sequence of transmission instants or, equivalently, the sequence of inter-transmission intervals and the set of protocols for which we can conclude that the NCS state (x, e) is stable in an appropriate sense. The origins of emulation-based NCS design in this sense begin with the pioneering work of Walsh *et al.* in [15] and [16] where NCS models in the form of (3.6) and its nonlinear counterpart were presented, together with conditions on the *maximum allowable transmission interval* (MATI) such that the resultant NCS was uniformly globally asymptotically or exponentially stable (UGAS, UGES) when using the RR or *maximum-error-first try-once-discard* (TOD) scheduling protocols. We defer a detailed discussion of these and other protocols until Section 3.3.2 and outline results in the spirit of those presented in [15, 16, 19].

Let $e \in \mathbb{R}^{n_e}$ and $x \in \mathbb{R}^{n_x}$. The following class of nonlinear systems was considered in [15]:

$$\begin{aligned} \dot{x} &= f(t, x, e), \\ \dot{e} &= g(t, x, e), \end{aligned} \tag{3.10}$$

with the shorthand notation:

$$\dot{z} = h(t, z), \tag{3.11}$$

where $z = (x^T \; e^T)^T$.

[4] Ordinarily and as in (3.9), the result of the transmission is to reset a component of error to zero, though we stress that for many of the results outlined in the chapter, this assumption is not necessary.

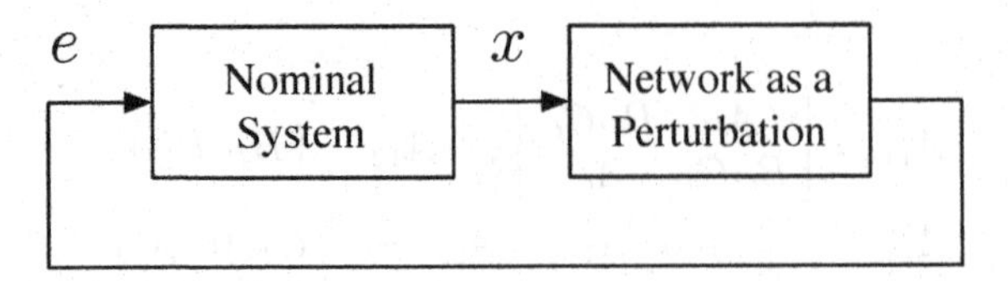

Fig. 3.5. Interconnection of signals in NCS dynamics

The Lipschitz constants for f, g and h are denoted, respectively, by k_f, k_g and k_h; that is, the right-hand side in (3.11) is assumed to be globally Lipschitz, uniformly in t. The class of linear systems (3.6) with the obvious shorthand:

$$\dot{z} = Az \tag{3.12}$$

was considered in [16, 19].

It is supposed in [15] that there exists a continuously differentiable Lyapunov function V such that the system (3.10) satisfies:

$$c_1|x|^2 \le V(t,x) \le c_2|x|^2 \quad \text{for all } x \in \mathbb{R}^{n_x} \text{ and } t \in \mathbb{R}, \tag{3.13}$$

$$\frac{\partial V}{\partial t} + \frac{\partial V}{\partial x} f(t,x,0) \le -c_3|x|^2 \quad \text{for almost all } x \in \mathbb{R}^{n_x} \text{ and } t \in \mathbb{R}, \tag{3.14}$$

$$\left|\frac{\partial V}{\partial x}\right| \le c_4\,|x|\,, \tag{3.15}$$

where c_1, c_2, c_3, c_4 are positive constants. A similar condition was used in [16, 19] for the linear system (3.6). Indeed, it was assumed that for some positive definite and symmetric matrix Q there exists a positive definite and symmetric matrix P that solves the Lyapunov matrix equation[5]:

$$A_{11}^T P + P A_{11} = -Q\,. \tag{3.16}$$

It is obvious that (3.16) implies that (3.13)–(3.15) are satisfied for the linear system (3.6), $V(x) = x^T P x$ and

$$c_1 = \lambda_{\min}(P);\ c_2 = \lambda_{\max}(P);\ c_3 = \lambda_{\min}(Q);\ c_4 = 2\lambda_{\max}(P), \tag{3.17}$$

where $\lambda_{\min}(\cdot)$ and $\lambda_{\min}(\cdot)$ denote the minimum and maximum singular value of a symmetric matrix, respectively. For linear systems, we can let

$$k_h = k_f = k_g = |A|\,. \tag{3.18}$$

A bound on MATI that guarantees the stability of the linear system (3.6) with the RR and TOD protocols was obtained in [16, 19]. We denote bounds computed in [16, 19], respectively, as τ_*^{RR} and τ_*^{TOD} for the RR and TOD

[5] The results in [16] are only presented for the special case $Q = I$. The result with general Q is presented in [19].

protocols. Similar bounds were obtained in [15] for nonlinear systems (3.10) with the RR and TOD protocols, where[6] the bounds obtained are also such that $\tau_*^{RR} = \tau_*^{TOD}$. The bounds in [15, 16, 19] can be expressed as:

$$\tau_*^{RR} = \tau_*^{TOD} = \frac{c_3}{M\ell(\ell+1)k_h k_f c_4}, \tag{3.19}$$

where the value of the constant M is different for linear and nonlinear systems and ℓ denotes the number of nodes that participate in scheduling. For nonlinear systems, we have

$$M = M_{NL} = 16\left(\frac{c_2}{c_1}\right)^{3/2}\left(\sqrt{\frac{c_2}{c_1}}+1\right), \tag{3.20}$$

established in [15]. Analogously in [16, 19], the following is obtained for linear systems

$$M = M_L = 8\sqrt{\frac{\lambda_{\max}(P)}{\lambda_{\min}(P)}}\left(\sqrt{\frac{\lambda_{\max}(P)}{\lambda_{\min}(P)}}+1\right), \tag{3.21}$$

where the meaning of all constants in (3.19) is explained through (3.17) and (3.18). These MATI bounds obtained in [15, 16, 19] do not differentiate between RR and TOD; that is $\tau_*^{TOD} = \tau_*^{RR}$.

In general, intuition suggests that MATI bounds should be protocol-dependent. Significant improvements upon these MATI bounds were made in [8] by efficiently capturing protocol properties through protocol-specific Lyapunov functions and characterizing the effects of transmission errors through L_p gains. Essentially, UGES and L_p input–output stability is with a MATI of:

$$0 < \tau < \frac{1}{L}\ln\left(1 + \frac{1-\theta}{\gamma/L+\theta}\right), \tag{3.22}$$

where $\theta \in [0,1)$ characterizes the ability of the protocol to reduce network error at transmission instants while $L > 0$ describes the speed of the network-error dynamics, and $\gamma > 0$ captures the effect of network-error on the behavior of the ideal system through an L_p gain. In particular, τ is protocol-dependent through θ – the better the protocol is at reducing network-error at transmission instants, the larger the MATI bound is, and hence, the less frequent transmissions have to be to guarantee stability of the NCS.

3.3 Modeling Networked Control Systems and Scheduling Protocols

The premise of networked control systems (NCS) is to spatially distribute a "traditional" control system across a number of *nodes* that will exchange

[6] Note that we do not use different notation for MATI bounds for linear and nonlinear systems, although they are different in general. This is because it always will be clear from the context which bound we mean.

data subject to the constraints of a shared data channel. These nodes include sensors, actuators and units that compute various control laws and the data channel is typically a wireless or wireline computer network, many examples of which can be found in [14].

Computer networks and communication systems present rich and sophisticated models of varying degrees of complexity, within stochastic and deterministic settings, and of various underlying physical communication media. For the vast majority of computer networks described in [14], the primary constraint on the exchange of data between nodes is that the respective channels are exclusive. This means that the attempt of more than one node to transmit data at a given time will result in data loss, i.e., a collision. Collisions can be prevented by arbitration of network access through the use of scheduling protocols that decide which node(s) can transmit and at what times.

The network models presented in this chapter aim to capture the essential aspects of control over networks in the context of several important settings:

(a) locally[7] arbitrated network access without packet dropouts;
(b) arbitrated network access with and without packet dropouts; and
(c) unarbitrated network access with and without packet dropouts.

Arbitration takes place through the use of a scheduling protocol adopted by every node in the network. A protocol can be thought of as a map

$$h\colon W \to \{1, \dots, \ell\} \tag{3.23}$$

that selects the node currently being allowed to transmit and an associated dynamical system that evolves the scheduler state variable $\omega \in W$. For spatially separated nodes, this generally means that each node must maintain a copy of the state ω that is evolved identically by the node (local knowledge with globally-known inputs), or, ω is known globally and updated in a distributed fashion. Such protocols are often referred to as contentionless protocols. For example, labeling the NCS nodes $\{a_1, a_2, \dots, a_\ell\}$, round-robin scheduling would entail apportioning the channel's time, $[0, \infty)$, into slots $\{s_1 := [t_0, t_1), s_2 := [t_1, t_2), \dots, \}$ such that node a_i is permitted to transmit during slot $s_{i+k\ell}$, $k = \{0, 1, \dots\}$. Depending on the context, this scheduling protocol is also known as time-division multiplexing or token ring and relies on each node being able to count transmissions. In this case,

$$\omega = \text{number of transmissions from some initial time.}$$

For networks with a large number of nodes, mobile nodes that are spatially separated across varying distances or networks with a varying number of nodes, it may be impractical or impossible to keep ω, the state information, synchronized across all nodes.

[7] By "locally" we mean that the arbitration process takes place without the exchange of global arbitration information prior to network access, e.g., a priority field.

The alternative is unarbitrated access in the sense that there is no global policy to enforce exclusive network access for a given node at a transmission instant. In particular, collisions may occur, and have to be detected and recovered from. The number that occur can often be reduced by employing various heuristics using data available to each node *locally*. Concrete and familiar examples of this approach include the family of carrier-sense multiple access protocols (CSMA) exemplified by Ethernet, p-persistent CSMA (Bluetooth, 802.11a/b/g) and variants of ALOHA. See [14] for an overview of these protocols and their operational characteristics.

Thus far, the discussion holds true for both computer and control networks. Where computer networks and control networks differ radically is in access patterns – ideally, a continuous-time control system would have nodes constantly transmitting sensor values and constantly receiving control values, in complete contrast to the usual assumption of access in short and irregular bursts for nodes in a computer network. Stated explicitly, *we assume continuous-time controllers and plant outputs are such that there will always be data to transmit when the network channel becomes idle.*

This assumption applies to all forms of network access in NCS, the key difference being that the unarbitrated network access does not enforce a particular choice of which link to transmit when the channel becomes idle whereas global arbitration would. We present a unified approach for the analysis of NCS both for ideal channels and in the presence of random packet dropouts and random inter-transmission times – effects that are essentially attributes of non-ideal or *stochastic* network channels.

We assume that every link in the NCS contests access to the network at either predetermined time-slots or times at which the network is sensed to be idle. This results in two potential sources of randomness:

(a) At any idle time or transmission slot, either some node j transmits successfully or a collision results or the transmitted packet is dropped. Denoting the probability that a packet is dropped or a collision occurs by p_0, we will always assume that the probabilities of successful transmission of links is identically equal to $(1-p_0)/\ell$ for an ℓ-link NCS without global arbitration. While this is not strictly necessary in our analyses, there is no reason to statically (off-line) favor any one link over another during contention by adjusting transmission-success probabilities. Contentionless protocols do, however, enforce a particular choice of which link to transmit in a given slot eliminating the possibility of a collision.
(b) Sensing the network as being idle, synchronizing to transmissions time-slots or else randomly waiting for a period of time after any of these events to reduce the likelihood of collisions are common features of network protocols. These uncertainties can be faithfully modeled with a stochastic (renewal) process. For the set of protocols we discuss, it is sufficient to restrict our attention to Poisson processes with some intensity λ or a

class of renewal processes where inter-transmission times are uniformly bounded, i.e., by the MATI.

3.3.1 Scheduling and a Hybrid System Model for NCS

We model the NCS as a so-called jump-continuous (hybrid) system, where jump times and the associated jump or reset maps are both potentially random but not necessarily so. Our NCS model incorporates the effects of exogenous perturbations w as first presented in [8]. As alluded to earlier, the model we present is general enough to examine several scheduling alternatives with and without packet dropouts when inter-transmission times are either uniformly bounded with a MATI or random.

Node data (controller and sensor values) are transmitted at (possibly) random transmission instants $\{t_0, t_1, \ldots, t_i\}, i \in \mathbb{N}$ and our NCS model is prescribed by the following dynamical and jump equations. In particular, for all $t \in [t_{i-1}, t_i]$:

$$\dot{x}_P = f_P(t, x_P, \hat{u}, w), \tag{3.24}$$

$$\dot{x}_C = f_C(t, x_C, \hat{y}, w), \tag{3.25}$$

$$u = g_C(t, x_C), \quad y = g_P(t, x_P), \tag{3.26}$$

$$\dot{\hat{y}} = 0, \quad \dot{\hat{u}} = 0^8, \quad \dot{\hat{e}} = 0, \tag{3.27}$$

and at each transmission instant t_i,

$$e(t_i^+) = Q_i(e(t_i))e(t_i)^9, \tag{3.28a}$$

or

$$\begin{aligned} e(t_i^+) &= Q_i(\hat{e}(t_i))e(t_i), \\ \hat{e}(t_i^+) &= \Lambda(i, (I - Q_i(\hat{e}(t_i)))e(t_i), \hat{e}(t_i)). \end{aligned} \tag{3.28b}$$

The effect of the protocol on the error is such that if the mth to nth nodes are successfully transmitted at transmission instant t_i the corresponding components of error, $e_n, \ldots, e_m$, experience a "jump". It may be the case that a single logical node (a "link") consists of several sensors or several actuators or both with the transmission of that link having the effect of setting multiple components of e to zero. It may also be the case that the network allows the transmission of more than one node at each transmission and our model allows for this extra degree of freedom. For transmission from mth node to nth node, we will always assume that $e_n(t_i^+), \ldots, e_m(t_i^+) = 0$ and hence, $Q_i(\cdot)e = [a_{kj}]e$, where $a_{kj} = 0$ for $k = j \in [n, m] \cup \{k \neq j\}$ and 1 elsewhere. We group the

[8] The assumption that $\dot{\hat{y}}$ and $\dot{\hat{u}}$ are zero simplifies the presentation and is not strictly necessary. Non-zero choices correspond to schemes that predict y and u between transmissions in an open-loop sense.

[9] Given $t \in \mathbb{R}$ and a piecewise continuous function $f \colon \mathbb{R} \to \mathbb{R}^n$, we use the notation $f(t^+) = \lim_{s \to t, s > t} f(s)$.

nodes that are transmitted together into logical links, associating a partition of size s_i, denoted by $\mathbf{e}_i = (e_{i1}, e_{i2}, \ldots, e_{is_i})$, of the error vector e such that we can write $e = (\mathbf{e}_1, \ldots, \mathbf{e}_\ell)$. We say that the NCS has ℓ links and $\sum_{i=1}^{\ell} s_i$ nodes. Note that this is purely a notational convenience and simplifies the description of scheduling protocols and the NCS itself.

The two alternative forms of the error jump-map (3.28a) and (3.28b) refer to two different situations with respect to the scheduler state ω in the abstract description of a scheduling protocol given in (3.23):

(a) $\omega \equiv (i, e)$ in (3.28a), where $Q_i(\cdot)$ may be a *random* jump map – in particular, Q_i may be the identity in the case where nothing was transmitted or a collision or dropout occurred.
(b) $\omega \equiv (i, \hat{e})$ in (3.28b), where $Q_i(\cdot)$ is an ordinary map and $\hat{e}$ is a state variable synchronously maintained and updated by all nodes.

In both cases, we refer to Q as the scheduling function and Λ as the decision-update function in (3.28b). The key difference between these two alternatives is the decision-vector $\hat{e}$. Special cases of $\hat{e}$-based scheduling were first considered in [18]. The model we introduced in [12] and described here formalizes the $\hat{e}$-based scheduling that was considered in [18] and it generalizes the NCS models considered in [8].

With respect to the available state-information, there are several alternatives as to what information the scheduler has available in making scheduling decisions prior to transmissions:

(a) (x, e, i) is known by all nodes;
(b) (e, i) is known by all nodes;
(c) i is known and any *broadcast* data becomes known *after transmissions*;[10]
(d) only i is known globally; or
(e) only local policies are adopted and no global information is used in scheduling.

These correspond to the following NCS scenarios:

(1) "Classical" control, that is, if (x, e, i) is known to all nodes prior to transmissions, transmissions would not be necessary as any of x, y, and u could be recovered.
(2) Each node can encode e into an arbitration field and participate in what is, in effect, a distributed scheduling decision, e.g., through binary countdown.
(3) Nodes only have i and local information available to make a scheduling decision, and once a transmission (broadcast) has taken place, are free to update their local information ($\hat{e}$) with the broadcast data.[11] To ensure

[10] This data can be used to evolve locally maintained state, e.g., $\hat{e}$.
[11] For reasons that shall become apparent, there is no loss of generality in assuming that the broadcast data is given by $(I - Q_i(\hat{e}(t_i)))e(t_i)$ – the component of error

that the nodes arrive at a unanimous decision, the update rule, and hence the local data is updated in the same fashion across all nodes.

(4) For situations (2)–(3) it is assumed that nodes can count the number of transmissions that have passed from some reference time and hence i is known. In this NCS scenario, no other data is known or maintained by nodes for scheduling purposes.

(5) Network access is, in effect, unarbitrated and access patterns are determined by local policy.

The maps prescribed by (3.28a) and (3.28b) are sufficiently general to capture the scenarios (2)–(5). We combine the controller and plant states into a vector $x = (x_P, x_C)$ and assuming g_P, g_C are continuous and a.e. (almost everywhere) C^1, for example, we can rewrite (3.24)–(3.28b) in a form more amenable to analysis:

$$\dot{x} = f(t, x, e, w), \quad t \in [t_{i-1}, t_i], \tag{3.29}$$

$$\dot{e} = g(t, x, e, w), \quad t \in [t_{i-1}, t_i], \tag{3.30}$$

and

$$e(t_i^+) = Q_i(e(t_i))e(t_i), \tag{3.31a}$$

or

$$\begin{aligned} e(t_i^+) &= Q_i(\hat{e}(t_i))e(t_i), \\ \hat{e}(t_i^+) &= \Lambda\big(i, (I - Q_i(\hat{e}(t_i)))e(t_i), \hat{e}(t_i)\big), \end{aligned} \tag{3.31b}$$

where $x \in \mathbb{R}^{n_x}, e \in \mathbb{R}^{n_e}, w \in \mathbb{R}^{n_w}$, and $\hat{e} \in \mathbb{R}^{n_{\hat{e}}}$.

Implicit in this definition is that there are no (pure) propagation delays. Transmission at time t_i results in the instant reset of the relevant error component to zero. We appeal to the robustness properties verified by the class of systems considered to assert that the results in this chapter remain true for sufficiently small delays.

With respect to (3.24)–(3.28b) and (3.29)–(3.31b), we further assume that the sequence of (attempted) transmission times $\{t_i\}_{i \in \mathbb{N}}$ is such that $t_{i+1} - t_i$ is exponentially distributed for all i or satisfy $\epsilon < t_{j+1} - t_j \leq \tau$ for all $j \geq 0$ where $\tau > 0$ and $\epsilon > 0$.[12] The constant τ is the *maximum allowable transmission interval* (MATI).

3.3.2 NCS Scheduling Protocol Properties

We have previously described protocols in a general setting as maps that effect errors at transmission instants. We now aim to identify general protocol

that was reset to zero at the ith transmission instant and hence, appears as the only input in (3.28b).

[12] This ensures that Zeno solutions cannot occur. Zeno behavior occurs in hybrid systems when there are an infinite number of discrete transitions in a finite period of time.

properties that appropriately characterize protocol behaviors and that are able to parametrize NCS stability under appropriate conditions. Recall that by "protocol" we refer to both the maps of the form (3.31a) and (3.31b) as well as an associated sequence of transmission times $\{t_i\}_{i=0}^{\infty}$, where $t_{i+1} - t_i$ is either uniformly bounded or exponentially distributed.

We introduce several protocol properties that are phrased in terms of membership in the class of Lyapunov UGES (uniformly globally exponentially stable) protocols, the class of PE_T (persistently exciting) protocols, the class of almost surely Lyapunov UGES protocols and the class of almost surely (a.s.) covering protocols.

3.3.3 Lyapunov UGES and a.s. UGES Scheduling Protocols

Let $\mathbf{E}[\cdot]$ and $\mathbf{P}\{\cdot\}$ denote the expectation and probability operators and let $X \sim \text{Exp}(\lambda)$ denote that X is an exponentially-distributed random variable with $\mathbf{E}[X] = 1/\lambda$. For purely deterministic maps and ignoring the dynamics introduced by (3.30), we can regard (3.31a) as a discrete-time system that captures the behavior of the scheduling protocol. The system is given by:

$$e^+ = Q_i(e)e\,. \tag{3.32}$$

Maps of this form were used to capture the behavior of the protocol in [8] on an ideal network. Describing the protocol in this fashion allows one to speak of uniformly globally asymptotically and exponentially stable (UGAS and UGES) scheduling protocols whenever the associated discrete-time system (3.32) is UGAS or UGES. Beyond taxonomy, the notion of UGES and UGAS protocols and the construction of smooth Lyapunov functions for the associated UGAS and UGES discrete-time systems is central to the stability analysis approach developed in [8] and [9].

NCS employing UGES and UGAS protocols on non-ideal network channels are still subject to packet losses and varying inter-transmission times. By assigning a probability, p_0, to the event that the channel drops a packet, we model the behavior of the protocol on non-ideal channels in this section with jump maps of the form

$$\tilde{Q}_i(e)e = q_i Q_i(e) + (1 - q_i)e, \tag{3.33}$$

where q_i is an i.i.d.[13] sequence of Bernoulli random variables that model the dropout process of channel with $\mathbf{P}\{q_i = 1\} = 1 - p_0$. Depending on the specific system, the sequence of arrival times (transmission instants) $\{t_i\}_{i\in\mathbb{N}}$ are either random and defined inductively by:

$$t_0 = \tau_0,$$

where $\tau_0 \sim \text{Exp}(\lambda)$ and for each $i > 0$,

[13] Independently identically distributed.

$$t_i = t_{i-1} + \tau_i,$$

$\tau_i \sim \text{Exp}(\lambda)$, where the sequence $\{\tau_i\}$ is i.i.d., or inter-transmission times are uniformly (deterministically) bounded by a MATI.

As in (3.32), it becomes natural to define the associated auxiliary discrete-time system for (3.33):

$$e^+ = q_i Q_i(e)e + (1 - q_i)e, \quad i \in \mathbb{N}, \tag{3.34}$$

where the sequence $\{q_i\}$ is defined as in (3.33).

We introduce the following definition with respect to System (3.34).

Definition 3.1 (Almost surely Lyapunov UGES protocols). Let $W : \mathbb{N} \times \mathbb{R}^{n_e} \to \mathbb{R}_{\geq 0}$ be given and suppose that κ_i is a sequence of nonnegative i.i.d. random variables and $a_1, a_2 > 0$ such that the following conditions hold for the discrete-time system (3.34) for all $i \in \mathbb{N}$ and all $e \in \mathbb{R}^{n_e}$:

$$a_1|e| \leq W(i, e) \leq a_2|e| \tag{3.35}$$

$$W(i+1, \tilde{Q}_i(e)e) \leq \kappa_i W(i, e) \tag{3.36}$$

$$\mathbf{E}[\kappa_i] < 1\,. \tag{3.37}$$

Then we say that (3.34) (equivalently, the contentionless protocol) is *almost surely uniformly globally exponentially stable (a.s. UGES) with Lyapunov function W.* □

Before discussing implications of this definition, we present a motivating example:

Example 3.2 (Try-once-discard). The TOD protocol was introduced in [16] and can be expressed with a model of the form (3.34) where

$$Q_i(e) = (I - \Psi(e))$$

and $\Psi(e) = \text{diag}\{\psi_1(e)I_{l_1}, \ldots, \psi_\ell(e)I_{l_\ell}\}$, with I_{l_j} identity matrices of dimension l_j and

$$\psi_j(e) = \begin{cases} 1, \text{ if } j = \min(\arg\max_j |\mathbf{e}_j|) \\ 0, \text{ otherwise.} \end{cases} \tag{3.38}$$

That is, TOD picks out the node with the largest magnitude of error for transmission. It was shown that TOD preserves stability properties of the network-free system in [15] (linear systems) and [8] (nonlinear systems with disturbance) for sufficiently small MATI. As in [8, Proposition 5], we set $W(i, e) = |e|$ and claim that TOD is a.s. Lyapunov UGES whenever the probability of a dropout, p_0 is such that

$$p_0 + (1 - p_0)\sqrt{\frac{\ell - 1}{\ell}} < 1. \tag{3.39}$$

□

The inequality (3.39) is a particular example of a more general condition that ensures that any Lyapunov UGES protocol in the sense of [8] is an a.s. Lyapunov UGES for sufficiently low probability of dropout. We first recall the definition of a Lyapunov UGES protocol:

Definition 3.2 (Lyapunov UGES protocols). A protocol (3.34) on an ideal channel ($p_0 = 0 \Rightarrow q_i = 1$) is said to be Lyapunov UGES in the sense of [8] if there exists $W\colon \mathbb{N} \times \mathbb{R}^{n_e} \to \mathbb{R}_{\geq 0}$, $a_1, a_2 > 0$, and $0 \leq \theta < 1$ such that for all $i \in \mathbb{N}$ and all $e \in \mathbb{R}^{n_e}$:

$$a_1|e| \leq W(i,e) \leq a_2|e| \tag{3.40}$$

$$W(i+1, Q_i(e)e) \leq \theta W(i,e)\,. \tag{3.41}$$

This definition admits the following proposition:

Proposition 3.1. Suppose that the protocol (3.34) on an ideal channel ($p_0 = 0 \Rightarrow q_i = 1$) is Lyapunov UGES. Then (3.34) is a.s. Lyapunov UGES on a non-ideal channel ($p_0 \geq 0$) if

$$p_0 + (1 - p_0)\theta < 1. \tag{3.42}$$

Remark 3.1. The rationale for the introduction of the class of a.s. Lyapunov UGES protocols is to provide an analysis framework for Lyapunov UGES protocols capable of handling random packet dropouts – any Lyapunov UGES protocol is automatically an a.s. Lyapunov UGES protocol for sufficiently low p_0. In the case where inter-transmission times are uniformly bounded by a MATI and $p_0 = 0$, we recover the usual definition of Lyapunov UGES protocols as in Definition 3.2. □

Remark 3.2. The definition of Lyapunov UGAS and hence, a.s. Lyapunov UGAS protocols is analogous and we refer the reader to [9] for details and results. □

3.3.4 PE_T Scheduling Protocols

Intuition suggests that schemes such as TOD should perform better than RR, as the node with the greatest error is transmitted at each transmission instant. TOD is certainly implementable in variants of CAN[14] as the error can be encoded into an arbitration field[15] in a frame but no such arbitration is possible for wireless channels and, indeed, many wireline channels and hence, it is often unreasonable to assume knowledge of the entire error vector e in these contexts.

Several variants of TOD were introduced in [18] that "estimate" the error vector and were shown to outperform RR in simulations. Stability results

[14] Control Area Network.

[15] Specifically, through binary countdown – see [14] and [17] for details.

are also provided for linear systems that lead to conservative estimates on performance bounds. One model of NCS that accommodates these variants was proposed in [12] that is a special case of (3.29)–(3.31b).

The variants of TOD presented in [18] as well as the RR scheduling protocol satisfy the following property: *there is a fixed (finite) number of transmissions T such that all nodes of the NCS have transmitted within T transmissions.* This T is related to the notion of a node's "silent-time" in [18]. This property is the point of departure of this section, and for reasons that will become apparent, we call protocols that satisfy this property *uniformly persistently exciting scheduling protocols*, or simply, PE protocols. Whenever T is known, we say that the protocol is PE_T. Round-robin is the first example of a PE_T protocol.

Example 3.3 (Round-robin). Round-robin scheduling is employed in the token ring and token bus network protocols as well as (once) being the ubiquitous scheduling protocol of time-sharing operating systems. Each link of the network is assigned a unique index and links are "visited" in the order of index. Consider an ℓ-link NCS. In terms of NCS scheduling, the discrete-time system is a linear time-varying system where the protocol map has no dependence on state:

$$e^+ = (I - \Delta(i))e, \tag{3.43}$$

where $\Delta(i) = \text{diag}\{\delta_1(i)I_{s_1}, \ldots, \delta_N(i)I_{s_\ell}\}$, and

$$\delta_k(i) = \begin{cases} 1, & \text{if } k - 1 = i \bmod \ell \\ 0, & \text{otherwise.} \end{cases} \tag{3.44}$$

It was established in [8] that RR is a Lyapunov UGES protocol and that it preserves stability properties of the network-free system for high enough transmission rates. As the protocol does not depend on NCS state it makes RR easily implementable and is PE_T with $T = \ell$. □

PE in the sense we have described is verified by many network technologies. Ethernet and 802.11 are examples of CSMA/CD protocols where it is known (see [2], for instance) that for a finite number of users (links), the expected waiting time for a link is finite. We pursue a stochastic analogue of PE for such protocols in Section 3.3.5.

For a more formal characterization of the PE property, it can be shown that if we integrate the equations (3.30) and $\dot{\hat{e}} = 0$ on the interval $[t_{i-1}^+, t_i]$ and then apply the jump map (3.31b) at t_i, the NCS induces the following discrete-time system:

$$e^+ = (I - \Psi(i, \hat{e}))(e + d), \tag{3.45}$$

$$\hat{e}^+ = \Lambda(i, \Psi(i, \hat{e})(e + d), \hat{e}), \tag{3.46}$$

where d captures the inter-sample behavior of $e(\cdot)$. This idea of examining an induced discrete-time system to evaluate protocol properties was first used in [8] as outlined in Section 3.3.3 though used here with a key difference: for specific initializations $(k, e(k), \hat{e}(k))$ and specific (bounded) values of $d(j), j \geq k$ the solution of the system (3.45)–(3.46) coincides with that of (3.29)–(3.31b) at time instants $t_j^+, j \geq k$ which is not the case for (3.34). As we think of the inter-sample behavior d as a perturbation, our formal definition of PE will be stated as a property that is robust to bounded perturbations.

Definition 3.3. The protocol (3.45)–(3.46) is said to be (robustly) persistently exciting in T or PE_T if there exists $T \in (0, \infty)$ such that

$$\prod_{k=i}^{i+T-1} Q_k(\hat{e}(k))) = \mathbf{0}, \tag{3.47}$$

holds for every $k \in \mathbb{N}$ and any initial condition $e(i)$ and $\hat{e}(i)$ where we have written $\phi_{\hat{e}}(k)$ in place of $\phi_{\hat{e}}(k, i, e(i), \hat{e}(i), d_{[k,i]})$ and all $d \in \ell_\infty$, where $\phi_{\hat{e}}(i) := \phi_{\hat{e}}(i, e, \hat{e}, d_{[k,i]})$ is the $\hat{e}$ component of the solution of the system (3.45)–(3.46). That is, the T-fold product of the jump map evaluated along any set of trajectories that can be generated by (3.45)–(3.46) from any set of initial condition is the zero matrix. □

The protocols below are typical of what has been proposed in NCS literature and what is used in practice. In what follows, we will always assume an *ℓ-link* NCS with the ith linking consisting of l_i nodes and an error vector $\mathbf{e}_i$. Two PE_T protocols are presented next though we note that the simplest example of a PE protocol is RR (Example 3.3).

Example 3.4 (Hybrid RR-TOD scheduling protocol). The hybrid RR-TOD scheduling protocol enforces PE in a time-periodic manner. For a prescribed $M \in \mathbb{N}$, the protocol takes the form:

$$e^+ = (I - \Omega(i, \hat{e}))(e + d), \tag{3.48}$$

$$\hat{e}^+ = (I - \Omega(i, \hat{e}))\hat{e} + \Omega(i, \hat{e})(e + d), \tag{3.49}$$

$$\Omega(i, \hat{e}) := \begin{cases} \operatorname{diag}\{p_1(i) I_{s_1}, \ldots, p_N(i) I_{s_N}\}, & \text{if } \operatorname{mod}(i, M) = 0 \\ \operatorname{diag}\{\psi_1(\hat{e}) I_{s_1}, \ldots, \psi_N(\hat{e}) I_{s_N}\}, & \text{otherwise,} \end{cases}$$

where, $p_n(i) = 1$ when $\operatorname{mod}(i/M, N) = n - 1$ and $p_n(i) = 0$ otherwise with ψ_j defined in (3.38). The hybrid RR-TOD protocol is PE_T with $T = MN$. In particular, when $M = 1$, we obtain the simplest PE_T protocol: "classical" RR. □

Example 3.5 (Constant-penalty TOD). Constant-penalty TOD (CP-TOD) [18] uses the mechanism of "silent-time" to ensure that every link is eventually visited within a finite window of time: each link j has a counter r_j that is incremented at every transmission instant that it is *not* scheduled and reset to zero when it *is* scheduled. Irrespective of the underlying scheduling protocol, when a link's counter reaches a predetermined threshold, say M, it will be scheduled. This ensures that every link is scheduled within $\ell + M$ transmission instants[16].

The underlying scheduler in this example is TOD and corresponds to the constant-penalty TOD scheme in [18] with a penalty (vector) of Θ:

$$e^+ = (I - \Phi(r,\zeta))(e + d), \tag{3.50}$$

$$\zeta^+ = (I - \Phi(r,\zeta))(\zeta + \Theta) + \Phi(r,\zeta)(e + d), \tag{3.51}$$

$$r^+ = (I - \Phi(r,\zeta))(r + \mathbf{1}), \tag{3.52}$$

where $\mathbf{1} = [1 \ \dots \ 1]^T$, the scheduling function Φ is given by

$$\Phi(r,\zeta) = \mathrm{diag}\{\varphi_j(r,\zeta)I_{s_j}\}, \quad j \in [1,\dots,\ell]$$

and

$$\varphi_n(r,\zeta) = \begin{cases} 1, & \text{if } [n = \min\{m\colon r_m \geq M\}] \vee \\ & [n = \min\left(\arg\max_{1\leq j\leq N} |\zeta_j|\right) \\ & \wedge (r_m < M, \forall\, m \in \{1,\dots,N\})] \\ 0, & \text{otherwise} \end{cases} \tag{3.53}$$

$a \wedge b$ and $a \vee b$ denote the logical conjunction and logical disjunction of two conditions a and b, respectively. The role of estimating e is played by ζ and through the term $\Phi(r,\zeta)e$, ζ is updated with $\mathbf{e}_j$ whenever the jth link is transmitted. For those links that are not transmitted, the estimated error is incremented by a fixed penalty Θ that might capture the worst-case growth of error (in the absence of disturbance) for a given MATI. In addition to performing this *ad hoc* estimation, the scheduling protocol counts the number of transmission instants that a link has not been visited for, the link's silent time, and schedules links that have exceeded a predetermined threshold for silent-time. In this way, if ζ is degenerating into an arbitrarily bad estimate of e, all links will continue to be visited within a fixed-length, finite window of transmission instants through the mechanism of forcing a finite silent-time for each link. In a loose sense, the protocol's behavior will "often" be qualitatively similar to that of RR, a protocol that has been shown to lead to L_p stability of the NCS with appropriate conditions. □

3.3.5 a.s. Covering Protocols

By a random protocol, we mean a sequence of random transmission times together with i.i.d. random jump maps Q_i that are independent of e with

[16] The silent-time protocols described in [18] have the links measure *continuous time* as opposed to counting the number of transmission instants elapsed (discrete-time) and set the silent-time threshold in terms of an integer multiple of MATI, say $M\mathcal{T}$. Since, for all $i \in \mathbb{N}$, $M\mathcal{T} \geq M(t_{s_{i+1}} - t_{s_{i+1}})$, our silent-time threshold will be smaller for the same M but the protocol will behave in precisely the same manner as when using the verbatim definition of silent-time given in [18].

reference to (3.31a). That is, Q_i are i.i.d. random matrices taking values in the finite set $\mathcal{M}_{n_e} = \{M_0, M_1, \ldots, M_\ell\}$, where $M_0 = I_{n_e}$ and M_j is such that

$$\begin{aligned} M_j e &= M_j(\mathbf{e}_1, \ldots, \mathbf{e}_j, \ldots, \mathbf{e}_\ell) \\ &= (\mathbf{e}_1, \ldots, \mathbf{e}_{j-1}, \mathbf{0}, \mathbf{e}_{j+1}, \ldots, \mathbf{e}_\ell). \end{aligned}$$

We make this definition more precise shortly. The intuition behind this model is that at a transmission time t_i, either some link j will acquire the channel and have its component of e set to zero, that is,

$$\mathbf{e}_j(t_i^+) = 0, \quad \mathbf{e}_k(t_i^+) = \mathbf{e}_k(t_i), \quad k \neq j.$$

Hence $Q_i = M_j$ or else more than one node attempted to transmit resulting in a collision with e remaining unchanged ($Q_i = M_0$). Due to random "back-off" times, and wait times inserted into medium access protocols, transmission times are potentially random. Collectively, these issues are the same issues presented in multi-user access in computer and mobile voice networks though the network access patterns are somewhat different. See [14] for an overview.

Definition 3.4. For an ℓ-link NCS, abstractly, we define a random protocol as a discrete Markov chain Q_i subordinated by a renewal process[17] $N(t)$ such that

(1) $Q_i \in \mathcal{M}_{n_e}$ are i.i.d. random $n_e \times n_e$ with associated link and collision probabilities given by

$$\mathbf{P}\{Q_i = M_k\} = p_k.$$

(2) The sequence of arrival times $\{t_i\}_{i\in\mathbb{N}}$ is defined inductively by:

$$t_0 = \tau_0,$$

where $\tau_0 \sim \text{Exp}(\lambda)$ and for each $i > 0$,

$$t_i = t_{i-1} + \tau_i,$$

$\tau_i \sim \text{Exp}(\lambda)$, where the sequence $\{\tau_i\}$ is i.i.d. We set

$$N(t) = \begin{cases} 0, & t \in [0, t_0), \\ k, & t \in [t_{k-1}, t_k). \end{cases}$$

Hence, $N(t)$ is a Poisson process with intensity λ. □

Essentially, the τ_is denote the wait time after the arrival of a packet (before a new transmission begins). When not otherwise stated, we will henceforth assume that $\mathbf{P}\{Q_i = M_k\} = \mathbf{P}\{Q_i = M_j\} = (1 - p_0)/\ell$, $k, j \neq 0$, i.e., each

[17] More precisely, the process of interest is in fact a marked point process. See [10] for an exposition.

link is equally likely to transmit successfully. This assumption is not strictly necessary for our analyses, however, any other distribution of probabilities results in a *static* choice of priorities among links where one link may be favored over another during contention. There may be examples of NCS that would benefit from such an adjustment of relative link priorities offline in terms of required transmission rates or greater robustness of stability but as these choices are made offline and not in response to the evolution of the NCS state online, we believe that the scope of exploiting this degree of freedom is limited.

We pursue here a stochastic analogue of the PE_T property described in Section 3.3.4.

Definition 3.5 (Cover time). Consider a random protocol in the sense of Definition 3.4 for an ℓ-link protocol and define

$$T_0 = \min\{j \geq 1 \colon \{M_1, \ldots, M_\ell\} \subset \{Q_0, \ldots, Q_{j-1}\}\}$$

and, inductively for $i > 0$,

$$T_i = \min\left\{j \geq 0 \colon \{M_1, \ldots, M_\ell\} \subset \{Q_{T_{i-1}}, \ldots, Q_{T_{i-1}+j-1}\}\right\}.$$

We refer to T_i as the ith cover time, and collectively the cover time process. It is clear from our definition of Q_i that T_i is a stationary process. □

Definition 3.6 (Covering sequence). Let $\tau_i = t_{i+1} - t_i$, as in Definition 3.4, that is, τ_i are inter-arrival times. We say that

$$C(j,k) = \{(Q_j, \tau_j), \ldots, (Q_k, \tau_k)\}, \quad k \geq j,$$

is a covering sequence if and only if $\{M_1, \ldots, M_\ell\} \subset C_{(1)}(j,k)$.[18] It is easy to see that cover times are simply the lengths of consecutive disjoint covering sequences. □

Remark 3.3. From our definition of random protocols, the distribution of T_n is given by the solution to the (weighted) coupon collectors problem. When $p_i = p_j$, $i, j \neq 0$, we have the following closed form expression for the expectation:

$$\mathbf{E}[T] = \ell H_\ell/(1 - p_0), \tag{3.54}$$

where H_ℓ is the ℓth harmonic number and we have dropped the time index n since T_n is stationary. We also have the bound for the distribution, $\mathbf{P}\{T_n \geq \beta\ell\ln\ell/(1-p_0)\} \leq \ell^{-(\beta-1)}/(1-p_0)$, for any $\beta > 1$. Intuitively, $T_n = \mathbf{E}[T]$ "most of the time" and $\mathbf{P}\{T_n < \infty\} = 1$. □

Our abstract definition of a contention protocol is a model for the contention protocols discussed earlier and to that end we present two natural examples in this setting.

[18] The notation $C_{(1)}(j,k)$ refers to the covering sequence of matrices Q_i with no reference to inter-transmission times τ_i, i.e., $\{Q_j, \ldots, Q_k\}$.

Definition 3.7 (Almost surely finite cover time). We say that a protocol is *a.s. covering* or has an *a.s. finite cover time* if in Definition 3.5

$$(\forall i \in \mathbb{N}) \quad \mathbf{P}\{T_i < \infty\} = 1. \qquad \square$$

Note that from the preceding discussion, this property is verified by all contention protocols in the sense of Definition 3.4.

3.3.6 Slotted p-Persistent CSMA

What has been referred to as "scheduling" and the associated scheduling protocols by [13] is generally known as medium access in the communications literature. Carrier sense multiple access with collision detection (CSMA/CD) is by far the most widely used medium access protocol by virtue of the sheer volume of Ethernet and Ethernet-like networking devices shipped and manufactured each year.

CSMA/CD is a simple protocol: Links listen for transmissions on the channel. A link wanting to transmit acquires the channel when it senses that the channel is idle. When more than one link senses that the channel is idle and begins transmission, a collision occurs. At this point, all transmissions are immediately aborted. There are several variants of CSMA/CD that prescribe how transmissions are rescheduled and how links initially acquire the channel.

With slotted p-persistent CSMA, rather than have links transmit whenever the channel is idle, links are only permitted to transmit at prescribed transmission slots that occur every $t_s > 0$ seconds in slotted protocols. At the start of slot s_k, links $S = \{i, \ldots, j\}$ intending to transmit acquire the channel with a probability of p. If a collision occurs, links S^c are permitted to transmit in the next slot and links S reschedule their transmissions at slots $\{s_{k+d_i}, \ldots, s_{k+d_j}\}$.

As alluded to earlier, the primary reason that CSMA protocols and indeed, all contention protocols work in practice is that the access patterns of computer and voice networks are "bursty" in nature. The assumption is that a link will occasionally transmit a burst of information and remain otherwise idle. Transmissions are expected to eventually succeed as links are "infrequently" contending for the channel.

The situation is quite different for control networks with the implication that medium access patterns are constant rather than bursty and for slotted p-persistent CSMA, we assume that every slot will be in contention. Another key difference between computer networks and NCS is in the treatment of collisions and dropouts. NCS should not buffer failed transmissions of controller or sensor values but rather, attempt to transmit the latest values when a slot is free. As the maximum number of links contending slots is constant for every slot, there is no reason for a link to delay transmission for more than one slot after a collision.

With these assumptions, consider an ℓ-link NCS with the p-persistent CSMA protocol. The probability $\mathbf{P}\{Q_i = M_j\}$ that a *particular* link j transmits successfully during the ith slot is given by

$$\mathbf{P}\{Q_i = M_j\} = p(1-p)^{\ell-1}.$$

It is clear that $\mathbf{P}\{Q_i = M_j\}$ is maximized when $p = 1/\ell$. We will henceforth set $p = 1/\ell$ and have that

$$\mathbf{P}\{Q_i = M_j\} = \frac{1}{\ell}\left(1 - \frac{1}{\ell}\right)^{\ell-1} = \frac{(\ell-1)^{\ell-1}}{\ell^\ell}.$$

Notice that in this "optimal" case,

$$\mathbf{P}\{Q_i = M_j\} = \mathbf{P}\{Q_i = M_k\} = (\ell-1)^{\ell-1}/\ell^\ell \text{ for } j, k \neq 0$$

and the probability of a collision is given by $\mathbf{P}\{Q_i = M_0\} = 1-(\ell-1)^{\ell-1}/\ell^{\ell-1}$. Finally, we assume that slots occur every $t_s > 0$ seconds and hence, p-persistent CSMA is a contention protocol in the sense of Definition 3.4 where inter-arrival times τ_i are deterministic.

3.3.7 CSMA with Random Waits

Whereas the use of fixed slots tends to improve throughput and reduce collisions with computer networks, e.g., slotted versus pure ALOHA, the contention by every link at every slot forces transmissions to happen in lock-step with NCS network access patterns with the potential for a collision at every slot.

Suppose that instead of immediately acquiring the channel with probability p after sensing the channel to be idle or after a new slot arrives, links instead wait for a random amount of time before transmitting. In particular, if a particular link j waits for a random time $\eta_j' \sim \text{Exp}(\lambda/\ell)$, then, $\mathbf{P}\{Q_i = M_j\} = (1-p_0)/\ell, j \neq 0$. The actual wait time before any particular transmission will be given by

$$\tau = \min\{\eta_1', \ldots, \eta_\ell'\};$$

that is, the link that waits the least gets to transmit first. Hence, $\tau \sim \text{Exp}(\lambda)$. Assuming the wait times are i.i.d. for each link, this is the prototypical example of what we mean by a stochastic protocol and a stochastic channel.

In the presence of transmission errors, p_0 is generally nonzero and conceptually, p-persistent CSMA and CSMA with random waits are essentially the same apart from the fact that the transmission process is truly random with the latter. While CSMA with random waits can be thought of as a protocol in its own right when the random waits are enforced explicitly in the implementation, it can also be thought of as a model of medium access with

NCS access patterns while using a class of CSMA wireless protocols. Delays in signal detection, multi-path effects and varying processor loads mean that links are only prepared to transmit after some delay upon sensing the channel being idle and although the cumulative effects of these delays may not be exponentially distributed, the principle remains the same.

3.4 NCS Stability

The notion of *robustness* of various stability properties plays a fundamental role in practical design and implementation of control systems as evidenced by the extensive literature discussing, for example, input-to-state stability (ISS), H_2, H_∞ design and variants of robust stability. To that end, [8] and [9] have examined L_p and input-to-state stability of NCS, respectively and it was shown in [13] that persistently exciting scheduling protocols lead to L_p stable NCS when appropriate conditions are imposed on transmission rates and the nominal system and similar results were provided for UGES and UGAS protocols in [8] and [9], respectively. While the proof techniques and settings are substantially different, the novel use of various small-gain theorems is a unifying theme throughout these results and a powerful tool for quantifying robustness. See [7, Chapter 5.4] for an introduction to the notion of input/output stability gain and [6] for general ISS small-gain results.

We outline several NCS stability results in the ensuing sections and refer the reader to [8], [9] and [13] where the results are stated and proved in greater generality. Finally, while these results are ISS or input–output stability (IOS) type results, whenever exogenous perturbations are removed, UGES and UGAS can be recovered under additional mild technical assumptions. See [8, Section II-B], for instance.

We first recall the definition of L_p stability and detectability for a system Σ_z with jumps:

$$\Sigma_z\colon \dot{z} = f(t, z, w), \quad t \in [t_i, t_{i+1}], \tag{3.55}$$

output $y(t) = g(t, z)$ and with jump equation

$$z(t_i^+) = h(i, z(t_i)). \tag{3.56}$$

Let $f\colon \mathbb{R} \to \mathbb{R}^n$ be a (Lebesgue) measurable function and define $\|f\|_p := \left(\int_{\mathbb{R}} |f(s)|^p ds\right)^{1/p}$ for $1 \leq p < \infty$ and define $\|f\|_\infty := \operatorname{ess.sup}_{t\in\mathbb{R}} |f(t)|$. We say that $f \in L_p$ for $p \in [1, \infty]$ whenever $\|f\|_p < \infty$. Let $f\colon \mathbb{R} \to \mathbb{R}^n$ and let $[a, b] \subset \mathbb{R}$. We use the notation

$$\|f[a, b]\|_p := \left(\int_{[a,b]} |f(s)|^p ds\right)^{1/p}$$

to denote the L_p norm of f when restricted to the interval $[a, b]$.

Definition 3.8. Let $p \in [1, \infty]$ and $\gamma \geq 0$ be given. We say that Σ_z is L_p stable from w to y with gain γ if $\exists\, K \geq 0$: $\|y[t_0,t]\|_p \leq K|z_0| + \gamma\|w[t_0,t]\|_p$.□

Definition 3.9. Let $p, q \in [1, \infty]$ and $\gamma \geq 0$ be given. The state z of Σ_z is said to be L_p to L_q detectable from output y with gain γ if $\exists\, K \geq 0$: $\|z[t_0,t]\|_q \leq K|z_0| + \gamma\|y[t_0,t]\|_p + \gamma\|w[t_0,t]\|_p$. □

An exposition of these ideas as they pertain to NCS can be found in [8, Section II-B].

3.4.1 L_p Stability of NCS with Lyapunov UGES Protocols

A more general version of the following result was first presented in [8] and asserts that Lyapunov UGES scheduling protocols preserve L_p stability of the network-free system under appropriate conditions and for small enough values of MATI.

Theorem 3.1. Consider NCS (3.29)–(3.31b) and suppose that:

(i) the NCS scheduling protocol (3.31a) is Lyapunov UGES with Lyapunov function W that is locally Lipschitz in e and uniformly in i, and there exists $L \geq 0$ such that:

$$\left\langle \frac{\partial W(i,e)}{\partial e}, g(t,x,e,w) \right\rangle \leq LW(i,e) + |\tilde{y}|, \tag{3.57}$$

for almost all $e \in \mathbb{R}^{n_e}$, for all $(x,w) \in \mathbb{R}^{n_x} \times \mathbb{R}^{n_w}$, for all $t \in (t_i, t_{i+1})$, for all $i \in \mathbb{N}$, where $\tilde{y}\colon \mathbb{R}^{n_e} \times \mathbb{R}^{n_w} \to \mathbb{R}$ is a continuous function of (x,w);

(ii) System (3.29) is L_p stable from (W,w) to $\tilde{y}$ with gain γ for some $p \in [1,\infty]$; (x,w) is L_p to L_p detectable from $\tilde{y}$; (e,w) is L_p to L_p detectable from W; and

(iii) MATI satisfies $\tau \in (\epsilon, \tau^*)$, $\epsilon \in (0, \tau^*)$, where

$$\tau^* = \frac{1}{L}\ln\left(\frac{L+\gamma}{\theta L + \gamma}\right), \tag{3.58}$$

and θ comes from (3.41).

Then, the NCS is L_p-stable from w to (x,e) with linear gain. □

Remark 3.4. Within the framework of hybrid systems presented in [4], results analogous to Theorem 3.1 are developed in [1] where τ^* is given by

$$\tau^* = \begin{cases} \dfrac{1}{Lr}\arctan\left(\dfrac{r(1-\theta)}{2\dfrac{\theta}{1+\theta}\left(\dfrac{\gamma}{L}-1\right)+1+\theta}\right), & \gamma > L, \\[2ex] \dfrac{1-\theta}{L(1+\theta)}, & \gamma = L, \\[2ex] \dfrac{1}{Lr}\operatorname{arctanh}\left(\dfrac{r(1-\theta)}{2\dfrac{\theta}{1+\theta}\left(\dfrac{\gamma}{L}-1\right)+1+\theta}\right), & \gamma < L, \end{cases} \tag{3.59}$$

where

$$r = \sqrt{\left|\left(\frac{\gamma}{L}\right)^2 - 1\right|}\,. \tag{3.60}$$

This bound is shown to improve upon (3.58) in [1] when verifying UGES. The results therein are stated for UGAS, UGES and semi-global practical ISS and can, in principle, be extended to apply to L_p IOS. □

3.4.2 L_p Stability of NCS with PE_T Protocols

The following theorem asserts that PE protocols lead to L_p stability of the NCS for sufficiently small MATI. While we do not provide a closed-form expression for MATI bounds, the bounds are readily obtained in examples by numerically solving for τ^* in (3.59). Note that we only consider stability of e and x. The decision-vector, if used in the protocol being analyzed, may fail to verify any stability properties but as $\hat{e}$ has no physical significance as a state vector whose evolution is governed by the protocol, this is generally not an issue. Let $\mathcal{A}_n$ denote the set of all $n \times n$ matrices and let $\mathcal{A}_n^+$ denote the subset of all matrices that are positive semi-definite, symmetric and have positive entries and let $\mathbb{R}_+^n$ denote the nonnegative orthant.

Theorem 3.2. Consider NCS (3.29)–(3.31b) and suppose that:

(i) the NCS scheduling protocol (3.31b) is uniformly persistently exciting in time T and there exists $A \in \mathcal{A}_{n_e}^+$ and a continuous $\tilde{y}\colon \mathbb{R}^{n_x} \times \mathbb{R}^{n_w} \to \mathbb{R}_+^{n_e}$ so that the error dynamics (3.30) satisfy[19]

$$\overline{g}(t, x, e, w) \preceq A\overline{e} + \tilde{y}(x, w) \tag{3.61}$$

for all $(x, e, w) \in \mathbb{R}^{n_x} \times \mathbb{R}^{n_e} \times \mathbb{R}^{n_w}$, for all $t \in (t_i, t_{i+1})$, for all $i \in \mathbb{N}$, where $\tilde{y} = G(x) + H(w)$;

(ii) System (3.29) is L_p stable from (e, w) to $G(x)$ with gain γ for some $p \in [1, \infty]$; (x, w) is L_p to L_p detectable from $\tilde{y}$;

(iii) and MATI satisfies $\tau \in (\epsilon, \tau^*)$, $\epsilon \in (0, \tau^*)$, where $\tau^* = \dfrac{\ln(z)}{|A|T}$ and z solves

$$z(|A| + \gamma T) - \gamma T z^{1-1/T} - 2|A| = 0, \tag{3.62}$$

where A comes from (3.61).

Then, the NCS is L_p-stable from w to (x, e) with linear gain. □

[19] Let $x = (x_1, \dots, x_n)$ and $y = (y_1, \dots, y_n) \in \mathbb{R}^n$. The *vector* partial order $\preceq$ is given by $x \preceq y \iff (x_1 \leq y_1) \wedge \cdots \wedge (x_n \leq y_n)$ and $\overline{e}$ and $\overline{g}$ are given by $\overline{e} := (|e_1|, \dots, |e_{n_e}|)^T$ and $t \overset{\overline{g}}{\mapsto} \overline{g(t)}$, respectively. That is, $\overline{e}$ is the vector that results from taking the absolute value of each scalar component of e and $\overline{g}$ operates analogously on the image of g.

Remark 3.5. Suppose that $g(t, x, e, w) = Bx + Ce + Dw$ and let $A = [a_{ij}]$, where $a_{ij} = \max\{|c_{ij}|, |c_{ji}|\}$ and $\tilde{y}(x, w) = \overline{Bx + Dw}$. We immediately have that A and $\tilde{y}(x, w)$ satisfy Condition 2 of Theorem 3.2 and $\|\tilde{y}(x, w)\|_p = \|Bx + Dw\|_p \leq \|Bx\|_p + \|Dw\|_p$. Whenever g satisfies a linear growth bound of the form $|g(t, x, e, w)| \leq L(|x| + |e| + |w|)$, it is straightforward to construct an appropriate A and $\tilde{y}$. □

Remark 3.6. Suppose that the network-free system is L_p stable from w to x with gain γ and the NCS satisfies the hypotheses of Theorem 3.2. Then for any $\gamma^* > \gamma$, it is possible to show that there exists a MATI τ such that the NCS is L_p stable from w to x with gain γ^*. This corollary of Theorem 3.2 is particularly useful in the design of optimal/robust controllers. □

3.4.3 L_p Stability of NCS with Random Protocols

The following result analyzes the input–output L_p stability (IOS) of NCS (in expectation), the essence of which is that outputs (or state) of an NCS verify a robustness property with respect to exogenous disturbances. We stress that it is only the network protocol and channel that induces randomness in our models and that the exogenous disturbances are L_p signals as in [8] and [13].

Although link cover times and inter-transmission are now random, and hence, not uniform, if the network-free system is L_p stable, the NCS remains so with any contention protocol, in the sense of our definition, whenever attempted transmissions occur "fast enough." By "fast enough" we mean that there exists a choice of intensity λ of the transmission process parameterized by properties of the protocol and the NCS dynamics such that the NCS is L_p stable-in-expectation from disturbance to NCS state with a finite expected gain.

Intuitively, and despite the presence of collisions, random packet dropouts and random inter-arrival times, it seems natural to expect that the stability of the NCS (3.24)–(3.28a) for high enough "average" transmission rates and in light of the a.s. cover times of contention protocols and in analogy with persistently exciting scheduling protocols, this stability ought to be robust in an L_p sense. In fact, if we relax our notion of "L_p stability" to "L_p stability-in-expectation," we can prove a positive result in that direction. The definition of these properties is obtained, essentially, by using expected norms $\mathbf{E}\|\cdot\|$ in lieu of $\|\cdot\|$ in Definitions 3.8 and 3.9. We stress that, as developed in this chapter, these notions only apply to hybrid systems of the form (3.55) and (3.56), i.e., we insist that w is "essentially" an L_p signal and not a Lèvy process (cf. [5]) specifically because we are concerned with robustness of stability in the sense of, e.g., [3], whereas a Lèvy process characterization of disturbances may be more appropriate in modeling sensor noise and quantization phenomena.

While the following results are stated for the delay and inter-arrival processes presented in Definition 3.4, it is straightforward to extend them to a more general class of processes.

Theorem 3.3. Consider an ℓ-link NCS (3.29)–(3.31b) and suppose that:

(i) the NCS employs a contention scheduling protocol with i.i.d. cover times T_i and the inter-arrival process is Poisson with intensity λ and also suppose that the NCS error dynamics satisfy

$$\overline{g}(t, x, e, w) \preceq A\overline{e} + \tilde{y}(x, w) \tag{3.63}$$

for all $(x, e, w) \in \mathbb{R}^{n_x} \times \mathbb{R}^{n_e} \times \mathbb{R}^{n_w}$ and almost all t, where A is a nonnegative symmetric $n_e \times n_e$ matrix with nonnegative entries and $\tilde{y} = G(x) + H(w)$;

(ii) System (3.29) is L_p stable-in-expectation from (e, w) to $G(x)$ with expected gain γ for some $p \in [1, \infty]$; (3.30) is L_p to L_p detectable-in-expectation from $\tilde{y}$.

Then, there exists $\lambda < \infty$ depending on $(\ell, |A|, \gamma, \mathbf{E}[T], p_0)$ such that the NCS is L_p stable-in-expectation from w to (x, e) with a finite linear expected gain $1/(1 - \gamma\gamma^*)$. Specifically, λ solves $\gamma^*\gamma < 1$ with

$$\gamma^* = \frac{\mathbf{E}[T](1 + \rho)}{(\lambda - |A|)(1 - \rho)},$$

where,

$$\rho = (\alpha(1 - p_0))^\ell \prod_{k=1}^{\ell} \frac{\ell - (k-1)}{\ell(1 - p_0\alpha) - (k-1)(1 - p_0)\alpha} - 1,$$

and $\alpha = \dfrac{\lambda}{\lambda - |A|}$ and $\lambda > \dfrac{|A|}{1 - p_0}$. □

Remark 3.7. While no bounds for λ are given, the requisite intensity can be found numerically. □

3.4.4 L_p Stability of NCS with a.s. Lyapunov Protocols

The following result is a natural extension to Theorem 3.1 for channels that have a non-zero probability of packet dropout and is intended to be used in much the same way as the latter result. While [8] presents sufficient conditions for L_p stability in the presence of deterministically characterized packet dropouts for Lyapunov UGES protocols, we believe the following result is a more natural treatment of dropouts and the conditions are directly verifiable.

Theorem 3.4. Consider NCS (3.29)–(3.31b) and suppose that:

(i) the NCS scheduling protocol (3.31a) is a.e. Lyapunov U GES with Lyapunov function W that is locally Lipschitz in e and uniformly in i, and there exists $L \geq 0$ such that:

$$\left\langle \frac{\partial W(i,e)}{\partial e}, g(t,x,e,w) \right\rangle \leq LW(i,e) + |\tilde{y}|, \tag{3.64}$$

for almost all $e \in \mathbb{R}^{n_e}$, for all $(x,w) \in \mathbb{R}^{n_x} \times \mathbb{R}^{n_w}$, all $t \in (t_i, t_{i+1})$, for all $i \in \mathbb{N}$, where $\tilde{y}\colon \mathbb{R}^{n_e} \times \mathbb{R}^{n_w} \to \mathbb{R}$ is a continuous function of (x,w);

(ii) System (3.29) is L_p stable from (W,w) to $\tilde{y}$ with finite expected gain γ for some $p \in [1,\infty]$; (x,w) is L_p to L_p detectable from $\tilde{y}$ with finite expected gain; e is L_p detectable from W with finite expected gain;

(iii) the channel packet dropout probability is given by $p_O \geq 0$ and (3.36) is satisfied with an i.i.d. sequence $\{\kappa_i\}$ such that the intensity of the inter-transmission process λ satisfies

$$\lambda > \frac{\gamma + L}{1 - \mathbf{E}[\kappa]}. \tag{3.65}$$

Then, the NCS is L_p-stable from w to (x,e) with finite expected linear gain.
□

Remark 3.8. As the motivation for studying a.s. Lyapunov UGES comes from the use of Lyapunov UGES protocols on non-ideal channels, we can restate several of the conditions of Theorem 3.4 in light of Proposition 3.1. Let θ be as in (3.41) and let the probability of packet dropout p_O satisfy (3.42). The requisite intensity in (3.65) becomes

$$\lambda > \frac{\gamma + L}{(1 - p_O)(1 - \theta)}. \tag{3.66}$$

□

Remark 3.9. As in [8] and [13], in both this and the preceding section, several generalizations and specializations of the stability results are possible. With additional technical assumptions on the NCS dynamics, one can conclude uniform global exponential stability (in expectation) and the assumptions on the various reset maps can be relaxed so as to infer ISS-like properties in lieu of L_p stability as discussed in [9]. If we forgo the detectability assumptions in the hypotheses of Theorems 3.3 and 3.4 we can only infer input-to-output stability-in-expectation. □

3.5 Case Studies and Comparisons

The aim of this section is to examine the various results presented in this chapter and compare them to results presented in the literature. For simplicity we will focus on the following linear time-invariant systems where the simplified equations for an ℓ-link NCS are given by (3.6) together with jump equations (3.31a) or (3.31b).

Example 3.6 (Batch Reactor). The linearized model of an unstable batch reactor is a two-input-two-output NCS that can be written as:

$$\dot{x}_P = A_P x_P + B_P u, \quad y = C_P x_P,$$

where $C_P = \begin{bmatrix} 1 & 0 & 1 & -1 \\ 0 & 1 & 0 & 0 \end{bmatrix}$,

$$A_P = \begin{bmatrix} 1.38 & -0.2077 & 6.715 & -5.676 \\ -0.5814 & -4.29 & 0 & 0.675 \\ 1.067 & 4.273 & -6.654 & 5.893 \\ 0.048 & 4.273 & 1.343 & -2.104 \end{bmatrix}, \quad B_P = \begin{bmatrix} 0 & 0 \\ 5.679 & 0 \\ 1.136 & -3.146 \\ 1.136 & 0 \end{bmatrix}.$$

The system is controlled by a PI controller with a state-space realization prescribed by

$$\dot{x}_C = A_C x_C + B_C y, \quad u = C_C x_C + D_C y,$$

and

$$A_C = \begin{bmatrix} 0 & 0 \\ 0 & 0 \end{bmatrix}, \quad B_C = \begin{bmatrix} 0 & 1 \\ 1 & 0 \end{bmatrix}, \quad C_C = -\begin{bmatrix} 2 & 0 \\ 0 & 8 \end{bmatrix}, \quad D_C = -\begin{bmatrix} 0 & 2 \\ -5 & 0 \end{bmatrix}.$$

Assuming that only the outputs are transmitted via the network, we have a two-link NCS ($\ell = 2, l_1 = l_2 = 1$) with error and state equations prescribed by (3.6) where

$$A_{11} = \begin{bmatrix} A_P + B_P D_C C_P & B_P C_C \\ B_C C_P & A_C \end{bmatrix}, \quad A_{12} = \begin{bmatrix} B_P D_C \\ B_C \end{bmatrix},$$
$$A_{21} = -\begin{bmatrix} C_P & 0 \end{bmatrix} A_{11}, \quad A_{22} = -\begin{bmatrix} C_P & 0 \end{bmatrix} A_{12}.$$

The error equation is given by

$$\dot{e} = A_{22} e + A_{21} x. \tag{3.67}$$

□

This example is used as the benchmark in comparing the inter-transmission bounds with the stability analysis frameworks outlined in this chapter and in [16, 18, 19].

3.5.1 Comparison of Analytical Inter-transmission Bounds

Prior to making numerical comparisons with respect to the bounds obtained for Example 3.6, we provide a brief summary of the analytical bounds in Table 3.1 as they apply in general. The various constants used are defined and explained in the respective referenced sections and details can be found in the respective sources cited in the table. These are analytical bounds that

Table 3.1. Summary of inter-transmission bounds for various classes of protocols

MATI – Section 3.2.2, [16, 18, 19] (Worst Case Analysis)	
RR & TOD Protocol	$\tau^{WC}_{3.2.2} = \dfrac{c_3}{M\ell(\ell+1)k_h k_f c_4}$
Silent-time Protocols silent-time T	$\tau^{ST}_{3.2.2} = \min\left\{\dfrac{\ln(2)}{k_h T}, \dfrac{S}{8}, \dfrac{S}{16c_1\sqrt{c_1/c_2}k_h}\right\}$, where $S = [k_h\sqrt{c_1/c_2}\sum_{i=1}^{\ell}(i+T-\ell)]^{-1}$
MATI – Section 3.4.1, [8] (Lyapunov UGES Analysis)	
RR Protocol	$\tau^{RR}_{3.4.1} = \dfrac{1}{k_h\sqrt{\ell}}\ln\left(\dfrac{\sqrt{\ell}(k_h+\gamma)}{k_h\sqrt{\ell-1}+\gamma\sqrt{\ell}}\right)$
TOD Protocol	$\tau^{TOD}_{3.4.1} = \dfrac{1}{L}\ln\left(\dfrac{L\sqrt{\ell}+\gamma\sqrt{\ell}}{L\sqrt{\ell-1}+\gamma\sqrt{\ell}}\right)$
MATI – Section 3.4.2, [13] (PE_T Analysis)	
PE_T Protocols (including RR)	$\tau^{PE}_{3.4.2} = \dfrac{\ln(z)}{\|A\|T}$, where z solves $z(\|A\|+\gamma T) - \gamma T z^{1-1/T} - 2\|A\| = 0$
Reciprocal-Intensity – Section 3.4.3, [11] (a.s. Cover Time Analysis)	
Stochastic Protocols $\mathbf{P}\{\text{dropout}\} = p_0$	$\tau^{STO}_{3.4.3} < \dfrac{(1-p_0)}{\|A\|}$, solved numerically via Theorem 3.3
Reciprocal-Intensity – Section 3.4.4, [11] (a.s. Lyapunov UGES Analysis)	
RR Protocol	$\tau^{RR}_{3.4.4} = \dfrac{(1-p_0)(\sqrt{\ell}-\sqrt{\ell-1})}{\ell(\gamma+k_h)}$
TOD Protocol	$\tau^{TOD}_{3.4.4} = \dfrac{(1-p_0)(\sqrt{\ell}-\sqrt{\ell-1})}{\sqrt{\ell}(\gamma+L)}$

guarantee stability. For all bounds presented, stability is in the sense of L_p (in-expectation) except for those derived in [16, 18, 19], where UGES is the applicable notion of stability.

Table 3.2 compares a selection of these MATI bounds as they apply to TOD and RR. It is shown in [13, Section VI-C] that for LTI systems employing RR scheduling, MATI bounds obtained within the framework outlined in Section 3.4.2 are asymptotically larger by a factor of $O(\ell^{1/2})$ than the MATI obtained in [8] which are, in turn, shown to be analytically superior to the bounds in [15] for both TOD and RR. As indicated in Remark 3.4, for protocols that are Lyapunov UGES or UGAS, [1] may offer improved MATI bounds over [8] and, for the batch reactor example, these were demonstrated to be an improvement of approximately 10%.

Table 3.2. Summary of analytic comparisons for NCS without dropouts for an ℓ-link NCS with constants as in (3.13)–(3.15)

Linear Systems	
RR Protocol	$\frac{\tau_{3.4.1}^{RR}}{\tau_{3.2.2}^{WC}} \geq 2\frac{\ell+1}{\sqrt{\ell}}\sqrt{\frac{c_2}{c_1}}\left(\sqrt{\frac{c_2}{c_1}}+1\right)$ $\frac{\tau_{3.4.2}^{PE}}{\tau_{3.4.1}^{RR}} \geq O(\ell^{1/2})$ as $\ell \to \infty$
TOD Protocol	$\frac{\tau_{3.4.1}^{TOD}}{\tau_{3.2.2}^{WC}} \geq 2(\ell+1)\sqrt{\frac{c_2}{c_1}}\left(\sqrt{\frac{c_2}{c_1}}+1\right)$
Nonlinear Systems	
RR Protocol	$\frac{\tau_{3.4.1}^{RR}}{\tau_{3.2.2}^{WC}} \geq 8\frac{\ell+1}{\sqrt{\ell}}\left(\frac{c_2}{c_1}\right)^{\frac{3}{2}}\left(\sqrt{\frac{c_2}{c_1}}+1\right)$
TOD Protocol	$\frac{\tau_{3.4.1}^{TOD}}{\tau_{3.2.2}^{WC}} \geq 8(\ell+1)\left(\frac{c_2}{c_1}\right)^{\frac{3}{2}}\left(\sqrt{\frac{c_2}{c_1}}+1\right)$

3.5.2 Comparison of Numerical Inter-transmission Bounds ($p_0 = 0$)

For simplicity, and since L_p stability results are not provided in [18], we will largely restrict our discussion without exogenous disturbances and examine bounds that verify UGES and related properties. Much of the focus will be on RR scheduling as it is the only scheduling protocol that can be mutually treated by the analysis frameworks in this chapter, [8] and [18] but several other protocols will be examined as well.

We present and compare various results for the batch reactor example, Example 3.6, following [8], [13], [15]. The comparison results are summarized below:

(a) The MATI bounds are shown in Table 3.3 with the bounds computed via the PE framework larger than those obtained using the results of [18] by a factor of 10^7 and larger than the bound obtained by the results of [8] by a factor of 1.5. The bounds $\tau_{3.4.2}^{PE}$ and $\tau_{3.2.2}^{WC}$ apply to any PE_T protocol for the original two-link system. The bound $\tau_{3.4.1}^{RR}$ only applies to RR ($T = \ell = 2$).

(b) When using RR, $\tau_{3.4.2}^{PE}$ that achieves UGES is equivalent to a network throughput of 84 kbps (assuming 128 byte frames), achievable on current 802.11g and 802.11b wireless networks and $\tau_{3.4.4}^{RR}$ requires an effective network throughput of approximately 125 kbps.

(c) We formally fix the constants used to compute the respective bounds and plot $\tau_{3.4.2}^{PE}$ and $\tau_{3.4.4}^{RR}$ with $T = \ell \in [1, 1000]$ in Fig. 3.6 to examine the behavior of the bounds as the number of links grow. We also fix $\ell = 2$ and allow $T \geq 2$ to vary for $\tau_{3.2.2}^{ST}$ and $\tau_{3.4.2}^{PE}$. Despite the relatively modest improvements for the nominal two-link system using RR, the differences are significant on the $\log_{10}(T) \times \log_{10}(\tau^*)$ scale used in Fig. 3.6 when we formally increase T or, equivalently, the number of links.

Simulations and alternative techniques for calculating MATI are a key test of the practicality of the MATI bounds and stability results produced in this chapter and in the literature. For linear systems with equidistant transmission times employing RR scheduling, an actual analytic MATI bound can be computed as discussed in [8, Section VII-A]. For general protocols, however,

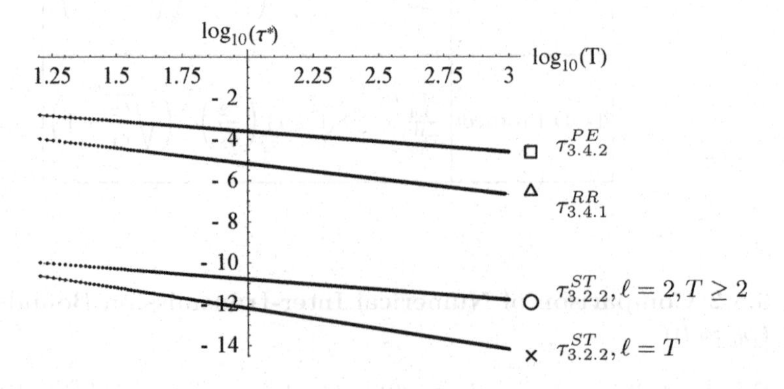

Fig. 3.6. Batch reactor MATI bounds comparison for PE_T protocols, $T \in [1, 1000]$

Table 3.3. MATI bounds achieving UGES for Example 3.6 with PE_T and Lyapunov UGES protocols

	$T=2$	$T=6$	$T=50$
$\tau_{3.4.2}^{PE}$	0.0123	0.004	4.75×10^{-4}
$\tau_{3.4.4}^{RR}$	0.0082	N/A	N/A
$\tau_{3.2.2}^{WC}$	1.05×10^{-9}	2.86×10^{-10}	3.18×10^{-11}
$\tau_{3.4.4}^{TOD}$	0.01	N/A	N/A
$\tau_{3.4.2}^{PE}/\tau_{3.2.2}^{WC}$	1.18×10^{7}	1.40×10^{7}	1.49×10^{7}
$\tau_{3.4.2}^{PE}/\tau_{3.4.4}^{RR}$	1.50	N/A	N/A
$\tau_{3.4.2}^{PE}/\tau_{3.4.4}^{TOD}$	1.23	N/A	N/A

simulations are the only resort, and as such, no firm conclusions can be drawn vis-a-vis the theoretical bounds for arbitrary NCS.

3.5.3 Comparison of Numerical Inter-transmission Bounds ($p_O > 0$)

Finally, we examine Example 3.6 for channels where $p_O > 0$. In particular, we look at the CSMA protocol described in Section 3.3.7 and hence,

$$\mathbf{E}[T] = 2 \cdot H_2/(1-p_O) = 3/(1-p_O). \tag{3.68}$$

By Theorem 3.3, the batch reactor system will be L_p stable-in-expectation from w to x if

$$\frac{\mathbf{E}[T](1+\rho)}{(\lambda - |A|)(1-\rho)}\gamma < 1, \tag{3.69}$$

where γ is the L_p gain of x subsystem from the input e to an "auxiliary" output $\tilde{y} = A_{21}x$.

By solving for λ numerically in (3.69), subject to the constraint $\lambda > |A|/(1-p_O)$, we are able to establish expected transmission rate bounds as a function of p_O that ensure L_p stability of the batch reactor system. The batch reactor system with the CSMA protocol was also simulated using expected transmission rates of $[1, \infty)$ transmissions per second for $p_O \in [0.1, 0.8]$. A bisection heuristic was used to find the intensities that resulted in stability with the ensemble average of multiple simulations with fixed initial conditions to yield the simulation-derived intensity bound.

The expected transmission rate bounds and expected inter-transmission times are shown in Table 3.4 as a function of dropout/collision probability p_O and plotted in Fig. 3.7. Simulation-derived bounds are also listed in Table 3.4.

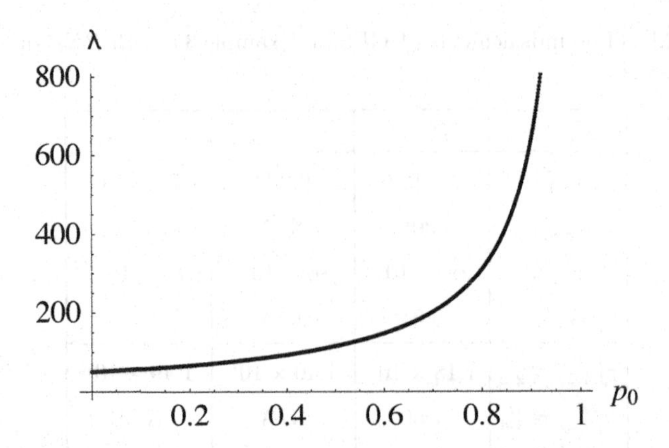

Fig. 3.7. Batch reactor expected transmission rate bounds for stochastic protocols as a function of dropout/collision probability p_O with identical initial conditions

For the initial condition used, the bounds obtained via Theorem 3.3 are within a factor of 4 of simulation-based bounds, and for example, demonstrate that with a 50% probability of dropout/collision, the network must deliver approximately 922 kbps (116 × 8 bits) of network throughput to maintain L_p stability. This is well within the realm of ordinary Ethernet and 802.11 wireless technology.

We can also consider the example within the context of a.s. Lyapunov UGES. Suppose that the TOD scheduling is employed. From Table 3.1, the requisite intensity for the conditions of Theorem 3.4 to be verified is

Table 3.4. Transmission rate and inter-transmission time bounds λ and $\tau^{STO}_{3.4.3} = 1/\lambda$ are derived via Theorem 3.3; λ^* and $\tau^{STO*}_{3.4.3} = 1/\lambda^*$ are derived via simulation

p_O	λ	$\tau^{STO}_{3.4.3} = 1/\lambda$ (s)	λ^*	$\tau^{STO*}_{3.4.3} = 1/\lambda^*$ (s)
0	50.19	0.02	14.77	0.0677
0.1	57.46	0.017	16.05	0.0623
0.2	66.52	0.015	18.38	0.0544
0.3	78.15	0.013	21.37	0.0468
0.4	93.63	0.011	25.00	0.0400
0.5	115.27	0.0087	31.65	0.0316
0.6	147.71	0.0068	37.74	0.0265
0.7	201.74	0.0049	61.35	0.0163
0.8	309.74	0.0032	145.77	0.00686

$$\lambda > \frac{108.07}{1 - p_0}.$$

For an ideal channel ($p_0 = 0$), this corresponds to a transmission at least once every 9.25 ms compared to a MATI of 0.01 s for the deterministic results presented in [8] – a factor of 1.08 improvement in favor of the deterministic results. The notion of MATI implies that every inter-transmission time is *uniformly bounded* whereas the intensity (or reciprocal) is an "average MATI" – individual inter-transmission times can individually exceed or fall short of the average. Notably, both values fall short of the figure obtained for the CSMA protocol of 0.02 s. As the characterization of dropouts in [8] is markedly different from that presented here, we do not pursue a comparison for $p_0 > 0$. We can, however, compare CSMA and TOD in the presence of dropouts as presented in this section and we see that the trend is continued for $p_0 > 0$, e.g., the requisite intensity for $p_0 = 0.5$ is over 216 for TOD and less than 116 for CSMA.

3.6 Conclusions

This chapter presented several general frameworks for emulation-based design of a general nonlinear control systems with disturbances that rely upon properties of the network-free system and various properties of the scheduling protocol used. Our guiding philosophy in the approach is the following qualitative statement that intuition suggests: *for high enough transmission rates, a scheduling protocol that is guaranteed to reduce the network-induced error within a finite amount of time ought to preserve stability properties of the network-free system.* In particular, this is the case for (a.s.) Lyapunov UGES and UGAS protocols as well as PE_T and a.s. covering protocols.

Quantitatively, the results outlined provide the sharpest bounds for MATI and expected transmission rate currently known in the literature for the classes of systems and protocols analyzed, and in some cases, are the only known results for certain classes of systems and scheduling protocols.

We qualify this observation by noting that the various protocol properties, namely, PE_T, Lyapunov UGES and UGES and their stochastic analogs are not necessarily the finest characterization possible of any particular protocol. This is reflected in the disparity between theoretical MATI and transmission rate values and those obtained by simulations. For example, it is known that for LTI systems employing RR with equidistant inter-transmission times, analytic MATI bounds that achieve UGES can be computed and, indeed, are as sharp or sharper than those obtained by any result in this chapter. The aim of this chapter, however, was to present emulation-type results and design procedures for the largest class of systems for which results are currently known and which are useful in practice. To that end, we believe that this work serves as a useful starting point from which there is still much scope for improvement.

References

1. Carnevale D, Teel AR, Nešić D (2007) A Lyapunov proof of an improved maximum allowable transfer interval for networked control systems. IEEE Trans Automat Control 52(5):892–897
2. Coyle E, Liu B (1985) A matrix representation of CSMA/CD networks. IEEE Trans Commun 33(1):53–64
3. Dullerud GE, Paganini F (2000) A course in robust control theory. Springer, New York
4. Goebel R, Teel AR (2006) Solutions to hybrid inclusions via set and graphical convergence with stability theory applications. Automatica 42(4):573–587
5. Hespanha JP, Teel AR (2006) Stochastic impulsive systems driven by renewal processes. In: Proc 17th International Symposium on Mathematical Theory of Networks and Systems, Kyoto, Japan
6. Jiang ZP, Teel AR, Praly L (1994) Small gain theorem for ISS systems and applications. Math Control Signals Syst 7:95–120
7. Khalil HK (2002) Nonlinear systems, 3rd edition. Prentice Hall, Englewood Cliffs, NJ
8. Nešić D, Teel AR (2004) Input–output stability properties of networked control systems. IEEE Trans Automatic Control 49:1650–1667
9. Nešić D, Teel AR (2004) Input-to-state stability of networked control system. Automatica 40:2121–2128
10. Sigman K (1995) Stationary marked point processes: an intuitive approach. Chapman & Hall, New York
11. Tabbara M, Nešić D (2007) Input–output stability of networked control systems with stochastic protocols and channels. IEEE Trans Automatic Control (to appear)
12. Tabbara M, Nešić D, Teel AR (2005) Input–output stability of wireless networked control systems. In: Proceedings of 44th IEEE Conference on Decision and Control, Seville, Spain, 209–214
13. Tabbara M, Nešić D, Teel AR (2007) Stability of wireless and wireline networked control systems. IEEE Trans Automatic Control (to appear)
14. Tanenbaum AS (2003) Computer networks, 4th edition. Prentice Hall, Englewood Cliffs, NJ
15. Walsh GC, Beldiman O, Bushnell LG (2001) Asymptotic behaviour of nonlinear networked control systems. IEEE Trans Automatic Control 46:1093–1097
16. Walsh GC, Ye H, Bushnell LG (1999) Stability analysis of networked control systems. In: Proceedings American Control Conference, San Diego, USA, 2876–2880
17. Walsh GC, Ye H (2001) Scheduling of networked control systems. IEEE Control Syst Mag 21:57–65,
18. Ye H, Walsh GC, Bushnell LG (2001) Real-time mixed-traffic wireless networks. IEEE Trans Ind Electron 48:883–890
19. Zhang W, Branicky MS, Phillips SM (2001) Stability of networked control systems. IEEE Control Syst Mag 21:84–99

4

Analysis and Design of Networked Predictive Control Systems

Guo-Ping Liu

University of Glamorgan, Pontypridd, CF37 1DL, UK gpliu@glam.ac.uk

Abstract. This chapter considers the analysis and design of networked control systems with random communication time delay, which is known to highly degrade the control performance of the control system. It introduces a novel control strategy of networked control systems, which is termed networked predictive control. The stability of the closed-loop networked predictive control system is analysed. The analytical criteria are obtained for both fixed and random communication time delays. The off-line and real-time simulation of the networked predictive control systems is detailed. Also, this control strategy is applied to a servo control system through the Ethernet. Various simulation and experimental results demonstrate the operation of the networked predictive control systems.

Keywords. Networked predictive control, networked system stability, random communication delay.

4.1 Introduction

With the emergence of high speed network technology that allows a cluster of devices to be linked together economically to form distributed networks which are capable of remote data transmission and data exchanges, distributed control systems based on networks are increasing rapidly in various applications (IEEE, 2002). Owing to the use of networks, the complexity and cost of distributed control systems are reduced greatly and the maintenance of the systems becomes much easier (Zhang *et al.*, 2001). Because of these attractive benefits, many industrial companies and institutions have shown great interest in applying various networks to remote control systems and manufacturing automation.

A feedback control system wherein the control loop is closed through a real-time network is known as a networked control system (NCS), which includes fieldbus control systems constructed on the base of bus technology (e.g., DeviceNet, ControlNet and LonWorks) and Internet based control systems using general computer networks. NCS is a completely distributed real-time feedback control system that is an integration of sensors, controllers, actuators and communication networks. It provides data transmission between devices in order that users at different sites can realize resource sharing and coordinating manipulation.

As there are more and more applications of networked control systems in industry, such as traffic, communication, aviation and spaceflight, more attention in this area has been paid to design and analysis of NCS (Zhivoglyadov, 2003). Generally speaking, there are three types of NCS methods: Type 1 – scheduling methods that guarantee network QoS (quality of service); Type 2 – control methods that guarantee system QoP (quality of performance); and Type 3 – integrated scheduling and control methods that consider both QoS and QoP. For Type 1, the following scheduling methods have been developed: scheduling method MEF (Maximum-Error-First) based on the MATI (Maximal-Allowable-Transfer-Interval) (Walsh *et al.*, 1999), and a sampling time scheduling method of network bandwidth allocation and sampling period decision for multi-loop NCSs by virtue of the notion "window", namely the service window of each transmission data in the network (Hong, 1995). For Type 2, there are many control methods developed for NCS, for example, augmented deterministic discrete-time model method (Halevi, 1988), queuing method (Luck and Ray, 1994), optimal stochastic control method (Nilsson, 1998), perturbation method (Walsh, 1999), fuzzy logic modulation method (Almutairi, 2001), event-based method (Tarn and Xi, 1998) and predictive control (Liu *et al.*, 2004). For Type 3, the following problems have been studied: the optimal sampling period selection problem for a set of digital controllers (Seto *et al.*, 1996), the sampling period optimization problem under the schedulability constraints (Ryu and Hong, 1997), and the NCS analysis and simulation problem solved by two MATLAB®-based toolboxes: Jitterbug and TrueTime (http://www.control.lth.se/~anton, 2003). Internet based control has also been considered for practical applications, for example, Internet-based process control (Yang *et al.*, 2003), Internet based control systems as a control device (Cushing, 2000), Internet robots (Taylor, 2000) and Internet based multimedia education (Nemoto *et al.*, 1997).

Although various control approaches have been developed for networked control systems, an approach to actively compensate for the random network delay is not available. This chapter first introduces a new control strategy to compensate for the network delay in networked control systems in an active way, which is named as the networked predictive control. Then, the stability analysis of closed-loop networked predictive control systems is discussed. After this, it details the off-line and real-time simulation and implementation of networked predictive control systems. Finally this strategy is applied to control a servo control system through the Ethernet.

4.2 Networked Predictive Control

Since there is an unknown network transmission delay, a networked predictive control scheme is proposed by Liu *et al.* (2004, 2005, 2006). It consists of two main parts: the control prediction generator (CPG) and the network delay compensator (NDC). The former is designed to generate a set of future control predictions. The latter is used to compensate for the unknown random network delay. To make use of the network advantage of transmitting data packages, a set of consecutive control predictions in the forward channel are packed and transmitted through the network at time t. So, this networked predictive control system (NPCS) structure is shown in Fig. 4.1.

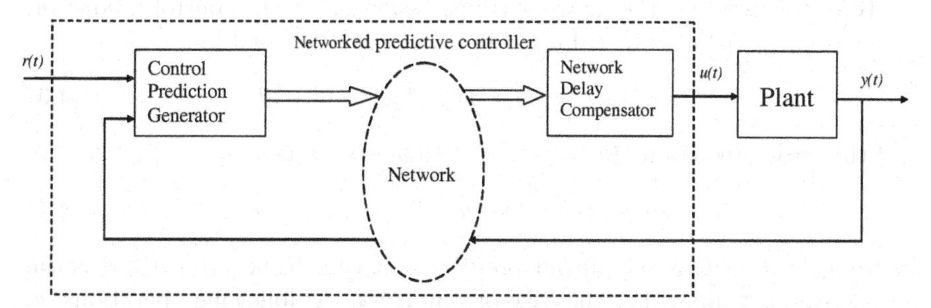

Fig. 4.1. The networked predictive control system

4.2.1 Design of the Control Prediction Generator

For the sake of simplicity, the following assumptions are made:

(a) The network time delay k in the forward channel is random but bounded by $\bar{k}$.
(b) The network time delay f in the backward channel is constant.
(c) The number of consecutive data package drops in the forward channel of the network is not greater than N_c.
(d) The data transmitted through the network are with a time stamp.

Let $\Re[z^{-1}, p]$ denote the set of polynomials in the indeterminate z^{-1} with coefficients in the field of real numbers and with the order p in a set of non-negative integer numbers. For example, the polynomial $A_k(z^{-1}) \in \Re[z^{-1}, n]$ is given by

$$A_k(z^{-1}) = a_{k,0} + a_{k,1}z^{-1} + a_{k,2}z^{-2} + \cdots + a_{k,n}z^{-n}.$$

Consider a SISO (single-input single-output) discrete-time plant described by the following:

$$A(z^{-1})y(t+d) = B(z^{-1})u(t) \tag{4.1}$$

where $y(t)$ and $u(t)$ are the output and control input of the plant, d is the time delay, and $A(z^{-1}) \in \Re[z^{-1}, n]$ and $B(z^{-1}) \in \Re[z^{-1}, m]$ are the system polynomials. If there is no network transmission delay, many control design methods are available for plant (4.1), for example, PID, LQG, MPC, etc. Here, it assumes that the controller of the system without network delay is given by

$$C(z^{-1})u(t) = D(z^{-1})e(t+d) \tag{4.2}$$

where the polynomials $C(z^{-1}) \in \Re[z^{-1}, n_c]$ and $D(z^{-1}) \in \Re[z^{-1}, n_d]$ and $e(t+d) = r(t+d) - \hat{y}(t+d|t)$ is the error between the future reference $r(t+d)$ and the output prediction $\hat{y}(t+d|t)$.

To compensate for the network transmission delay, the control prediction sequence $u(t+i|t)$ at time t, for $i = 0, 1, 2, \ldots$, is generated by

$$C(z^{-1})u(t+i|t) = D(z^{-1})e(t+d+i|t) \tag{4.3}$$

and the error prediction $e(t+d+i|t)$ at time t is defined as

$$e(t+d+i|t) = r(t+d+i) - \hat{y}(t+d+i|t) \tag{4.4}$$

where $\hat{y}(t+d+i|t)$ is the output prediction at time t and $r(t+d+i)$ is the future reference input. For the sake of simplicity, the following operations on predictions are used:

$$z^{-1}v(i|j) = v(i-1|j), \quad \text{if} \quad i > j > 0 \tag{4.5}$$

$$z^{-1}v(i|i) = v(i-1|i-1) \tag{4.6}$$

where $v(\cdot|\cdot)$ represents the control prediction $u(\cdot|\cdot)$ and output prediction $\hat{y}(\cdot|\cdot)$. For example,

$$z^{-3}v(t+2|t) = z^{-2}v(t+1|t) = z^{-1}v(t|t) = v(t-1|t-1).$$

For $i = 0, 1, 2, \ldots, N$, there exists the following Diophantine equation:

$$A(z^{-1})E_i(z^{-1}) + z^{-i-f}F_i(z^{-1}) = 1 \tag{4.7}$$

where the polynomials $E_i(z^{-1}) \in \Re[z^{-1}, i+f-1]$ and $F_i(z^{-1}) \in \Re[z^{-1}, n-1]$. It is clear from Assumption (b) that the past outputs up to time $t-f$ are available on the control prediction generator side. Combining the above and the controlled plant yields the following output predictions at t:

$$\begin{aligned}
\hat{y}(t+d|t) &= F_d(z^{-1})y(t-f) + B(z^{-1})E_d(z^{-1})u(t|t) \\
\hat{y}(t+d+1|t) &= F_{d+1}(z^{-1})y(t-f) + B(z^{-1})E_{d+1}(z^{-1})u(t+1|t) \\
&\vdots \\
\hat{y}(t+d+N|t) &= F_{d+N}(z^{-1})y(t-f) + B(z^{-1})E_{d+N}(z^{-1})u(t+N|t)
\end{aligned} \tag{4.8}$$

which can be rewritten as

$$\begin{bmatrix} \hat{y}(t+d|t) \\ \hat{y}(t+d+1|t) \\ \vdots \\ \hat{y}(t+d+N|t) \end{bmatrix} = \begin{bmatrix} F_d(z^{-1}) \\ F_{d+1}(z^{-1}) \\ \vdots \\ F_{d+N}(z^{-1}) \end{bmatrix} y(t-f) + \begin{bmatrix} B(z^{-1})E_d(z^{-1})u(t|t) \\ B(z^{-1})E_{d+1}(z^{-1})u(t+1|t) \\ \vdots \\ B(z^{-1})E_{d+N}(z^{-1})u(t+N|t) \end{bmatrix}. \tag{4.9}$$

Considering Assumptions (a), (b), and (c), there should be $N \geq f+\bar{k}+N_c$ so that the issue of the random network delay and data dropout can be solved.

The second term on the right side of the above can be separated into two parts: the first part contains the control sequence before time t and the second part the future control prediction sequence. So, let

$$\begin{bmatrix} B(z^{-1})E_d(z^{-1})u(t|t) \\ B(z^{-1})E_{d+1}(z^{-1})u(t+1|t) \\ \vdots \\ B(z^{-1})E_{d+N}(z^{-1})u(t+N|t) \end{bmatrix} = \begin{bmatrix} G_d(z^{-1}) \\ G_{d+1}(z^{-1}) \\ \vdots \\ G_{d+N}(z^{-1}) \end{bmatrix} u(t-1|t-1) + M_1 \begin{bmatrix} u(t|t) \\ u(t+1|t) \\ \vdots \\ u(t+N|t) \end{bmatrix} \tag{4.10}$$

where the polynomial $G_k(z^{-1}) \in \Re[z^{-1}, m+f+d-2]$ and the matrix $M_1 \in \Re^{(N+1)\times(N+1)}$. Thus,

$$\hat{Y}(t+d|t) = F(z^{-1})y(t-f) + G(z^{-1})u(t-1|t-1) + M_1 U(t|t) \tag{4.11}$$

where

$$\hat{Y}(t+d|t) = \left[\hat{y}(t+d|t), \hat{y}(t+d+1|t), \cdots, \hat{y}(t+d+N|t)\right]^T \tag{4.12}$$

$$U(t|t) = \left[u(t|t), u(t+1|t), \cdots, u(t+N|t)\right]^T \tag{4.13}$$

$$G(z^{-1}) = \left[G_d(z^{-1}), G_{d+1}(z^{-1}), \cdots, G_{d+N}(z^{-1})\right]^T \tag{4.14}$$

$$F(z^{-1}) = \left[F_d(z^{-1}), F_{d+1}(z^{-1}), \cdots, F_{d+N}(z^{-1})\right]^T. \tag{4.15}$$

From the controller designed for the system without network delay, it is clear that the future control sequence can be expressed by

$$C(z^{-1})U(t|t) = D(z^{-1})\left(R(t+d) - \hat{Y}(t+d|t)\right) \tag{4.16}$$

where $R(t+d) = \left[r(t+d), r(t+d+1), \cdots, r(t+d+N)\right]^T$.

The term $C(z^{-1})U(t|t)$ can also be separated into two parts: the first part contains the control sequence before time t and the second part the predicted future control sequence. Then, let

$$C(z^{-1})U(t|t) = H(z^{-1})u(t-1|t-1) + LU(t|t) \tag{4.17}$$

where $H(z^{-1}) = \left[H_0(z^{-1}), H_1(z^{-1}), \cdots, H_N(z^{-1})\right]^T$, the polynomial

$$H_i(z^{-1}) \in \Re[z^{-1}, \max\{n_c - i - 1, 0\}]$$

and the matrix $L \in \Re^{(N+1)\times(N+1)}$. Combining (4.11), (4.16) and (4.17) gives

$$\begin{aligned} H(z^{-1})u(t-1|t-1) + LU(t|t) &= D(z^{-1})R(t+d) - D(z^{-1})F(z^{-1})y(t-f) \\ &\quad - D(z^{-1})G(z^{-1})u(t-1|t-1) - D(z^{-1})M_1U(t|t) \end{aligned} \tag{4.18}$$

Let

$$\Gamma(z^{-1})u(t-1|t-1) + MU(t|t) = D(z^{-1})\left(G(z^{-1})u(t-1|t-1) + M_1U(t|t)\right) \tag{4.19}$$

where $\Gamma(z^{-1}) = \left[\Gamma_0(z^{-1}),\ \Gamma_1(z^{-1}), \cdots, \Gamma_N(z^{-1})\right]^T$, the polynomial

$$\Gamma_i(z^{-1}) \in \Re[z^{-1}, \max\{n_d + m + f + d - 2, 0\}]$$

and the matrix $M \in \Re^{(N+1)\times(N+1)}$. It is assumed that matrix $L+M$ is not singular, which can be achieved through the design of polynomials $C(z^{-1})$ and $D(z^{-1})$. As a result,

$$\begin{aligned} U(t|t) = (L+M)^{-1}\big[&D(z^{-1})R(t+d) - D(z^{-1})F(z^{-1})y(t-f) \\ &- \left(\Gamma(z^{-1}) + H(z^{-1})\right)u(t-1|t-1)\big]. \end{aligned} \tag{4.20}$$

Therefore, the control prediction sequence can be determined by the following predictive controller:

$$\begin{aligned} \begin{bmatrix} u(t|t) \\ u(t+1|t) \\ \vdots \\ u(t+N|t) \end{bmatrix} &= \begin{bmatrix} P_0(z^{-1}) \\ P_1(z^{-1}) \\ \vdots \\ P_N(z^{-1}) \end{bmatrix} r(t+d+N) - \begin{bmatrix} Q_0(z^{-1}) \\ Q_1(z^{-1}) \\ \vdots \\ Q_N(z^{-1}) \end{bmatrix} y(t-f) \\ &\quad - \begin{bmatrix} S_0(z^{-1}) \\ S_1(z^{-1}) \\ \vdots \\ S_N(z^{-1}) \end{bmatrix} u(t-1|t-1) \end{aligned} \tag{4.21}$$

where

$$[P_0(z^{-1})\ P_1(z^{-1})\cdots P_N(z^{-1})]^T = (L+M)^{-1} \times [z^{-N}\ z^{-N+1}\cdots 1]^T D(z^{-1})$$
$$[Q_0(z^{-1})\ Q_1(z^{-1})\cdots Q_N(z^{-1})]^T = (L+M)^{-1}F(z^{-1})D(z^{-1})$$
$$[S_0(z^{-1})\ S_1(z^{-1})\cdots S_N(z^{-1})]^T = (L+M)^{-1}\left(\Gamma(z^{-1})+H(z^{-1})\right)$$

and the polynomial $P_i(z^{-1}) \in \Re[z^{-1}, n_d+N]$, $Q_i(z^{-1}) \in \Re[z^{-1}, n_d+n-1]$, and $S_i(z^{-1}) \in \Re[z^{-1}, \max\{n_c-i-1,\ n_d+m+f+d-2,\ 0\}]$.

4.2.2 Design of the Network Delay Compensator

In order to compensate for the network transmission delay, a network delay compensator is proposed. A very important characteristic of communication networks is that they can transmit a set of data at the same time. Thus, it is assumed that all control predictions at time t are packed and sent to the plant side through the network. The network delay compensator chooses the latest control value from the control prediction sequences available on the plant side. For example, if the following predictive control sequences are received on the plant side:

$$\begin{bmatrix} u(t-k_1|t-k_1) \\ u(t-k_1+1|t-k_1) \\ \vdots \\ u(t|t-k_1) \\ \vdots \\ u(t+N-k_1|t-k_1) \end{bmatrix}, \begin{bmatrix} u(t-k_2|t-k_2) \\ u(t-k_2+1|t-k_2) \\ \vdots \\ u(t|t-k_2) \\ \vdots \\ u(t+N-k_2|t-k_2) \end{bmatrix}, \begin{bmatrix} u(t-k_t|t-k_t) \\ u(t-k_t+1|t-k_t) \\ \vdots \\ u(t|t-k_t) \\ \vdots \\ u(t+N-k_t|t-k_t) \end{bmatrix} \tag{4.22}$$

where the control values $u(t|t-k_i)$ for $i = 1, 2, \ldots, t$, are available to be chosen as the control input of the plant at time t, the output of the network delay compensator will be

$$u(t) = u(t|t-k) \tag{4.23}$$

where $k = \min\{k_1, k_2, \cdots, k_t\}$, and $u(t)$ is the latest predictive control value for time t. Actually, it is clear from Assumption (d) that only one control prediction sequence needs to be kept in the network delay compensator and updated by the latest one received from the control prediction generator.

4.2.3 Algorithm of Networked Predictive Control

Following the above subsections, an algorithm of the networked predictive control scheme is proposed as follows:

Step 1: Design a controller for the system without network transmission delay to satisfy the requirements using conventional control methods, for example, PID, LQG, model predictve control, etc.

Step 2: Calculate the output sequence of the control prediction generator using (4.21).

Step 3: Transmit the output sequence of the control prediction generator to the controlled plant through a network each time.

Step 4: Apply the network delay compensator to choose the control input for the plant using (4.23).

4.3 Stability of Networked Predictive Control Systems

The stability of a closed-loop system is the most important issue in the design of control systems. This section considers the stability of networked control systems for two cases: the first one is the case of fixed network transmission delay and the second one is the case of random network transmission delay.

4.3.1 Fixed Network Transmission Delay

It is assumed that the network transmission delays k and f in the forward and backward channels are constant. From the control prediction sequence derived in the previous section, it can be obtained that

$$u(t|t) = P_0(z^{-1})r(t+d+N) - Q_0(z^{-1})y(t-f) - S_0(z^{-1})u(t-1|t-1). \quad (4.24)$$

Then

$$u(t|t) = \frac{P_0(z^{-1})r(t+d+N) - Q_0(z^{-1})y(t-f)}{1+S_0(z^{-1})z^{-1}}. \quad (4.25)$$

Using (4.21) and (4.25), the k-step-ahead predictive control at time t is expressed by

$$\begin{aligned} u(t+k|t) &= P_k(z^{-1})r(t+d+N) - Q_k(z^{-1})y(t-f) - S_k(z^{-1})u(t-1|t-1) \\ &= \frac{P_k(z^{-1})+P_k(z^{-1})S_0(z^{-1})z^{-1}-P_0(z^{-1})S_k(z^{-1})z^{-1}}{1+S_0(z^{-1})z^{-1}}r(t+d+N) \\ &\quad - \frac{Q_k(z^{-1})+Q_k(z^{-1})S_0(z^{-1})z^{-1}-Q_0(z^{-1})S_k(z^{-1})z^{-1}}{1+S_0(z^{-1})z^{-1}}y(t-f). \end{aligned} \quad (4.26)$$

As the network transmission delay is assumed to be fixed (say k), the transmission delay compensator is taken as

$$u(t+k) = u(t+k|t). \tag{4.27}$$

Thus, the closed-loop system becomes

$$\begin{aligned} A(z^{-1})y(t+d+k) &= B(z^{-1})u(t+k) = B(z^{-1})u(t+k|t) \\ &= B(z^{-1})\frac{P_k(z^{-1})+P_k(z^{-1})S_0(z^{-1})z^{-1}-P_0(z^{-1})S_k(z^{-1})z^{-1}}{1+S_0(z^{-1})z^{-1}}r(t+d+N) \\ &\quad - B(z^{-1})\frac{Q_k(z^{-1})+Q_k(z^{-1})S_0(z^{-1})z^{-1}-Q_0(z^{-1})S_k(z^{-1})z^{-1}}{1+S_0(z^{-1})z^{-1}}y(t-f). \end{aligned} \tag{4.28}$$

The closed-loop characteristic equation is

$$\begin{aligned} A(z^{-1})\left(1+S_0(z^{-1})z^{-1}\right) + z^{-d-f-k}B(z^{-1})\big(Q_k(z^{-1}) + Q_k(z^{-1})S_0(z^{-1})z^{-1} \\ - Q_0(z^{-1})S_k(z^{-1})z^{-1}\big) = 0. \end{aligned} \tag{4.29}$$

Therefore, the stability criterion of the closed-loop networked predictve control system with constant network delay is that the system is stable if and only if the roots of the above polynomial are within the unit circle.

4.3.2 Random Network Communication Time Delay

Without losing the generality of the stability analysis, it is assumed that the reference input $R(t)$ is zero. Let

$$\begin{aligned} Q(z^{-1}) &= \left[Q_0(z^{-1})\ Q_1(z^{-1}) \cdots Q_N(z^{-1})\right]^T \\ S(z^{-1}) &= \left[S_0(z^{-1})\ S_1(z^{-1}) \cdots S_N(z^{-1})\right]^T. \end{aligned}$$

Let G_1 and F_1 be the coefficient matrices of polynomial vectors $S(z^{-1})$ and $Q(z^{-1})$, respectively, $\bar{n}$ and $\bar{m}$ be the highest order of the polynomials in vectors $S(z^{-1})$ and $Q(z^{-1})$, respectively, $I_{n\times m}$ denote an $n\times m$ unit matrix and $0_{n\times m}$ denote an $n\times m$ zero matrix. Then, it is clear that (4.21) can be rewritten in the following form:

$$\bar{U}(t) = G_1\tilde{U}(t-1) + F_1\tilde{Y}(t-f) \tag{4.30}$$

where

$$\begin{aligned} \bar{U}(t) &\triangleq \left[u(t|t)\ u(t+1|t) \cdots u(t+N-1|t)\right]^T, \\ \tilde{Y}(t-f) &\triangleq \left[y(t-f)\ y(t-f-1) \cdots y(t-f-\bar{m})\right]^T, \\ \tilde{U}(t-1) &\triangleq \left[u(t-1|t-1)\ u(t-2|t-2) \cdots u(t-\bar{n}-1|t-\bar{n}-1)\right]^T, \\ G_1 &\in \Re^{(N+1)\times(\bar{n}+1)},\quad F_1 \in \Re^{(N+1)\times(\bar{m}+1)}. \end{aligned}$$

Since the control sequences are transmitted to the plant side via a communication network, several control prediction sequences may arrive at the plant side at the same time with different time delays $k_1, k_2, \cdots, k_p$. Let the smallest delay at time t be $k = \min\{k_1, k_2, \cdots, k_p\}$. Then, the latest control prediction sequence on the plant side is

$$\begin{aligned}\bar{U}(t-k) &= G_1 \tilde{U}(t-1-k) + F_1 \tilde{Y}(t-f-k) \\ &= \left[0_{(N+1)\times k}\; G_1\; 0_{(N+1)\times(\bar{k}-k)}\right] \hat{U}(t-1) \\ &\quad + \left[0_{(N+1)\times(f+k)}\; F_1\; 0_{(N+1)\times(\bar{k}-k)}\right] Y(t-1)\end{aligned} \tag{4.31}$$

where

$$\begin{aligned}\hat{U}(t) &= \left[u(t|t)\; u(t-1|t-1) \cdots u(t-\bar{n}-\bar{k}|t-\bar{n}-\bar{k})\right] \\ Y(t) &= \left[y(t)\; y(t-1) \cdots y(t-f-\bar{m}-\bar{k})\right].\end{aligned}$$

According to (4.23), the control input of the plant is the $(k+1)$th element in vector $\bar{U}(t-k)$, that is,

$$\begin{aligned}u(t) &= u(t|t-k) \\ &= \left[0_{1\times k}\; 1\; 0_{1\times(N-k)}\right] \bar{U}(t-k) \\ &= \left[0_{1\times k}\; 1\; 0_{1\times(N-k)}\right] \left(\left[0_{(N+1)\times k}\; G_1\; 0_{(N+1)\times(\bar{k}-k)}\right] \hat{U}(t-1)\right. \\ &\quad \left. + \left[0_{(N+1)\times(f+k)}\; F_1\; 0_{(N+1)\times(\bar{k}-k)}\right] Y(t-1)\right) \\ &= c(k)\hat{U}(t-1) + d(k)Y(t-1)\end{aligned} \tag{4.32}$$

where

$$\begin{aligned}c(k) &= \left[0_{1\times k}\; 1\; 0_{1\times(N-k)}\right] \left[0_{(N+1)\times k}\; G_1\; 0_{(N+1)\times(\bar{k}-k)}\right] \\ d(k) &= \left[0_{1\times k}\; 1\; 0_{1\times(N-k)}\right] \left[0_{(N+1)\times(f+k)}\; F_1\; 0_{(N+1)\times(\bar{k}-k)}\right].\end{aligned}$$

Thus, based on (4.32), the control vector on the plant side can be expressed by

$$U(t) = EU(t-1) + C(k)\hat{U}(t-1) + D(k)Y(t-1) \tag{4.33}$$

where

$$U(t) \triangleq \left[u(t)\; u(t-1) \cdots u(t-m-d+1)\right]^T$$

$$C(k) = \begin{bmatrix} c(k) \\ 0_{(m+d-1)\times(\bar{n}+\bar{k}+1)} \end{bmatrix}$$

$$D(k) = \begin{bmatrix} d(k) \\ 0_{(m+d-1)\times(\bar{m}+f+\bar{k}+1)} \end{bmatrix}$$

$$E = \begin{bmatrix} 0_{1\times(m+d-1)} & 0_{1\times 1} \\ I_{(m+d-1)\times(m+d-1)} & 0_{(m+d-1)\times 1} \end{bmatrix}.$$

Actually, it is clear from (4.1) that the output vector of the plant can be described by

$$Y(t) = A_1 Y(t-1) + B_1 U(t-1) \tag{4.34}$$

where

$$A_1 = \begin{bmatrix} \begin{bmatrix} -a_1 & -a_2 & \cdots & -a_n & 0_{1\times(\bar{m}+f+\bar{k}+1-n)} \end{bmatrix} \\ \begin{bmatrix} I_{(\bar{m}+f+\bar{k})\times(\bar{m}+f+\bar{k})} & 0_{(\bar{m}+f+\bar{k})\times 1} \end{bmatrix} \end{bmatrix} \in \Re^{(\bar{m}+f+\bar{k}+1)\times(\bar{m}+f+\bar{k}+1)}$$

$$B_1 = \begin{bmatrix} \begin{bmatrix} 0_{1\times(d-1)} & b_0 & b_1 & \cdots & b_m \end{bmatrix} \\ 0_{(\bar{m}+f+\bar{k})\times(m+d)} \end{bmatrix} \in \Re^{(\bar{m}+f+\bar{k}+1)\times(m+d)}.$$

In addition, since $u(t|t)$ is the first row of $\bar{U}(t)$ in (4.17), it can be calculated by

$$u(t|t) = \begin{bmatrix} 1 & 0 & \cdots & 0 \end{bmatrix} G_1 \tilde{U}(t-1) + \begin{bmatrix} 1 & 0 & \cdots & 0 \end{bmatrix} F_1 \tilde{Y}(t-f). \tag{4.35}$$

Let

$$\begin{bmatrix} \bar{g}_0 & \bar{g}_1 & \cdots & \bar{g}_{\bar{n}} \end{bmatrix} \triangleq \begin{bmatrix} 1 & 0 & \cdots & 0 \end{bmatrix} G_1$$
$$\begin{bmatrix} \bar{f}_0 & \bar{f}_1 & \cdots & \bar{f}_{\bar{m}} \end{bmatrix} \triangleq \begin{bmatrix} 1 & 0 & \cdots & 0 \end{bmatrix} F_1.$$

Using (4.35), the vector $\hat{U}(t)$ can be constructed by

$$\hat{U}(t) = G_2 \hat{U}(t-1) + F_2 Y(t-1) \tag{4.36}$$

where

$$G_2 = \begin{bmatrix} \begin{bmatrix} \bar{g}_0 & \bar{g}_1 & \cdots & \bar{g}_{\bar{n}} & 0_{1\times\bar{k}} \end{bmatrix} \\ \begin{bmatrix} I_{(\bar{n}+\bar{k})\times(\bar{n}+\bar{k})} & 0_{(\bar{n}+\bar{k})\times 1} \end{bmatrix} \end{bmatrix} \in \Re^{(\bar{n}+\bar{k}+1)\times(\bar{n}+\bar{k}+1)}$$

$$F_2 = \begin{bmatrix} \begin{bmatrix} 0_{1\times(f-1)} & \bar{f}_0 & \bar{f}_1 & \cdots & \bar{f}_{\bar{m}} & 0_{1\times\bar{k}} \end{bmatrix} \\ 0_{(\bar{n}+\bar{k})\times(\bar{m}+f+\bar{k}+1)} \end{bmatrix} \in \Re^{(\bar{n}+\bar{k}+1)\times(\bar{m}+f+\bar{k}+1)}.$$

As a result, combining (4.33), (4.34) and (4.36) yields the following closed-loop system:

$$X_t = \Lambda(k) X_{t-1} \tag{4.37}$$

where

$$X_t = \begin{bmatrix} Y(t) \\ U(t) \\ \hat{U}(t) \end{bmatrix},$$

$$\Lambda(k) = \begin{bmatrix} A_1 & B_1 & 0_{(\bar{m}+f+\bar{k}+1)\times(\bar{n}+\bar{k}+1)} \\ D(k) & E & C(k) \\ F_2 & 0_{(\bar{n}+\bar{k}+1)\times(m+d)} & G_2 \end{bmatrix}.$$

As time delay k changes randomly between 0 and the upper bound $\bar{k}$, the above system is a switched system. Thus, the following theorem provides a sufficient condition for the closed-loop networked predictive control system.

Theorem 4.1. If there exists a positive definite matrix P such that

$$\Lambda^T(k) P \Lambda(k) - P < 0 \tag{4.38}$$

for all $k \in \{0, 1, \cdots, \bar{k}\}$, then the closed-loop system (4.37) is stable for all random delays.

Proof. Let the Lyapunov function be

$$V_t = X_t^T P X_t. \tag{4.39}$$

Then

$$V_{t+1} - V_t = X_{t+1}^T P X_{t+1} - X_t^T P X_t = X_t^T (\Lambda^T(k) P \Lambda(k) - P) X_t. \tag{4.40}$$

If (4.38) is satisfied, then $V_{t+1} - V_t < 0$ for any $X_t \neq 0$. Therefore the closed-loop system is stable for random time delay k in the forward channel and constant time delay f in the feedback channel. □

4.4 Simulation of Networked Predictive Control Systems

As the MATLAB/Simulink® simulation environment provides various powerful tools for control system design, the simulation of networked predictive control systems is carried out using MATLAB® and Simulink®. This section illustrates the simulation strategy of NPCS using a particular control system – a servo control system.

4.4.1 Estimation of Network Transmission Delay

In networked control systems, one important issue is the network transmission delay. Here, the following assumptions are made:

(a) The network delays in the forward channel and feedback channel are the same.
(b) The network delays do not change very rapidly.

In the networked predictive control system, a signature signal (e.g., a sine wave signal) with time stamp is used to measure the network delay. This signal is continuously looped in the whole networked control system, which starts from the plant side, goes through the feedback channel and comes back from the forward channel. Using the current signal value, the total network delay in both the forward and feedback channels can be calculated. So, the forward and feedback time delays are half the total network delay, which can be calculated on the plant side and controller side, respectively.

4.4.2 Off-line Simulation

To simulate the network delay, a set of unit-delay blocks are connected in a series path and one of their outputs will be randomly switched to the network delay compensator on the actuator side if it is not transmitted before. So, an

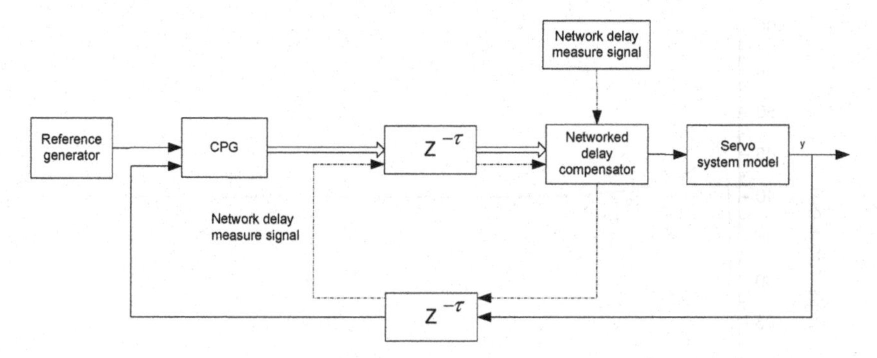

Fig. 4.2. Off-line simulation of NPCS

off-line simulation structure is presented for the networked predictive control system, as shown in Fig. 4.2.

To show the operation of the off-line simulation of NPCS, the following model of a servo control system is considered:

$$G(z^{-1}) = \frac{1.2782z^{-1} - 0.0087}{0.6581z^{-2} - 1.6617z^{-1} + 1}. \tag{4.41}$$

The above transfer function model is estimated from the measured input–output (i.e., voltage–angle) data of the servo control test rig using the least squares algorithm.

Two cases for the network delay are simulated: one is constant delay and the other is random delay. The step responses of the networked predictive control system for the cases of 1-step, 2-step and 3-step constant network delays in both forward and feedback channels are shown in Fig. 4.3. It is clear from the results that the control performance of the closed-loop system for the three different network delays is the same and their plots overlap each other. This means that the networked predictive control scheme can compensate for the network delay actively.

For the case of a random network delay, a random sequence is generated to simulate the network delay. The response of the closed-loop NPCS with random delay is given in Fig. 4.4. Clearly, the NPCS with random delay also has very similar control performance to the one without time delay.

4.4.3 Real-time Simulation

The real-time simulation was carried out, where the control program runs in a real-time embedded microprocessor system and the plant to be controlled is a mathematical model. A real-time simulation structure for the networked predictive control system is designed, which is shown in Fig. 4.5. It is composed of the controller part and the simulated plant part, which run in two separate embedded microprocessor systems that are linked through Ethernet, i.e., the networked control board (NCB) and the networked implement board (NIB).

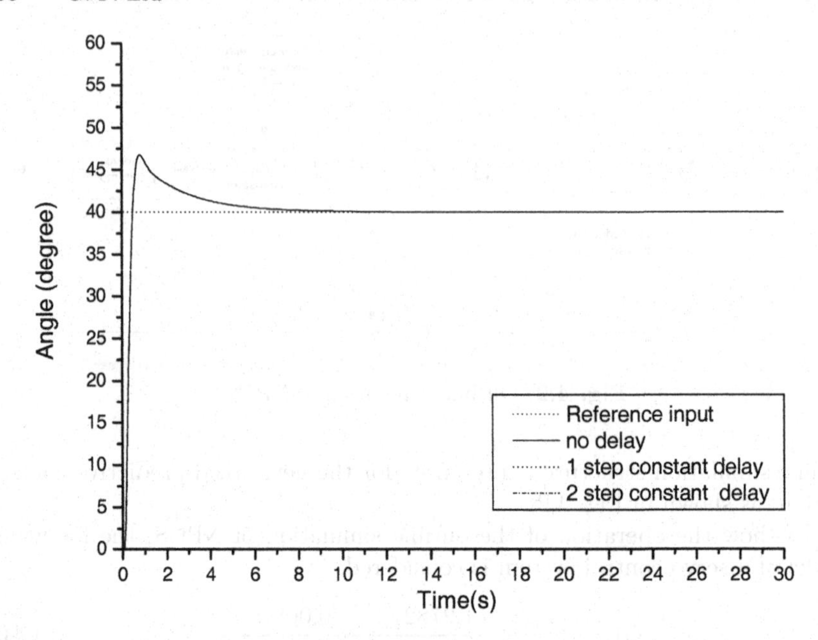

Fig. 4.3. Step responses of the NPCS with various simulated constant network delay

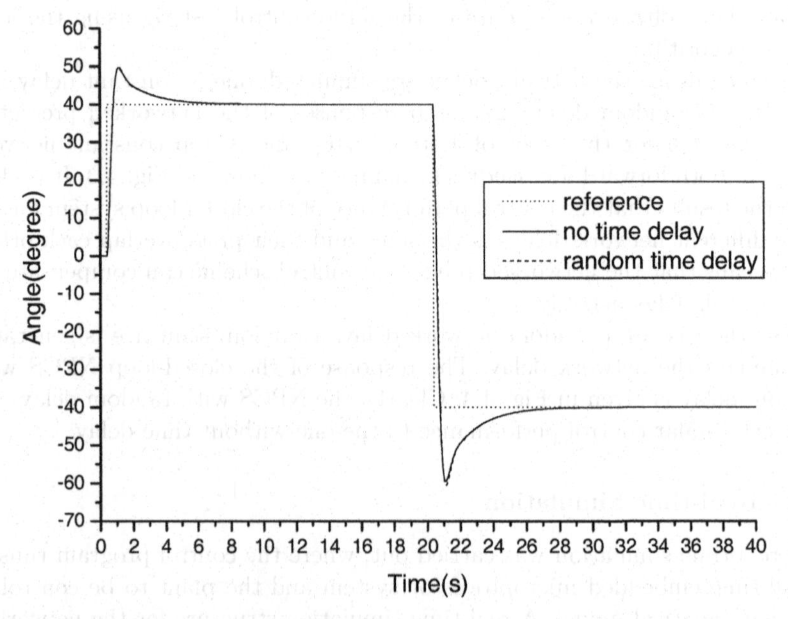

Fig. 4.4. Responses of the NPCS with simulated random network delay

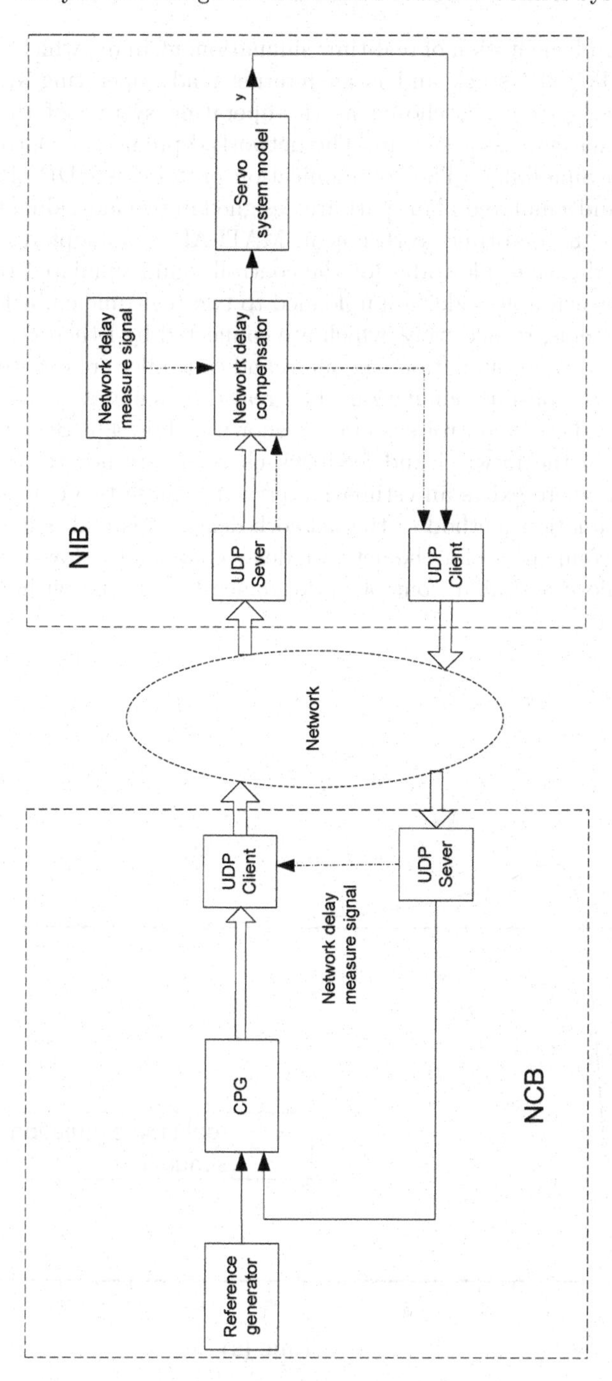

Fig. 4.5. Block diagram of real-time simulation

For the implementation of real-time simulation, uClinux, which is equipped with a full TCP/IP stack and is an internet-ready operating system (OS) for embedded systems, is chosen as the operating system of the real-time embedded microprocessor system. The networked predictive control strategy is realized in Simulink®. The communication protocol is UDP/IP. The controller part and simulated plant part are designed in two individual Simulink® blocks. Then the real-time workshop in MATLAB® is adopted to generate two individual executable codes for the controller and simulated plant parts. Finally, the executable codes are uploaded to two real-time embedded microprocessor systems, respectively, which are connected by Ethernet.

For the real-time simulation, the plant and controller are exactly the same as those for the off-line simulation. The real-time simulation results of the networked predictive control system are shown in Fig. 4.6. Because the network delays in the forward and feedback channels are not the same in the real network, there exists an estimation error for those two delays using the proposed estimation method of the network delay. This makes it difficult for the NPC to compensate for the network delay completely. However, the NPC can still achieve a similar control performance to the one without network delay.

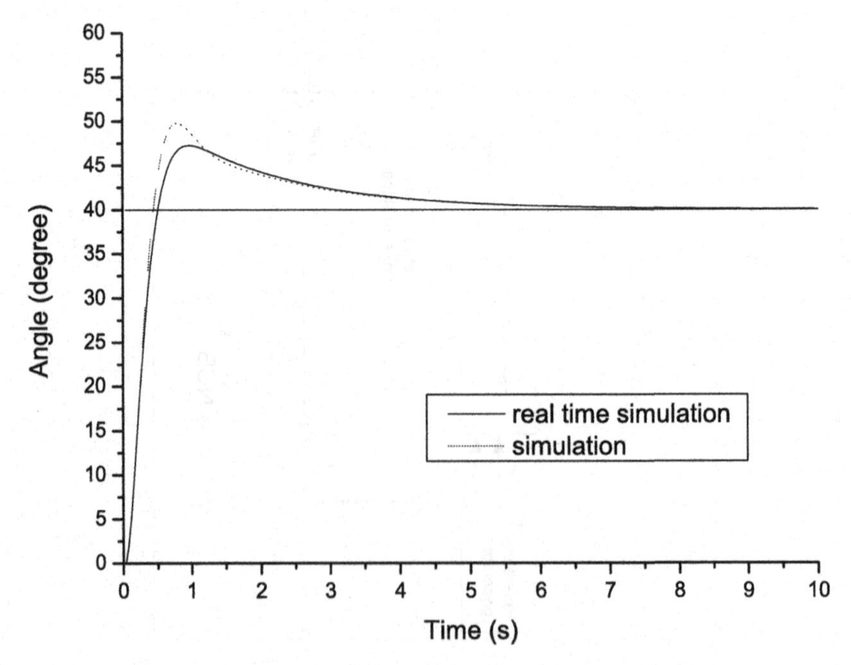

Fig. 4.6. Step responses of NPCS in real-time simulation

4.5 Implementation of Networked Predictive Control Systems

In order to implement networked predictive control systems in practice, some specific software and hardware were designed. They mainly include the software of supervisory, device driver library and interface library, and the hardware of embedded microprocessor systems and control system test rig.

4.5.1 Software of Networked Control Systems

MATLAB/Simulink® software package provides the user with a convenient way to model, simulate and analyse a control system through a visual graphic interface. Real-time workshop (RTW) can generate optimized ANSI C code from control system blocks in Simulink®. uClinux is a concise operating system for embedded systems. How to make full use of the characteristics of Simulink®/RTW and uClinux for NPCS is introduced below.

Supervisory Software

The supervisory software of networked control systems through the Internet can be divided into two parts, one on the client side and the other on the server side, which is composed of a client/server architecture. The client side mainly provides an interface for the user and the server side includes the control and data acquisition programs to fulfil the control task. The overall system architecture is depicted in Fig. 4.7.

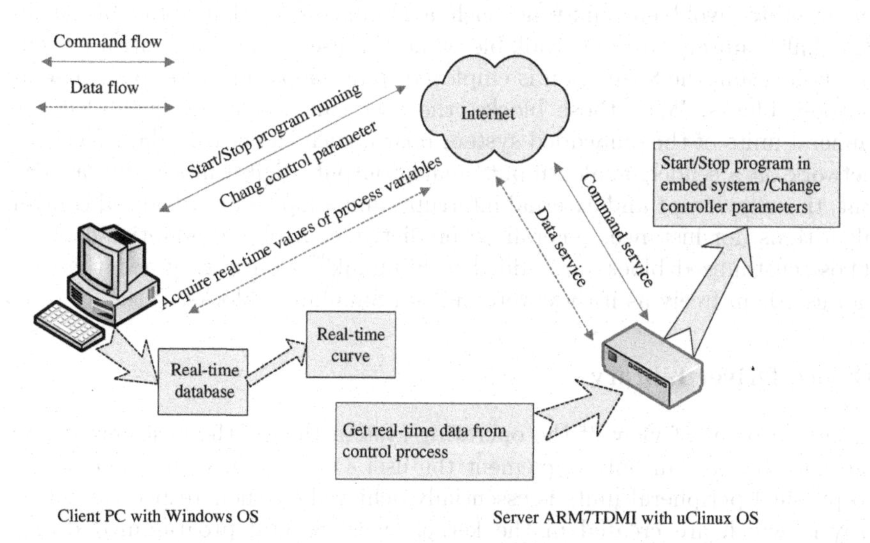

Fig. 4.7. Structure of the supervisory software

The software on the client side is programmed in Visual C++ language. It includes two functions; one is the control function and the other is the monitoring function. The control function primarily responses to the interaction from the user. With this function, the user can send commands and change the parameters of the controller via the Internet. The client first establishes the connection with the server (here, the server runs in an embedded platform) through the TCP/IP protocol. It can not only send messages to the server but also receive messages from the server. As to the monitoring function, the client side can save all acquirable parameters and/or process variables to a real-time database, and at the same time can display the real-time trend of the corresponding variables, which enables the user to know the current status of controlled processes. The curve displays are refreshed automatically at a fixed interval.

The server side has two embedded microprocessor systems that run the control prediction generator algorithm and the network delay compensator algorithm, respectively. The server program is designed in C language and must use a C library subject to uClinux, i.e., uclibc. The server mainly deals with the commands from the client, transfers the received requests, then takes relevant actions, such as starting/stopping the program running, sending the real-time data of the required process variables to the client, or changing the parameters of the controller on-line in real time. As a result, the client can either receive the data from the server or obtain an instantaneous response results for the changes of the control parameters.

Interface Library for Simulink®

In order to avoid encountering such a circumstance that some blocks in Simulink® library are not available while the user creates a control system block diagram, the S-function is employed to develop and mask some general purpose blocks. With those blocks, the user can access the expanded peripheral units of the embedded system board, such as network data receiver, network data sender, analog input, analog output, digital input, digital output, timer interrupt and external interrupt, and adopt some advanced control algorithms, for instance, generalized predictive control and adaptive control. Those customized blocks are added to Simulink® as a library and the user can use them freely as if they were built-in Simulink® blocks.

Device Driver Library

From the point of view of the operating system theory, the customized I/O interface blocks can only implement the user's specific program. The access to physical peripheral units is essentially achieved by their respective device drivers which are created in the kernel layer. Several programmed device drivers in the uClinux kernel are developed, which form the device driver

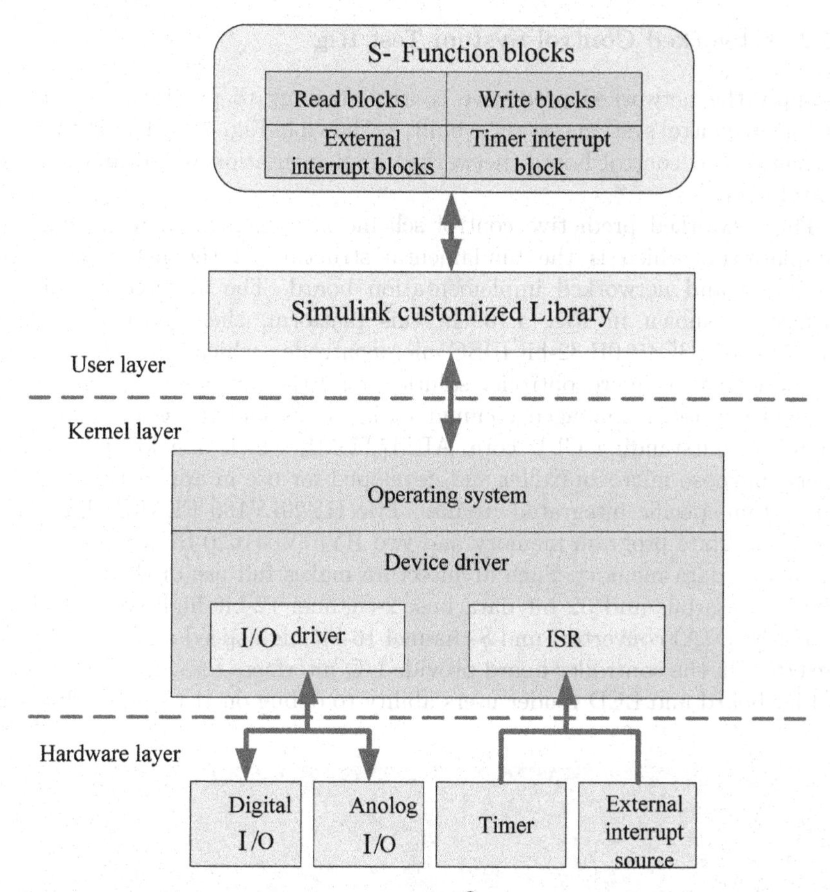

Fig. 4.8. Relationship between the Simulink® customized library and the device driver

library that renders users the ability to manipulate I/O interfaces. The relationship between the customized Simulink® library in the user layer and the device driver library in the kernel layer is shown in Fig. 4.8.

Generation of Application Programs with Simulink®/RTW

As an embedded system has limited system sources, it does not have the ability to develop application programs itself. In order to generate executable codes for embedded systems, the master–slave mode is adopted, that is, programming, compiling and linking of an application program are performed on a host PC which usually has a Linux OS whereas the created executable codes are uploaded through a network to the target embedded system and finally run on it, which is called the cross-compilation.

4.5.2 Networked Control System Test Rig

To apply the networked predictive control strategy to practical systems, a networked control system test rig is built, as shown in Fig. 4.9. This rig consists of a networked control board, networked implementation board and a servo control plant.

The networked predictive control scheme is implemented in an embedded platform, which is the fundamental structure of the networked control board and networked implementation board. The architecture of the platform is shown in Fig. 4.10. In this platform, there is a Samsung's ARM7TDMI S3C4510B 32-bit RISC microcontroller, which is a cost-effective, high-performance microcontroller solution for Ethernet-based systems. It is designed for use in managed communication hubs and routers and is built around an outstanding CPU core, ARM7TDMI, which is a low-power and general purpose microcontroller and developed for use in application-specific and custom-specific integrated circuits. Two HY29LV160 FLASH chips provide 1M$\times$32bits program memory and two HY57V641620 SDRAM chips for 4M$\times$32bits data memory. Such architecture makes full use of the S3C4510B 32bit address bus and 32 bit data bus. 2-channel 12-bit high speed digital-to-analog (D/A) converters and 8-channel 16-bit high speed analog-to-digital converters in the controller board provide I/O interfaces for controlled plants. 4$\times$4 keyboard and LED render users ability to debug on the spot. uClinux is

Fig. 4.9. The networked servo control system

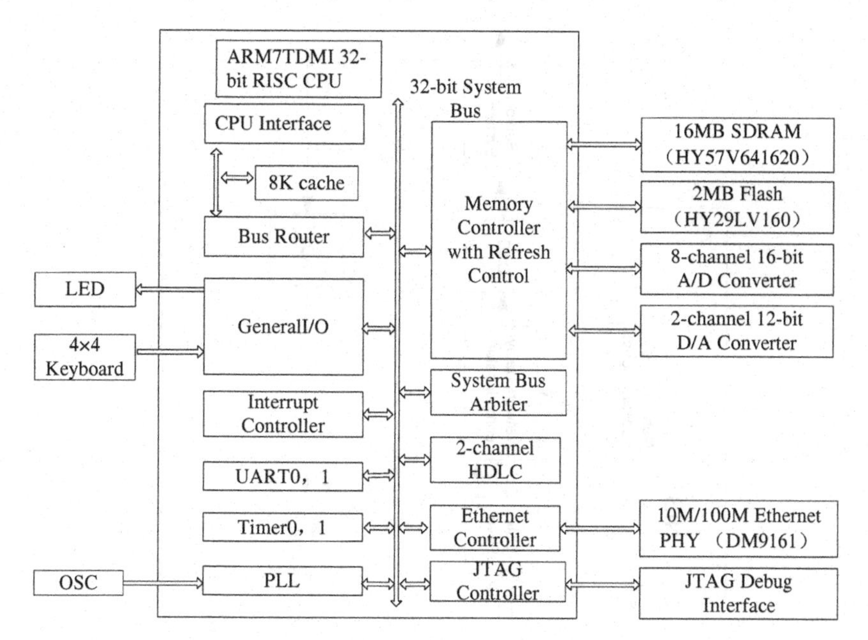

Fig. 4.10. Hardware architecture of the platform

a derivation of Linux 2.0 kernel intended for microcontrollers without memory management units (MMUs). It has a small kernel about 600 kB, while retaining the main advantages of the standard Linux, such as the excellent file system and powerful network capability. In view of ARM7TDMI with non-MMU, uClinux is naturally adopted as OS in embedded systems.

4.5.3 Practical Experiments

For the practical application, the block diagram of the networked predictive servo control system is shown in Fig. 4.11. The difference from the real-time simulation is that the plant to be controlled is a real servo control system. The network connecting the networked control board and the networked implementation board is the Intranet on the university campus.

The responses of the real closed-loop servo control system with a PI controller are shown in Fig. 4.12, where the NPC is not used. They indicate that the response of the closed loop in the case of network delay is different from that in the case of no network delay. It is clear that the network delay makes the control performance poor.

The control performance of the networked predictive control strategy is given in Fig. 4.13, using the same PI controller as above but including the network delay compensator. To have a long network time delay, extra artificial time-delay (e.g., 1-step, 2-step and 3-step delays) was added to the network. It is clearly shown from the experimental results for various network delay cases

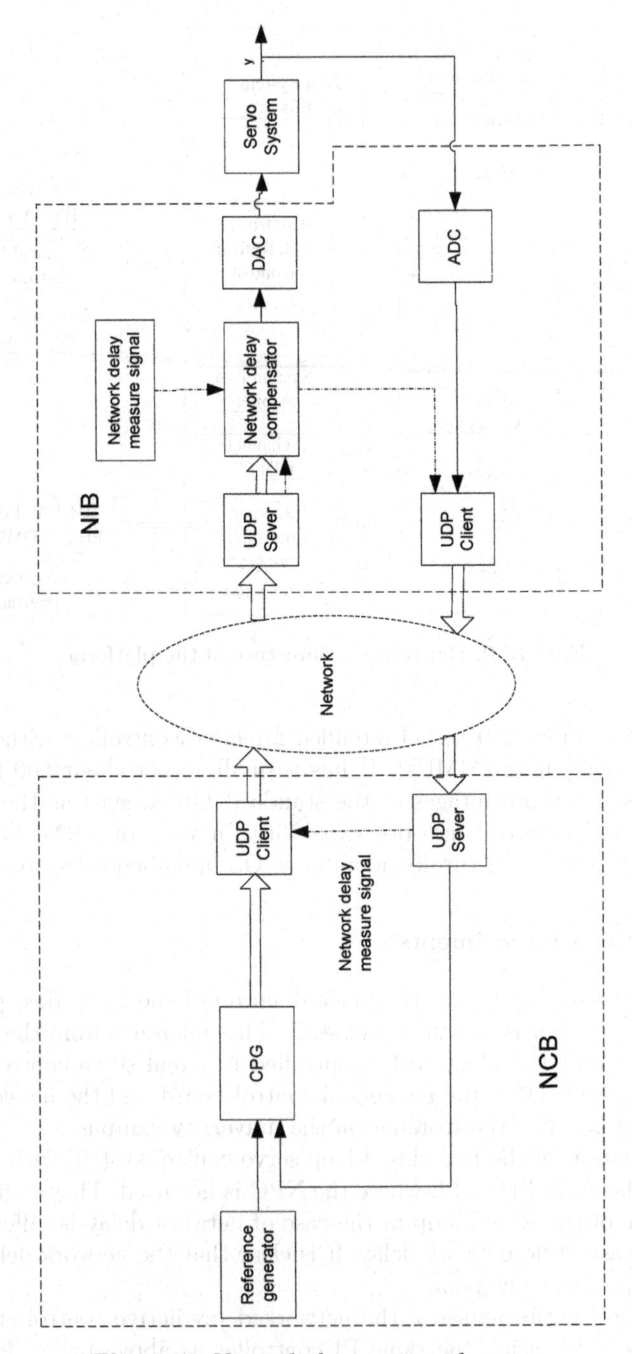

Fig. 4.11. Networked servo control system

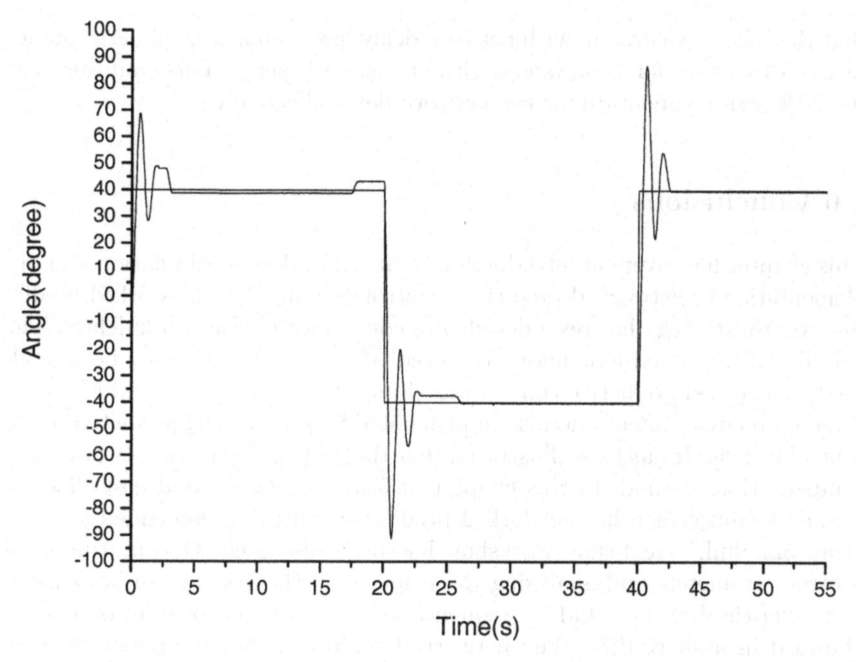

Fig. 4.12. Response of the closed-loop networked PI control system

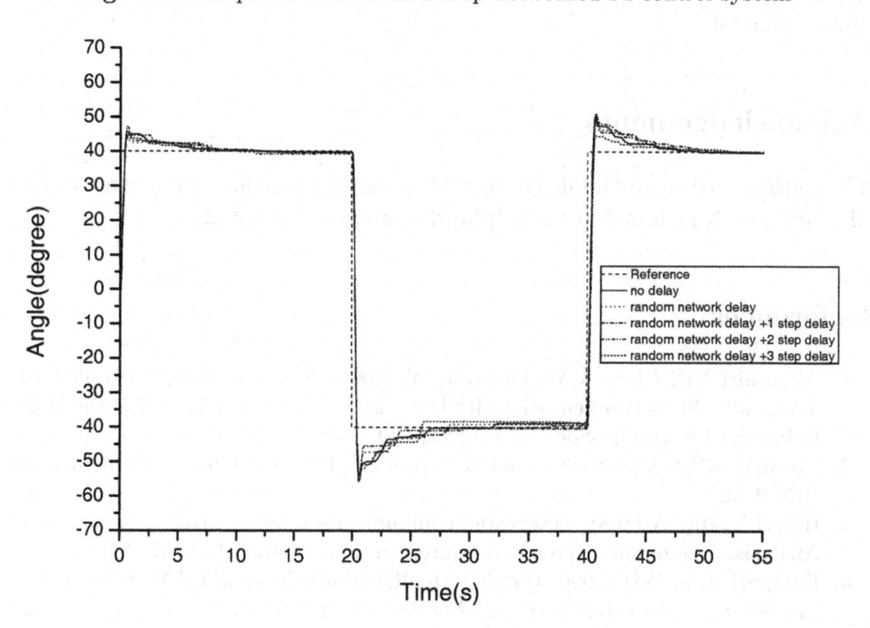

Fig. 4.13. Responses of the closed-loop NPCS

that the NPC for a system with network delay has similar control performance to the PI control for the system without network delay. This confirms that the NPC can compensate for the network delay effectively.

4.6 Conclusions

This chapter has given an introduction to analysis, design, simulation and implementation of networked predictive control systems. The networked predictive control strategy has been detailed to compensate for network delays. The stability of the closed-loop networked predictive control system has been analysed. Networked predictive control was simulated in the off-line and real-time simulation environment and also implemented in a networked predictive servo control test rig. It has been illustrated that the NPC is an active network delay compensation method. In this chapter, a fast, convenient, and cost-effective scheme for implementing networked predictive control in embedded systems using Simulink®/real-time workshop has been discussed. On-line tuning of control parameters and analysing the response of the system can be realized easily via the Internet, and consequently an optimal control solution can be obtained in a short time. The networked servo control experiment through the Intranet has successfully demonstrated the effectiveness of networked predictive control.

Acknowledgements

The author wishes to thank Dr. David Rees, Dr. Senchun Chai, Dr. Junxia Mu, and Dr. Xueyuan Nie for helpful discussions and comments.

References

1. Almutairi NB, Chow MY, Tipsuwan Y (2001) Network-based controlled DC motor with fuzzy compensation. In: Proc. 27th Annual Conference of the IEEE Industrial Electronics Society, Denver, CO, 3:1844–1849
2. Cushing M (2000) Process control across the Internet. Chemical Engineering 107:80–82
3. Halevi Y, Ray A (1988) Integrated communication and control systems: Part I Analysis. J Dynamic Systems Measurement and Control 110:367–373
4. Hong SH, Kim WH (2000) Bandwidth allocation scheme in CAN protocol. IEE Proceedings–Control Theory and Applications 147:37–44
5. Hong SH (1995) Scheduling algorithm of data sampling times in the integrated communication and control systems. IEEE Transactions on Control Systems Technology 3:225-230
6. IEEE (2002) Special issue on systems and control methods for communication Networks. IEEE Transactions on Automatic Control 47:877–1038

7. Liu GP, Mu J, Rees D (2004) Networked predictive control of systems with random communication delay. In: Proceedings of the UKACC Control, Bath, UK
8. Liu GP, Rees D, Chai SC, Nie XY (2005) Design, simulation and implementation of networked predictive control systems. Measurement & Control 38:17–21
9. Liu GP, Mu JX, Rees D, Chai SC (2006) Design and stability analysis of networked control systems with random communication time delay using the modified MPC. International Journal of Control 79:287–296
10. Luck R, Ray A (1994) Experimental verification of a delay compensation algorithm for integrated communication and control systems. International Journal of Control 59:1357–1372
11. Nemoto Y, Hamamoto N, Suzuki R, Ikegami T, Hashimoto Y, Ide T, Ohta K, Mansfield G, Kato N (1997) Construction and utilization experiment of multimedia education system using satellite ETS-V and the Internet. IEICE Transaction on Information and Systems 2:162–169
12. Nilsson J (1998) Real-time control systems with delays. PhD dissertation, Lund Institute of Technology, Sweden
13. Ryu M, Hong S (1988) Toward automatic synthesis of schedulable real-time controllers. Integrated Computer-Aided Engineering 5:261–277
14. Seto JP, Sha L, Shin KG (1996) On task schedulability in real-time control systems. In: Proceedings of the 17th IEEE Real-time Systems Symposium, Washington, DC, 12–21
15. Tarn TJ, Xi N (1998) Planning and control of internet-based teleoperation. In: Proceedings of SPIE: Telemanipulator and telepresence technologies V, Boston, MA, 3524:189–193
16. Taylor K, Dalton B (2000) Internet robots: A new robotics niche. IEEE Robotics and Automation Magazine 7:27–34
17. Walsh GC, Ye H, Bushnell L (1999) Stability analysis of networked control Systems. In: Proceedings of the 1999 American Control Conference, San Diego, CA, 4:2876–2880
18. Yang SH, Chen X, Edwards DW, Alty JL (2003) Design issues and implementation of Internet based process control. Control Engineering Practice 11:709–720
19. Zhang W, Branicky MS, Phillips SM (2001) Stability of networked control Systems. IEEE Control Systems Magazine 21:84–99
20. Zhivoglyadov PV, Middleton RH (2003) Networked control design for linear Systems. Automatica 39:743–750

5

Robust H_∞ Control and Filtering of Networked Control Systems

Dong Yue[1], Qing-Long Han[2], and James Lam[3]

[1] Nanjing Normal University, Nanjing, P. R. China medongy@njnu.edu.cn
[2] Central Queensland University, Rockhampton, Australia q.han@cqu.edu.au
[3] University of Hong Kong, Hong Kong, P. R. China james.lam@hku.hk

Abstract. This chapter is concerned with the design of robust H_∞ controllers and H_∞ filters for uncertain networked control systems (NCS) with the effects of both the network-induced delay and data dropout taken into consideration. A new analysis method for H_∞ performance of NCS is provided by introducing some slack matrix variables and employing the information of the lower bound of the network-induced delay. The criteria derived for the design of H_∞ controller and H_∞ filter are expressed as a set of linear matrix inequalities. Finally, numerical examples and simulation results are given to illustrate the effectiveness of the method.

Keywords. Networked control systems, stability, filtering, Lyapunov functional.

5.1 Introduction

As is well known, in modern industrial systems, sensors, controllers and plants are often connected over a network medium [1], and are called networked control systems (NCS). There are many advantages of NCS, such as low cost, reduced weight and power requirements, simple installation and maintenance, and high reliability. Thus, increasing research interest has recently been focused on the study of the stability, stabilization and signal estimation of NCS [1, 14, 20, 24, 29]. However, since the sampling data and output and controller signals are transmitted through a network, network-induced delays and data dropout in NCS are inevitable. For NCS with different scheduling protocols, the network-induced delay may be constant, time-varying, or even random [29]. Recently, the stability analysis and stabilization controller design for NCS have been investigated by researchers, with the effects of network-induced delay and/or data dropouts taken into account. In these studies, analysis and synthesis methods are provided based on discrete-time models

[4, 14], continuous-time models [20] or hybrid system models [29]. Considering the effects of external disturbance on the system, stability analysis and disturbance attenuation analysis are carried out in [9] based on a framework of discrete-time switched systems. However, only the case when there are no parameter uncertainties in the system is considered and no controller synthesis method is given in [9]. Moreover, the effects of controller-to-actuator delay is neglected in [9].

For the case when statistical information about external noise may not be known exactly and uncertainties may exist in the system model due to modelling errors, the signal estimate problem has been investigated based on an H_∞ filtering technique. Compared with the Kalman filter, the advantage of H_∞ filtering is that noise sources can be arbitrary signals with bounded energy, or bounded average power instead of being Gaussian. When considering the presence of time delay in the system state and/or output measurement, the H_∞ filter design problem of time delay systems has been studied based on a delay-independent approach [2, 21, 23, 25] or delay-dependent approach [3]. However, it is assumed in all these references that the time delay existing in the state or the output is either constant or slowly time-varying. Here, by "slowly time-varying" we mean that the derivative of the time-varying delay exists and the bound on the derivative is less than 1. It will be seen from the next section that the system over a network connection is essentially a system with fast time-varying delay. Therefore, the methods in the existing references [2, 3, 23, 25] cannot be used for H_∞ filtering purposes in such a class of systems.

In this chapter we are concerned with the design of a robust H_∞ controller and H_∞ filter for uncertain networked control systems. Both network-induced delay and data dropout are considered in the model. The network-induced delay considered in the model is composed of sensor-to-controller delay and controller-to-actuator delay, as well as computation delay. In our method, discretization of the system model and the assumption that the controller dynamics is continuous are *not needed* for design of the controller and filter. In contrast with methods based on discrete-time models, our method is formulated in the continuous-time domain; that is, the inter-sampling behavior is taken into account. Through introducing some slack matrix parameters, we first give some criteria for guaranteeing the H_∞ performance of the NCS and then present control and filter design conditions based on the criteria. Since the lower bound of the network-induced delay is employed to derive the criteria, considerably less conservative results can be obtained by using the criteria in this chapter, especially for the case where the lower bound of the network-induced delay is *nonzero*. The criteria for H_∞ performance analysis, H_∞ control synthesis and H_∞ filter design are derived based on a linear matrix inequality (LMI) approach. To illustrate the effectiveness of the methods, numerical examples and simulation results are given. Some of the results given in this chapter have appeared in [26, 28].

Notation: $\mathbb{R}^n$ denotes the n-dimensional Euclidean space, $\mathbb{R}^{n\times m}$ is the set of $n \times m$ real matrices, I is the identity matrix of appropriate dimensions, $\|\cdot\|$ stands for the Euclidean vector norm or the induced matrix 2-norm as appropriate. The notation $X > 0$ (respectively, $X \geq 0$), for $X \in \mathbb{R}^{n\times n}$ means that the matrix X is a real symmetric positive definite (respectively, positive semi-definite) matrix. $\lambda_{\max}(P)$ ($\lambda_{\min}(P)$) denotes the maximum (minimum) eigenvalue of a real symmetric matrix P. For an arbitrarily real matrix B and two real symmetric matrices A and C, $\begin{bmatrix} A & * \\ B & C \end{bmatrix}$ denotes a real symmetric matrix, where $*$ denotes the entries implied by symmetry.

5.2 Robust H_∞ Control of NCS

5.2.1 System Description and Preliminaries

Consider the following system with parameter uncertainties given by

$$\begin{cases} \dot{x}(t) = [A + \Delta A(t)]\, x(t) + [B + \Delta B(t)]\, u(t) + B_w w(t), \\ x(t_0) = x_0, \\ z(t) = Cx(t) + Du(t), \end{cases} \tag{5.1}$$

where $x(t) \in \mathbb{R}^n$, $u(t) \in \mathbb{R}^m$ and $z(t) \in \mathbb{R}^q$ are the state vector, control input vector and controlled output, respectively; $x_0 \in \mathbb{R}^n$ denotes the initial condition; A, B, B_w, C and D are some constant matrices of appropriate dimensions; $\Delta A(t)$ and $\Delta B(t)$ denote the parameter uncertainties satisfying the following condition:

$$\left[\Delta A(t)\ \Delta B(t)\right] = GF(t)\left[E_a\ E_b\right], \tag{5.2}$$

where G, E_a and E_b are constant matrices of appropriate dimensions and $F(t)$ is an unknown time-varying matrix, which is Lebesque measurable in t and satisfies $F^T(t)F(t) \leq I$. $w(t) \in L_2\,[t_0, \infty)$ denotes the external perturbation. Throughout this chapter, we assume that the system (5.1) is controlled through a network.

As pointed out in [1, 29], the presence of the network may often lead to signal transmitting delay and data dropout, which can degrade the performance of the closed-loop system. In the presence of the control network, under a linear control law, the control system (5.1) can be expressed as

$$\dot{x}(t) = [A + \Delta A(t)]\, x(t) + [B + \Delta B(t)]\, u(t) + B_w w(t), \tag{5.3}$$

$$z(t) = Cx(t) + Du(t),\ t \in [i_k h + \tau_k, i_{k+1}h + \tau_{k+1}) \tag{5.4}$$

$$u(t^+) = Kx(t - \tau_k),\ t \in \{i_k h + \tau_k, k = 1, 2, \ldots\}, \tag{5.5}$$

where h is the sampling period, i_k ($k = 1, 2, 3, \ldots$) are some integers and $\{i_1, i_2, i_3, \ldots\} \subset \{1, 2, 3, \ldots\}$. The time delay τ_k denotes the time from the instant $i_k h$ when sensor nodes sample sensor data from a plant to the instant when actuators transfer data to the plant. Obviously, $\cup_{k=1}^{\infty}\left[i_k h+\tau_k, i_{k+1}h+\tau_{k+1}\right) = \left[t_0, \infty\right), t_0 \geq 0$. In this chapter, we assume that $u(t) = 0$ before the first control signal reaches the plant.

Remark 5.1. In (5.3)–(5.5), $\{i_1, i_2, i_3, \ldots\}$ is a subset of $\{1, 2, 3, \ldots\}$. Moreover, it is not required that $i_{k+1} > i_k$. When $\{i_1, i_2, i_3, \ldots\} = \{1, 2, 3, \ldots\}$, it means that no packet dropout occurs in the transmission. If $i_{k+1} = i_k + 1$, it implies that $h + \tau_{k+1} > \tau_k$, which includes $\tau_k = \hat{\tau}$ and $\tau_k < h$ as special cases, where $\hat{\tau}$ is a constant. In addition, the effects of the network-induced delay, data packet dropout and external perturbation are simultaneously considered in (5.3)–(5.5). □

Assumption 5.1. The sensor is clock-driven, the controller and actuator are event-driven. □

Assumption 5.2. Two constants $\eta > 0$ and $\tau_m \geq 0$ exist such that

$$(i_{k+1} - i_k)h + \tau_{k+1} \leq \eta, \quad k = 1, 2, \ldots, \tag{5.6}$$

$$\tau_m \leq \tau_k, \quad k = 1, 2, \ldots, \tag{5.7}$$

where $\eta \geq \tau_m$. □

Remark 5.2. It is assumed in Assumption 5.1 that the controller and actuator are event-driven, such as a zero-order hold. This means that the controller and actuator will be updated when the new data packet comes. From the physical point of view, to guarantee the stability of NCS, only a finite amount of data dropout can be tolerated. Moreover, in a real NCS, a data packet that has expired in a limited transmission time will be lost based on the commonly used network protocols. Thus, τ_k should be bounded. In Condition (5.6), η can be used to reflect the allowable bound on the amount of data dropout and network-induced delays. It is known that the network-induced delay consists of waiting time delay, frame time delay and propagation delay. Since propagation delay always exists when data is transmitted through a network and can be monitored, it is reasonable to assume that τ_m exists and is a number larger than zero, i.e., $\tau_m > 0$. It is noted that the introduction of lower bound τ_m on the time delay is important, and may lead to less conservative result than would be obtained based on the assumption $\tau_m = 0$. This observation will be shown in numerical examples. □

Under Assumption 5.1, System (5.3)–(5.5) can be rewritten using an equivalent form as follows:

$$\dot{x}(t) = \left[A + \Delta A(t)\right] x(t) + \left[B + \Delta B(t)\right] Kx(i_k h) + B_w w(t), \tag{5.8}$$

$$x(t) = \Phi(t, t_0 - \eta)x(t_0 - \eta) \triangleq \phi(t), \; t \in \left[t_0 - \eta, t_0\right], \tag{5.9}$$

$$z(t) = Cx(t) + DKx(i_k h), \; t \in \left[i_k h + \tau_k, i_{k+1}h + \tau_{k+1}\right), \tag{5.10}$$

where $\Phi(t,t_0-\eta)$ is a solution of $\dot{\Phi}(t,t_0-\eta) = [A+\Delta A(t)]\,\Phi(t,t_0-\eta)$, $t\in[t_0-\eta,t_0]$.

Remark 5.3. Since $i_kh = t-(t-i_kh)$, define $\tau(t)=t-i_kh$, which denotes the time-varying delay in the control signal. Obviously,

$$\tau_k \le \tau(t) \le (i_{k+1}-i_k)\,h+\tau_{k+1}, \quad t\in[i_kh+\tau_k, i_{k+1}h+\tau_{k+1}). \qquad \Box$$

Remark 5.4. (5.8)–(5.10) can be used to express the mathematical model of networked control systems when the transmitted data is single-packet. For the multiple-packet transmission case, since the arrival time of sensor messages at the controller or the arrival time of controller messages at the actuator may be different, especially for the case when the sampling time of sensors is different, a buffer before the controller and actuator is needed. By employing the buffer technology on the network, the model (5.8)–(5.10) can also be used to express the NCS with multiple-packet transmission. □

Definition 5.1. System (5.8)–(5.10) is said to be robustly exponentially stable with an H_∞ norm bound γ, if the following hold:

(i) System (5.8)–(5.10) with $w(t)\equiv 0$ is robustly exponentially stable; that is, there exist constants $\alpha>0$ and $\beta>0$ such that

$$\|x(t)\| \le \alpha \sup_{t_0-\eta\le s\le t_0} \|\phi(s)\|\, e^{-\beta t},$$

$t\ge t_0$, for all admissible uncertainties $\Delta A(t)$ and $\Delta B(t)$; and

(ii) under the assumption of zero initial condition, the controlled output $z(t)$ satisfies $\|z(t)\|_2 \le \gamma\,\|w(t)\|_2$ for any nonzero $w(t)\in L_2\,[0,\infty)$. □

5.2.2 H_∞ Performance Analysis

Define

$$\tau_0 = \frac{\eta+\tau_m}{2}, \quad \delta = \frac{\eta-\tau_m}{2}.$$

For any matrices N_i, S_i and M_i $(i=1,2,3,4)$ of appropriate dimensions, it can be seen that

$$\left[x^T(t)N_1 + x^T(i_kh)N_2 + x^T(t-\tau_0)N_3 + \dot{x}^T(t)N_4\right] \times \left[x(t)-x(i_kh)-\int_{i_kh}^{t}\dot{x}(s)ds\right] = 0, \tag{5.11}$$

$$\left[x^T(t)S_1 + x^T(i_kh)S_2 + x^T(t-\tau_0)S_3 + \dot{x}^T(t)S_4\right] \times \left[x(i_kh)-x(t-\tau_0)-\int_{t-\tau_0}^{i_kh}\dot{x}(s)ds\right] = 0, \tag{5.12}$$

and

$$\begin{aligned}&\left[x^T(t)M_1 + x^T(i_kh)M_2 + x^T(t-\tau_0)M_3 + \dot{x}^T(t)M_4\right]\\ &\qquad \times \left[-\left[A+\Delta A(t)\right]x(t) - \left[B+\Delta B(t)\right]Kx(i_kh)\right.\\ &\qquad \left.-B_w w(t) + \dot{x}(t)\right] = 0.\end{aligned} \tag{5.13}$$

Next, based on a Lyapunov–Krasovskii functional and combining (5.11)–(5.13), we prove the following result.

Lemma 5.1. For given scalars τ_m, η and γ and a matrix K, if there exist matrices P_k $(k=1,2,3)$, $W_j>0$, $T_j>0$, $R_j>0$ $(j=1,2)$ and any matrices N_i, S_i and M_i $(i=1,2,3,4)$ of appropriate dimensions such that

$$\begin{bmatrix}\Xi_{11}+\mathrm{diag}\begin{bmatrix}W_1 & 0 & 0 & W_2\end{bmatrix} & * \\ \Xi_{21} & \Xi_{22}\end{bmatrix}<0, \tag{5.14}$$

$$\begin{bmatrix}P_1 & P_2\\ P_2^T & P_3\end{bmatrix}>0, \tag{5.15}$$

where

$$\Xi_{11} = \begin{bmatrix}\Gamma_{11} & * & * & *\\ \Gamma_{21} & \Gamma_{22} & * & *\\ \Gamma_{31} & \Gamma_{32} & \Gamma_{33} & *\\ \Gamma_{41} & \Gamma_{42} & \Gamma_{43} & \Gamma_{44}\end{bmatrix},$$

$$\Xi_{21} = \begin{bmatrix}\tau_0 P_3 & 0 & -\tau_0 P_3 & \tau_0 P_2^T\\ \eta N_1^T & \eta N_2^T & \eta N_3^T & \eta N_4^T\\ \delta S_1^T & \delta S_2^T & \delta S_3^T & \delta S_4^T\\ C & DK & 0 & 0\\ -B_w^T M_1^T & -B_w^T M_2^T & -B_w^T M_3^T & -B_w^T M_4^T\end{bmatrix},$$

$$\Xi_{22} = \mathrm{diag}\left(-\tau_0 T_2,\ -\eta R_1,\ -\delta R_2,\ -I,\ -\gamma^2 I\right),$$

$$\begin{aligned}\Gamma_{11} &= P_2 + P_2^T + T_1 + \tau_0 T_2 + N_1 + N_1^T\\ &\quad -M_1\left[A+\Delta A(t)\right] - \left[A+\Delta A(t)\right]^T M_1^T,\\ \Gamma_{21} &= N_2 - N_1^T + S_1^T - M_2\left[A+\Delta A(t)\right] - K^T\left[B+\Delta B(t)\right]^T M_1^T,\\ \Gamma_{31} &= N_3 - P_2^T - S_1^T - M_3\left[A+\Delta A(t)\right],\\ \Gamma_{41} &= M_1^T + N_4 + P_1 - M_4\left[A+\Delta A(t)\right],\\ \Gamma_{22} &= -N_2 - N_2^T + S_2 + S_2^T - M_2\left[B+\Delta B(t)\right]K\\ &\quad -K^T\left[B+\Delta B(t)\right]^T M_2^T,\\ \Gamma_{32} &= -N_3 + S_3 - S_2^T - M_3\left[B+\Delta B(t)\right]K,\\ \Gamma_{42} &= -N_4 + S_4 + M_2^T - M_4\left[B+\Delta B(t)\right]K,\\ \Gamma_{33} &= -T_1 - S_3 - S_3^T,\\ \Gamma_{43} &= -S_4 + M_3^T,\\ \Gamma_{44} &= M_4 + M_4^T + \eta R_1 + 2\delta R_2,\end{aligned}$$

then, System (5.8)–(5.10) with a control network satisfying Assumption 5.2 is robustly exponentially stable with an H_∞ norm bound γ.

Proof. Construct a Lyapunov–Krasovskii functional as

$$\begin{aligned}V(t) &= x^T(t)P_1x(t) + 2x^T(t)P_2\left(\int_{t-\tau_0}^{t} x(s)ds\right)\\&+\left(\int_{t-\tau_0}^{t} x(s)ds\right)P_3\left(\int_{t-\tau_0}^{t} x(s)ds\right)\\&+\int_{t-\tau_0}^{t} x^T(s)T_1x(s)ds+\int_{t-\tau_0}^{t}\int_{s}^{t} x^T(v)T_2x(v)dvds\\&+\int_{t-\eta}^{t}\int_{s}^{t}\dot{x}^T(v)R_1\dot{x}(v)dvds+2\delta\int_{t-\tau_0+\delta}^{t}\dot{x}^T(s)R_2\dot{x}(s)ds\\&+\int_{t-\tau_0-\delta}^{t-\tau_0+\delta}\int_{s}^{t-\tau_0+\delta}\dot{x}^T(v)R_2\dot{x}(v)dvds,\end{aligned}\tag{5.16}$$

where $\begin{bmatrix} P_1 & P_2 \\ P_2^T & P_3 \end{bmatrix} > 0$, $T_i > 0$ and $R_i > 0$ $(i = 1, 2)$. Taking the time derivative of $V(t)$ along the trajectory of (5.8) yields that, for $t \in [i_kh + \tau_k, i_{k+1}h + \tau_{k+1})$,

$$\begin{aligned}\dot{V}(t) &= 2x^T(t)P_1\dot{x}(t) + 2\dot{x}^T(t)P_2\int_{t-\tau_0}^{t} x(s)ds\\&+2x^T(t)P_2\left(x(t)-x(t-\tau_0)\right)\\&+2\left(x(t)-x(t-\tau_0)\right)P_3\int_{t-\tau_0}^{t} x(s)ds\\&+x^T(t)\left(T_1+\tau_0T_2\right)x(t)-x^T(t-\tau_0)T_1x(t-\tau_0)\\&-\int_{t-\tau_0}^{t} x^T(s)T_2x(s)ds+\dot{x}^T(t)\left(\eta R_1+2\delta R_2\right)\dot{x}(t)\\&-\int_{t-\eta}^{t}\dot{x}^T(s)R_1\dot{x}(s)ds-\int_{t-\tau_0-\delta}^{t-\tau_0+\delta}\dot{x}^T(s)R_2\dot{x}(s)ds\\&+2e^T(t)N\left[x(t)-x(i_kh)-\int_{i_kh}^{t}\dot{x}(s)ds\right]\\&+2e^T(t)S\left[x(i_kh)-x(t-\tau_0)-\int_{t-\tau_0}^{i_kh}\dot{x}(s)ds\right]\\&+2e^T(t)M\left[-\left[A+\Delta A(t)\right]x(t)\right.\\&\left.-\left[B+\Delta B(t)\right]Kx(i_kh)-B_ww(t)+\dot{x}(t)\right]\\&+\left[Cx(t)+DKx(i_kh)\right]^T\left[Cx(t)+DKx(i_kh)\right]\\&-\gamma^2w^T(t)w(t)-z^T(t)z(t)+\gamma^2w^T(t)w(t),\end{aligned}\tag{5.17}$$

where

$$
\begin{aligned}
e^T(t) &= \left[x^T(t)\; x^T(i_k h)\; x^T(t-\tau_0)\; \dot{x}^T(t) \right], \\
N^T &= \left[N_1^T\; N_2^T\; N_3^T\; N_4^T \right], \\
S^T &= \left[S_1^T\; S_2^T\; S_3^T\; S_4^T \right], \\
M^T &= \left[M_1^T\; M_2^T\; M_3^T\; M_4^T \right].
\end{aligned}
$$

Since $i_k h = t - (t - i_k h)$, define $\tau(t) = t - i_k h$. Then

$$
\tau_k \le \tau(t) \le (i_{k+1} - i_k)\, h + \tau_{k+1}, \quad t \in [i_k h + \tau_k, i_{k+1} h + \tau_{k+1}). \tag{5.18}
$$

It is easy to see that, for $t \in [i_k h + \tau_k, i_{k+1} h + \tau_{k+1})$,

$$
-2e^T(t)N \int_{i_k h}^{t} \dot{x}(s)ds \le \eta e^T(t) N R_1^{-1} N^T e(t) + \int_{i_k h}^{t} \dot{x}^T(s) R_1 \dot{x}(s)ds, \tag{5.19}
$$

and

$$
-2e^T(t)S \int_{t-\tau_0}^{i_k h} \dot{x}(s)ds \le \delta e^T(t) S R_2^{-1} S^T e(t) + \begin{cases} \displaystyle\int_{t-\tau_0}^{i_k h} \dot{x}^T(s) R_2 \dot{x}(s)ds, \; t < \tau_0 + i_k h, \\ \displaystyle\int_{i_k h}^{t-\tau_0} \dot{x}^T(s) R_2 \dot{x}(s)ds, \; t \ge \tau_0 + i_k h. \end{cases} \tag{5.20}
$$

From (5.18), one can easily show that, for $t \in [i_k h + \tau_k,\; i_{k+1} h + \tau_{k+1})$,

$$
\int_{i_k h}^{t} \dot{x}^T(s) R_1 \dot{x}(s)ds \le \int_{t-\eta}^{t} \dot{x}^T(s) R_1 \dot{x}(s)ds, \tag{5.21}
$$

and for $t < \tau_0 + i_k h$,

$$
\int_{t-\tau_0}^{i_k h} \dot{x}^T(s) R_2 \dot{x}(s)ds \le \int_{t-\tau_0-\delta}^{t-\tau_0+\delta} \dot{x}^T(s) R_2 \dot{x}(s)ds, \tag{5.22}
$$

and for $t \ge \tau_0 + i_k h$,

$$
\int_{i_k h}^{t-\tau_0} \dot{x}^T(s) R_2 \dot{x}(s)ds \le \int_{t-\tau_0-\delta}^{t-\tau_0+\delta} \dot{x}^T(s) R_2 \dot{x}(s)ds. \tag{5.23}
$$

Then, combining (5.17)–(5.23), we can show that, for
$t \in [i_k h + \tau_k, i_{k+1} h + \tau_{k+1})$,

$$
\dot{V}(t) \le \left[e^T(t)\; \int_{t-\tau_0}^{t} x^T(s)ds\; w^T(t) \right] \left\{ \begin{bmatrix} \Xi_{11} & * \\ \tilde{\Xi}_{21} & \tilde{\Xi}_{22} \end{bmatrix} + \eta \tilde{N} R_1^{-1} \tilde{N}^T + \delta \tilde{S} R_2^{-1} \tilde{S}^T + \tilde{C}\tilde{C}^T \right\} \begin{bmatrix} e(t) \\ \int_{t-\tau_0}^{t} x(s)ds \\ w(t) \end{bmatrix} \tag{5.24}
$$

where

$$\tilde{\Xi}_{21} = \begin{bmatrix} P_3 & 0 & -P_3 & P_2^T \\ -B_w^T M_1^T & -B_w^T M_2^T & -B_w^T M_3^T & -B_w^T M_4^T \end{bmatrix},$$

$$\tilde{\Xi}_{22} = \mathrm{diag}\left(-\frac{1}{\tau_0}T_2,\ -\gamma^2 I\right),\quad \tilde{N}^T = \begin{bmatrix} N^T & 0 & 0 \end{bmatrix},$$

$$\tilde{S}^T = \begin{bmatrix} S^T & 0 & 0 \end{bmatrix},\quad \tilde{C}^T = \begin{bmatrix} C & DK & 0 & 0 & 0 & 0 \end{bmatrix}.$$

By Schur complement and combining (5.14) and (5.24) we obtain for $t \in [i_k h + \tau_k, i_{k+1} h + \tau_{k+1})$,

$$\dot{V}(t) \leq -z^T(t)z(t) + \gamma^2 w^T(t)w(t). \tag{5.25}$$

Integrating both sides of (5.25) from $i_k h + \tau_k$ to $t \in [i_k h + \tau_k,\ i_{k+1} h + \tau_{k+1})$, we have

$$V(t) - V(i_k h + \tau_k) \leq -\int_{i_k h + \tau_k}^{t} z^T(s)z(s)ds + \int_{i_k h + \tau_k}^{t} \gamma^2 w^T(s)w(s)ds. \tag{5.26}$$

Since $\cup_{k=1}^{\infty} [i_k h + \tau_k, i_{k+1} h + \tau_{k+1}) = [t_0, \infty)$ and $V(t)$ is continuous in t (since $x(t)$ is continuous in t), from (5.26), we can see that

$$V(t) - V(t_0) \leq -\int_{t_0}^{t} z^T(s)z(s)ds + \int_{t_0}^{t} \gamma^2 w^T(s)w(s)ds. \tag{5.27}$$

Then, letting $t \to \infty$ and under zero initial condition, we can show from (5.27) that

$$\int_{t_0}^{\infty} z^T(s)z(s)ds \leq \gamma^2 \int_{t_0}^{\infty} w^T(s)w(s)ds, \tag{5.28}$$

thus, $\|z(t)\|_2 \leq \gamma \|w(t)\|_2$.

Next, we prove the exponential stability of System (5.8). In this case, the external perturbation $w(t)$ is assumed to be zero. Then, using the Lyapunov–Krasovskii functional (5.16), for $t \in [i_k h + \tau_k, i_{k+1} h + \tau_{k+1})$, one can obtain that

$$\dot{V}(t) \leq \begin{bmatrix} e^T(t) & \int_{t-\tau_0}^{t} x^T(s)ds \end{bmatrix} \left\{ \begin{bmatrix} \Xi_{11} & * \\ \hat{\Xi}_{21} & -\frac{1}{\tau_0}T_2 \end{bmatrix} + \eta \begin{bmatrix} N \\ 0 \end{bmatrix} R_1^{-1} \begin{bmatrix} N^T & 0 \end{bmatrix} + \delta \begin{bmatrix} S \\ 0 \end{bmatrix} R_2^{-1} \begin{bmatrix} S^T & 0 \end{bmatrix} \right\} \begin{bmatrix} e(t) \\ \int_{t-\tau_0}^{t} x(s)ds \end{bmatrix}, \tag{5.29}$$

where $\hat{\Xi}_{21} = \begin{bmatrix} P_3 & 0 & -P_3 & P_2^T \end{bmatrix}$.

Using (5.14) and Schur complements, we can conclude from (5.29) that for $t \in [i_k h + \tau_k, i_{k+1}h + \tau_{k+1})$,

$$\dot{V}(t) \leq -\lambda \|x(t)\|^2 - \lambda \|\dot{x}(t)\|^2, \tag{5.30}$$

where $\lambda = \min\{\lambda_{\min}(W_1), \lambda_{\min}(W_2)\}$.

Defining a new function as

$$W(t) = e^{\varepsilon t}V(t) \tag{5.31}$$

and taking its time derivative for $t \in [i_k h + \tau_k, i_{k+1}h + \tau_{k+1})$ yields

$$\begin{aligned}\dot{W}(t) &= \varepsilon e^{\varepsilon t}V(t) + e^{\varepsilon t}\dot{V}(t) \\ &\leq \varepsilon e^{\varepsilon t}V(t) - \lambda e^{\varepsilon t}\|x(t)\|^2 - \lambda e^{\varepsilon t}\|\dot{x}(t)\|^2. \end{aligned}\tag{5.32}$$

Integrating both sides of (5.32) from $i_k h + \tau_k$ to t, we obtain

$$\begin{aligned}W(t) - W(i_k h + \tau_k) \leq &\int_{i_k h+\tau_k}^{t} \varepsilon e^{\varepsilon s}V(s)ds - \lambda \int_{i_k h+\tau_k}^{t} e^{\varepsilon s}\|x(s)\|^2 ds \\ &-\lambda \int_{i_k h+\tau_k}^{t} e^{\varepsilon s}\|\dot{x}(s)\|^2 ds.\end{aligned}\tag{5.33}$$

Since $V(t)$ is continuous on $[t_0, \infty)$, $W(t)$ is also continuous on $[t_0, \infty)$. Then, from (5.33), it is easy to see that

$$\begin{aligned}W(t) - W(t_0) \leq &\int_{t_0}^{t} \varepsilon e^{\varepsilon s}V(s)ds - \lambda \int_{t_0}^{t} e^{\varepsilon s}\|x(s)\|^2 ds \\ &-\lambda \int_{t_0}^{t} e^{\varepsilon s}\|\dot{x}(s)\|^2 ds.\end{aligned}\tag{5.34}$$

By using a similar analysis method to that in [10], it can be seen from (5.16), (5.31) and (5.34) that, if $\varepsilon > 0$ is chosen small enough, a constant $\rho > 0$ can be found such that

$$V(t) \leq \rho \sup_{t_0-\eta\leq s\leq t_0} \|\phi(s)\|^2 e^{-\varepsilon t},\ t \geq t_0, \tag{5.35}$$

which can further imply from (5.16) that

$$\|x(t)\| \leq \sqrt{\lambda_{\min}^{-1}(P)\rho} \sup_{t_0-\eta\leq s\leq t_0} \|\phi(s)\| e^{-\varepsilon t/2},\ t \geq t_0, \tag{5.36}$$

where $P = \begin{bmatrix} P_1 & P_2 \\ P_2^T & P_3 \end{bmatrix}$. Then, by Definition 5.1, the result is established. □

Remark 5.5. If we do not consider the effects of network-induced delay and sampling rate on the system, (5.8) becomes

$$\dot{x}(t) = A_0(t)x(t) + B_w w(t), \tag{5.37}$$

where $A_0(t) = A + \Delta A(t) + BK + \Delta B(t)K$. In this case, $\tau_m = 0$ and $\eta = 0$. Thus, $V(t)$ in (5.16) reduces to

$$V(t) = x^T(t)P_1 x(t).$$

The derivative of $V(t)$, i.e., (5.17), becomes

$$\begin{aligned}\dot{V}(t) &= 2x^T(t)P_1\dot{x}(t) \\ &\quad +2e^T(t)M\left[-A_0(t)x(t) - B_w w(t) + \dot{x}(t)\right] \\ &\quad +\left[Cx(t) + DKx(i_k h)\right]^T\left[Cx(t) + DKx(i_k h)\right] \\ &\quad -\gamma^2 w^T(t)w(t) - z^T(t)z(t) + \gamma^2 w^T(t)w(t).\end{aligned} \tag{5.38}$$

Letting $M_1 = -P_1$ and $M_2 = 0$, we have from (5.38) that

$$\begin{aligned}\dot{V}(t) &= 2x^T(t)P_1\left[A_0(t)x(t) + B_w w(t)\right] \\ &\quad +\left[Cx(t) + DKx(i_k h)\right]^T\left[Cx(t) + DKx(i_k h)\right] \\ &\quad -\gamma^2 w^T(t)w(t) - z^T(t)z(t) + \gamma^2 w^T(t)w(t).\end{aligned} \tag{5.39}$$

Then, from (5.39), we can obtain the H_∞ performance criterion commonly used for the system without considering the effects of network-induced delay and sampling rate on the system. Since M_1 and M_2 in our criterion can be any matrices, it is easy to see that using our criterion we can obtain better (that is, less conservative) results than those by other methods even for the case where the effects of network-induced delay and sampling rate on the system are not considered. □

The parameter uncertainties $\Delta A(t)$ and $\Delta B(t)$ are contained in (5.14). Therefore, Lemma 5.1 cannot directly be used to determine the performance of the closed-loop System (5.8)–(5.10). The following result is given to provide a sufficient condition for guaranteeing the feasibility of (5.14) (its proof can easily be established by combining (5.14), (5.15) and (5.2), and hence is omitted).

Theorem 5.1. For given scalars τ_m, η and γ and a matrix K, if there exist matrices P_k $(k = 1,2,3)$, $T_j > 0$, $R_j > 0$ $(j = 1,2)$ and any matrices N_i, S_i and M_i $(i = 1,2,3,4)$ of appropriate dimensions and a scalar $\varepsilon > 0$ such that

$$\begin{bmatrix} \Psi_{11} & * \\ \Psi_{21} & \Psi_{22} \end{bmatrix} < 0, \tag{5.40}$$

$$\begin{bmatrix} P_1 & P_2 \\ P_2^T & P_3 \end{bmatrix} > 0, \tag{5.41}$$

where

$$\Psi_{11} = \begin{bmatrix} \Gamma'_{11} & * & * & * \\ \Gamma'_{21} & \Gamma'_{22} & * & * \\ \Gamma'_{31} & \Gamma'_{32} & \Gamma'_{33} & * \\ \Gamma'_{41} & \Gamma'_{42} & \Gamma'_{43} & \Gamma'_{44} \end{bmatrix},$$

$$\Psi_{21} = \begin{bmatrix} \tau_0 P_3 & 0 & -\tau_0 P_3 & \tau_0 P_2^T \\ \eta N_1^T & \eta N_2^T & \eta N_3^T & \eta N_4^T \\ \delta S_1^T & \delta S_2^T & \delta S_3^T & \delta S_4^T \\ C & DK & 0 & 0 \\ -B_w^T M_1^T & -B_w^T M_2^T & -B_w^T M_3^T & -B_w^T M_4^T \\ G^T M_1^T & G^T M_2^T & G^T M_3^T & G^T M_4^T \end{bmatrix},$$

$$\Psi_{22} = \operatorname{diag}\left(-\tau_0 T_2,\ -\eta R_1,\ -\delta R_2,\ -I,\ -\gamma^2 I,\ -\varepsilon I\right),$$

$$\begin{aligned}
\Gamma'_{11} &= P_2 + P_2^T + T_1 + \tau_0 T_2 + N_1 + N_1^T - M_1 A - A^T M_1^T + \varepsilon E_a^T E_a, \\
\Gamma'_{21} &= N_2 - N_1^T + S_1^T - M_2 A - K^T B^T M_1^T + \varepsilon K^T E_b^T E_a, \\
\Gamma'_{31} &= N_3 - P_2^T - S_1^T - M_3 A, \\
\Gamma'_{41} &= M_1^T + N_4 + P_1 - M_4 A, \\
\Gamma'_{22} &= -N_2 - N_2^T + S_2 + S_2^T - M_2 BK - K^T B^T M_2^T + \varepsilon K^T E_b^T E_b K, \\
\Gamma'_{32} &= -N_3 + S_3 - S_2^T - M_3 BK, \\
\Gamma'_{42} &= -N_4 + S_4 + M_2^T - M_4 BK, \\
\Gamma'_{33} &= -T_1 - S_3 - S_3^T, \\
\Gamma'_{43} &= -S_4 + M_3^T, \\
\Gamma'_{44} &= M_4 + M_4^T + \eta R_1 + 2\delta R_2,
\end{aligned}$$

then, System (5.8)–(5.10) with a control network satisfying Assumption 5.2 is robustly exponentially stable with an H_∞ norm bound γ. □

Remark 5.6. For given τ_m, η and γ and a matrix K, if the solvability of LMIs (5.40)–(5.41) is feasible, we know that as long as Assumption 5.2 is satisfied, System (5.8)–(5.10) is robustly exponentially stable with an H_∞ norm bound γ. If there is no data dropout in the network, (5.6) becomes

$$h + \tau_{k+1} \leq \eta. \tag{5.42}$$

For a chosen sampling rate h, we can find that the upper bound of time delay τ_{k+1} that guarantees the H_∞ performance of System (5.8)–(5.10) is $\eta - h$. Moreover, the allowable amount of data dropout that guarantees the H_∞ performance of System (5.8)–(5.10) can also be determined by (5.6). □

5.2.3 Robust H_∞ Controller Design

Based on Theorem 5.1, we are now in a position to design the feedback gain K, which can make the System (5.1) robustly exponentially stable with an H_∞ norm bound γ.

Theorem 5.2. For given scalars ρ_l $(l = 2, 3, 4)$, τ_m, η and γ, if there exist matrices $\tilde{P}_k$ $(k = 1, 2, 3)$, $\tilde{T}_j > 0$, $\tilde{R}_j > 0$ $(j = 1, 2)$, a nonsingular matrix X and any matrices $\tilde{N}_i$ and $\tilde{S}_i$ $(i = 1, 2, 3, 4)$ of appropriate dimensions and a scalar $\mu > 0$ such that

$$\begin{bmatrix} \Phi_{11} & * \\ \Phi_{21} & \Phi_{22} \end{bmatrix} < 0, \tag{5.43}$$

$$\begin{bmatrix} \tilde{P}_1 & \tilde{P}_2 \\ \tilde{P}_2^T & \tilde{P}_3 \end{bmatrix} > 0, \tag{5.44}$$

where

$$\Phi_{11} = \begin{bmatrix} \Sigma_{11} & * & * & * \\ \Sigma_{21} & \Sigma_{22} & * & * \\ \Sigma_{31} & \Sigma_{32} & \Sigma_{33} & * \\ \Sigma_{41} & \Sigma_{42} & \Sigma_{43} & \Sigma_{44} \end{bmatrix},$$

$$\Phi_{21} = \begin{bmatrix} \tau_0 \tilde{P}_3 & 0 & -\tau_0 \tilde{P}_3 & \tau_0 \tilde{P}_2^T \\ \eta \tilde{N}_1^T & \eta \tilde{N}_2^T & \eta \tilde{N}_3^T & \eta \tilde{N}_4^T \\ \delta \tilde{S}_1^T & \delta \tilde{S}_2^T & \delta \tilde{S}_3^T & \delta \tilde{S}_4^T \\ CX^T & DY & 0 & 0 \\ -B_w^T & -\rho_2 B_w^T & -\rho_3 B_w^T & -\rho_4 B_w^T \\ E_a X^T & E_b Y & 0 & 0 \end{bmatrix},$$

$$\Phi_{22} = \mathrm{diag}\left(-\tau_0 \tilde{T}_2,\ -\eta \tilde{R}_1,\ -\delta \tilde{R}_2,\ -I,\ -\gamma^2 I,\ -\mu I \right),$$

$$\begin{aligned}
\Sigma_{11} &= \tilde{P}_2 + \tilde{P}_2^T + \tilde{T}_1 + \tau_0 \tilde{T}_2 + \tilde{N}_1 + \tilde{N}_1^T - AX^T - XA^T + \mu GG^T, \\
\Sigma_{21} &= \tilde{N}_2 - \tilde{N}_1^T + \tilde{S}_1^T - \rho_2 AX^T - Y^T B^T + \mu \rho_2 GG^T, \\
\Sigma_{31} &= \tilde{N}_3 - \tilde{P}_2^T - \tilde{S}_1^T - \rho_3 AX^T + \mu \rho_3 GG^T, \\
\Sigma_{41} &= X + \tilde{N}_4 + \tilde{P}_1 - \rho_4 AX^T + \mu \rho_4 GG^T, \\
\Sigma_{22} &= -\tilde{N}_2 - \tilde{N}_2^T + \tilde{S}_2 + \tilde{S}_2^T - \rho_2 BY - \rho_2 Y^T B^T + \mu \rho_2^2 GG^T, \\
\Sigma_{32} &= -\tilde{N}_3 + \tilde{S}_3 - \tilde{S}_2^T - \rho_3 BY + \mu \rho_3 \rho_2 GG^T, \\
\Sigma_{42} &= -\tilde{N}_4 + \tilde{S}_4 + \rho_2 X - \rho_4 BY + \mu \rho_4 \rho_2 GG^T, \\
\Sigma_{33} &= -\tilde{T}_1 - \tilde{S}_3 - \tilde{S}_3^T + \mu \rho_3^2 GG^T, \\
\Sigma_{43} &= -\tilde{S}_4 + \rho_3 X + \mu \rho_4 \rho_3 GG^T, \\
\Sigma_{44} &= \rho_4 X + \rho_4 X^T + \eta \tilde{R}_1 + 2\delta \tilde{R}_2 + \mu \rho_4^2 GG^T,
\end{aligned}$$

then, under the controller $u(t) = Kx(t)$ with $K = YX^{-T}$, System (5.1) with a control network satisfying Assumption 5.2 is robustly exponentially stabilizable with an H_∞ norm bound γ.

Proof. Define $M_1 = M_0$, $M_2 = \rho_2 M_0$, $M_3 = \rho_3 M_0$, $M_4 = \rho_4 M_0$ and $\rho_4 \neq 0$ in (5.40) and refer it to as (5.40)$'$. Obviously, (5.40)$'$ implies that M_0 is nonsingular. Pre- and post-multiplying both sides of (5.40)$'$ with

$$\text{diag}(X, X, X, X, X, X, X, I, I, I)$$

and its transpose, and (5.41) with $\text{diag}(X, X)$ and its transpose, where $X = M_0^{-1}$, and introducing new variables $\tilde{P}_k = XP_kX^T$ $(k = 1,2,3)$, $\tilde{N}_i = XN_iX^T$, $\tilde{S}_i = XS_iX^T$ $(i = 1,2,3,4)$, $\tilde{T}_j = XT_jX^T$, $\tilde{R}_j = XR_jX^T$ $(j = 1,2)$, $Y = KX^T$ and $\mu = \varepsilon^{-1}$, we can obtain (5.43) and (5.44) by using Schur complements. It is easy to see that (5.43) and (5.44), respectively imply (5.40)′ and (5.41). Therefore, from Theorem 5.1, we complete the proof. □

5.2.4 Numerical Examples

Example 5.1. Consider the following system borrowed from [29]:

$$\dot{x}(t) = \begin{bmatrix} 0 & 1 \\ 0 & -0.1 \end{bmatrix} x(t) + \begin{bmatrix} 0 \\ 0.1 \end{bmatrix} u(t). \tag{5.45}$$

When considering the effect of external perturbation on the system, (5.45) can be expressed as

$$\dot{x}(t) = \begin{bmatrix} 0 & 1 \\ 0 & -0.1 \end{bmatrix} x(t) + \begin{bmatrix} 0 \\ 0.1 \end{bmatrix} u(t) + \begin{bmatrix} 0.1 \\ 0.1 \end{bmatrix} w(t), \tag{5.46}$$

$$z(t) = \begin{bmatrix} 0 & 1 \end{bmatrix} x(t) + 0.1u(t). \tag{5.47}$$

For this example, we will employ the same feedback controller as in [29], that is, $u(t) = \begin{bmatrix} -3.75 & -11.5 \end{bmatrix} x(t)$. This controller is designed without considering the presence of the network. Under an assumption that the controller dynamics in (5.45) is continuous and/or the communication medium is error-free [20], the maximum allowable transfer intervals (MATI), also called MADB [8], that guarantees the stability of System (5.45) controlled over a network are given in Table 5.1 which were obtained based on different methods. It has been pointed out in [27] that the method in [27] can be applicable to systems when the controller is computer-based and the control input signal arrives at the plant through a network. In this case, it has been found in [27] that the maximum allowable value of $\eta_{\max}$ is 0.8695. Choosing $\tau_m = 0$ and using Theorem 5.1, we compute that the maximum allowable value of $\eta_{\max}$ can be 0.8871. When the lower bound of the time delay τ_k exists, for brevity, we consider the constant delay case, that is, $\tau_k = \hat{\tau}$, and no data dropout (in

Table 5.1. MATIs based on different methods

Methods	MATI
Zhang *et al.*, 2001 [29]	4.5×10^{-4}
Park *et al.*, 2002 [16]	0.0538
Kim *et al.*, 2003 [8]	0.7805
Yue *et al.*, 2004 [27]	0.8695

this case, $\tau_m = \hat{\tau}$ and $\eta = \hat{\tau} + h$). From [27], we have $\eta_{\max} = 0.8695$. Then, the maximum allowable value of $\hat{\tau}_{\max}$ is $0.8695 - h$. For the case $h = 0.3$, $\hat{\tau}_{\max} = 0.5695$. Using Theorem 5.1 with $\tau_0 = \hat{\tau} + \frac{h}{2}$ and $\delta = \frac{h}{2}$, we find that $\hat{\tau}_{\max}$ is 0.6916 when $h = 0.3$. In this case, $\eta_{\max} = 0.9916$. Obviously, from these figures, it can be found that the present method can obtain a much less conservative result than that by the method in [27], especially for the case when a lower bound on time delay τ_k exists.

Next, we consider the H_∞ performance of System (5.46)–(5.47) under the given controller. For the case $\tau_m = 0$ we find that a minimum allowable value of $\gamma_{\min}$ is 6.82 for $\eta = 0.8695$. For the case $h = 0.3$ and constant time delay, that is, $\tau_k = \hat{\tau}$, a minimum allowable value of $\gamma_{\min}$ is 1.26 when $\hat{\tau} = 0.5695$.□

Example 5.2. Consider the following uncertain system controlled over a network:

$$\dot{x}(t) = \left(\begin{bmatrix} -1 & 0 & -0.5 \\ 1 & -0.5 & 0 \\ 0 & 0 & 0.5 \end{bmatrix} + \Delta A(t) \right) x(t)$$

$$+ \begin{bmatrix} 0 \\ 0 \\ 1 \end{bmatrix} u(t) + \begin{bmatrix} 1 \\ 1 \\ 1 \end{bmatrix} w(t), \tag{5.48}$$

$$z(t) = \begin{bmatrix} 1 \; 0 \; 1 \end{bmatrix} x(t) + 0.1u(t), \tag{5.49}$$

where $\|\Delta A(t)\| \leq 0.01$. For brevity, it is assumed in this example that $i_k = k$, that is, no data dropout occurs in the network. It is also assumed that $\tau_m = 0.1$, which is the lower bound of the network-induced delay. In the following discussion, the sampling rate is chosen as 0.2, that is, $h = 0.2$.

Using Theorem 5.2 with $\rho_2 = \rho_3 = 0.2$ and $\rho_4 = 2$, it is found that, for $\eta = 0.5$, the values of the minimum allowable value of $\gamma_{\min}$ is 1.9 with X and Y respectively given by

$$X = \begin{bmatrix} -4.8685 & -1.6898 & 4.5822 \\ 1.4794 & -130.3345 & -0.3929 \\ 2.4762 & 0.0790 & -4.7527 \end{bmatrix}, \quad Y = \begin{bmatrix} -3.7066 \\ -0.0765 \\ 5.2430 \end{bmatrix}^T.$$

Then, the feedback gain K can be computed as

$$K = [-0.5425 \quad -0.0014 \quad -1.3858].$$

By (5.6), it can be found that the allowable upper value of the network-induced delay is 0.3. Simulation results of System (5.48)–(5.49) with controller $u(t) = \begin{bmatrix} -0.5425 \; -0.0014 \; -1.3858 \end{bmatrix} x(t)$ is shown in Fig. 5.1. For this simulation,

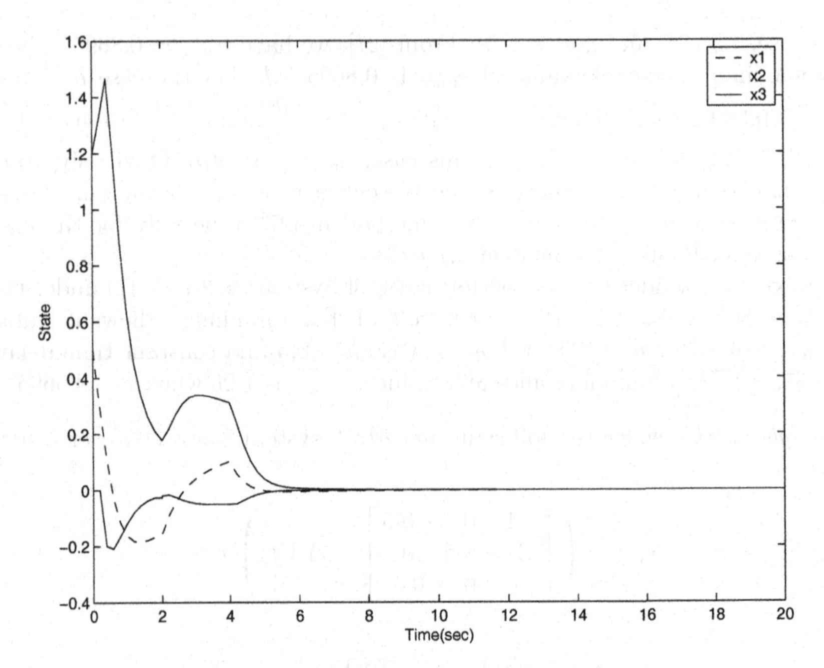

Fig. 5.1. Simulation of System (5.48)–(5.49)

initial values of the states are $x_1(0) = 0.5$, $x_2(0) = -0.5$ and $x_3(0) = 1.2$, the disturbance signal $w(t)$ is defined by

$$w(t) = \begin{cases} 0.3, & 2\text{s} \leq t \leq 4\text{s}, \\ 0, & \text{otherwise.} \end{cases}$$

□

5.3 Robust H_∞ Filter Design of NCS

5.3.1 Modeling a Network-based Filter

Consider an uncertain linear system

$$\dot{x}(t) = [A + \Delta A(t)]\, x(t) + Bw(t), \tag{5.50}$$

$$y(t) = [C_1 + \Delta C_1(t)]\, x(t) + Dw(t), \tag{5.51}$$

$$z(t) = Lx(t), \tag{5.52}$$

where $x(t) \in \mathbb{R}^n$ is the state, $y(t) \in \mathbb{R}^r$ is the measurement, $z(t) \in \mathbb{R}^q$ is the signal that will be estimated, and $w(t) \in \mathbb{R}^p$ is the disturbance input; A, B, C_1, D, and L are some constant matrices of appropriate dimensions; $\Delta A(t)$ and $\Delta C_1(t)$ denote the parameter uncertainties.

In this chapter, we consider the filter of the following form for the estimate of $z(t)$

$$\dot{x}_f(t) = A_f x_f(t) + B_f \hat{y}(t), \tag{5.53}$$
$$z_f(t) = L_f x_f(t), \tag{5.54}$$

where $x_f(t) \in \mathbb{R}^n$, $\hat{y}(t) \in \mathbb{R}^r$ are the inputs of the filter, A_f, B_f, and L_f will be determined later.

Remark 5.7. For a traditional filtering problem, the effect of the transmission network is neglected. In this case, $\hat{y}(t) = y(t)$. □

Different from the traditional filtering problem, the data of measurement $y(t)$ of System (5.50) is sampled and transmitted through a common network medium. Considering the effect of the common network on data transmission, (5.53) can be expressed as

$$\dot{x}_f(t) = A_f x_f(t) + B_f \left[C_1 + \Delta C_1(i_k h)\right] x(i_k h) + B_f D w(i_k h), \quad t \in \left[i_k h + \tau_k, i_{k+1} h + \tau_{k+1}\right), \ k = 1, 2, 3, \ldots \tag{5.55}$$

where h is the sampling period, i_k $(k = 1, 2, 3, \ldots)$ are as defined in (5.3)–(5.5).

As in Remark 5.3, defining $\tau(t) = t - i_k h$, $t \in \left[i_k h + \tau_k, i_{k+1} h + \tau_{k+1}\right)$, (5.55) can be further written as

$$\begin{aligned}\dot{x}_f(t) = {} & A_f x_f(t) + B_f \left[C_1 + \Delta C_1(t - \tau(t))\right] x(t - \tau(t)) \\ & + B_f D w(t - \tau(t)),\end{aligned} \tag{5.56}$$

where $\tau_k \le \tau(t) \le (i_{k+1} - i_k)h + \tau_{k+1}$ for $t \in \left[i_k h + \tau_k, i_{k+1} h + \tau_{k+1}\right)$, τ_k and $(i_{k+1} - i_k)h + \tau_{k+1}$ satisfy Assumption 5.2.

Remark 5.8. If the effect of the network is not considered, i.e., $\tau_k = 0$, then (5.56) reduces to a traditional sampled-data system. □

Define $\Delta \tilde{C}_1(t) = \Delta C_1(t - \tau(t))$ and $v(t) = w(t - \tau(t))$. Then (5.56) becomes

$$\dot{x}_f(t) = A_f x_f(t) + B_f \left[C_1 + \Delta \tilde{C}_1(t)\right] x(t - \tau(t)) + B_f D v(t). \tag{5.57}$$

Remark 5.9. It is found that the network-based filter is very different from the traditional filter. In the former case, there exist a network-induced delay and data dropout at the input of the filter. Moreover, the delay belongs to a class of fast time-varying delays. In other words, we cannot find a positive scalar less than 1, which is an upper bound of the derivative of $\tau(t)$. Therefore, for the design of filters (5.54) and (5.57), the methods in [2, 3, 23, 25] cannot be applied since the upper bound of the derivative of the time delay is required to be less than 1 in these references. □

In this section, we propose a design method for filter (5.57) and (5.54). Without loss of generality, consider the filtering problem for the following system:

$$\dot{x}(t) = [A + \Delta A(t)]\,x(t) + [A_1 + \Delta A_1(t)]\,x(t-\tau(t)) + Bw(t), \tag{5.58}$$

$$y(t) = [C + \Delta C(t)]\,x(t) + [C_1 + \Delta C_1(t)]\,x(t-\tau(t)) + Dv(t), \tag{5.59}$$

$$z(t) = Lx(t), \tag{5.60}$$

$$x(t) = \psi(t),\ t \in [t_0 - \eta, t_0], \tag{5.61}$$

where $\tau(t)$ is the time delay satisfying

$$\tau_m \le \tau(t) \le \eta, \tag{5.62}$$

τ_m and η are two known constants, and $\Delta A(t)$, $\Delta A_1(t)$, $\Delta C(t)$, and $\Delta C_1(t)$ denote the parameter uncertainties satisfying

$$\begin{bmatrix} \Delta A(t) & \Delta A_1(t) \\ \Delta C(t) & \Delta C_1(t) \end{bmatrix} = \begin{bmatrix} G_1 \\ G_2 \end{bmatrix} F(t) \begin{bmatrix} E_1 & E_2 \end{bmatrix}. \tag{5.63}$$

The filter that will be designed is of the following form:

$$\dot{x}_f(t) = A_f x_f(t) + B_f y(t), \tag{5.64}$$

$$z_f(t) = L_f x_f(t), \tag{5.65}$$

$$x_f(t) = 0,\ t \le t_0. \tag{5.66}$$

Define $\zeta(t) = \left[x^T(t)\ x_f^T(t)\right]^T$, $e(t) = z(t) - z_f(t)$ and $\beta(t) = \begin{bmatrix} w(t) \\ v(t) \end{bmatrix}$. Combining (5.58)–(5.61) and (5.64)–(5.66), we can obtain the filtering-error system

$$\dot{\zeta}(t) = \left[\tilde{A} + \Delta\tilde{A}(t)\right]\zeta(t) + \left[\tilde{A}_1 + \Delta\tilde{A}_1(t)\right] \times \zeta(t-\tau(t)) + \tilde{B}\beta(t), \tag{5.67}$$

$$e(t) = \tilde{L}\zeta(t), \tag{5.68}$$

$$\zeta(t) = \phi(t) \triangleq \begin{bmatrix} \psi(t) \\ 0 \end{bmatrix},\ t \in [t_0 - \eta, t_0], \tag{5.69}$$

where

$$\tilde{A} = \begin{bmatrix} A & 0 \\ B_f C & A_f \end{bmatrix},\ \tilde{A}_1 = \begin{bmatrix} A_1 & 0 \\ B_f C_1 & 0 \end{bmatrix},$$

$$\tilde{B} = \begin{bmatrix} B & 0 \\ 0 & B_f D \end{bmatrix},\ \tilde{L} = \begin{bmatrix} L & -L_f \end{bmatrix},$$

$$\Delta\tilde{A}(t) = \begin{bmatrix} \Delta A(t) & 0 \\ B_f \Delta C(t) & 0 \end{bmatrix},\ \Delta\tilde{A}_1(t) = \begin{bmatrix} \Delta A_1(t) & 0 \\ B_f \Delta C_1(t) & 0 \end{bmatrix}.$$

In light of (5.63), $\Delta\tilde{A}(t)$ and $\Delta\tilde{A}_1(t)$ can be expressed as

$$\Delta\tilde{A}(t) = \begin{bmatrix} G_1 \\ B_f G_2 \end{bmatrix} F(t) \begin{bmatrix} E_1 \; 0 \end{bmatrix} \triangleq G F(t) \tilde{E}_1,$$

$$\Delta\tilde{A}_1(t) = \begin{bmatrix} G_1 \\ B_f G_2 \end{bmatrix} F(t) \begin{bmatrix} E_2 \; 0 \end{bmatrix} \triangleq G F(t) \tilde{E}_2.$$

Remark 5.10. When $[A_1 + \Delta A_1(t)] = 0$, $[C + \Delta C(t)] = 0$, and $\tau(t) = t - i_k h$, for $t \in [i_k h + \tau_k, i_{k+1} h + \tau_{k+1})$, combining (5.64) and (5.59) yields (5.57), which becomes the design problem of network-based filter. More specifically, suppose that $\Delta A(t)$ and $\Delta C_1(t)$ in (5.50) and (5.51) can be expressed as

$$\Delta A(t) = G_a F(t) E_a, \; \Delta C_1(t) = G_c F(t) E_c.$$

For System (5.50)–(5.52) with the network-based filter (5.53)–(5.54), we can obtain the corresponding network-based filtering-error System (5.67)′–(5.69)′, which is derived from System (5.67)–(5.69) with the following system matrices

$$\tilde{A} = \begin{bmatrix} A & 0 \\ 0 & A_f \end{bmatrix}, \; \tilde{A}_1 = \begin{bmatrix} 0 & 0 \\ B_f C_1 & 0 \end{bmatrix},$$

$$\tilde{B} = \begin{bmatrix} B & 0 \\ 0 & B_f D \end{bmatrix}, \; \tilde{L} = \begin{bmatrix} L & -L_f \end{bmatrix},$$

$$\Delta\tilde{A}(t) = \begin{bmatrix} \Delta A(t) & 0 \\ 0 & 0 \end{bmatrix}, \; \Delta\tilde{A}_1(t) = \begin{bmatrix} 0 & 0 \\ B_f \Delta C_1(t) & 0 \end{bmatrix}, \tag{5.70}$$

and $\Delta\tilde{A}(t)$ and $\Delta\tilde{A}_1(t)$ can be expressed as

$$\Delta\tilde{A}(t) = \begin{bmatrix} G_1 \\ B_f G_2 \end{bmatrix} F(t) \begin{bmatrix} E_1 \; 0 \end{bmatrix} \triangleq G F(t) \tilde{E}_1,$$

$$\Delta\tilde{A}_1(t) = \begin{bmatrix} G_1 \\ B_f G_2 \end{bmatrix} F(t) \begin{bmatrix} E_2 \; 0 \end{bmatrix} \triangleq G F(t) \tilde{E}_2, \tag{5.71}$$

with $G_1 = \begin{bmatrix} G_a \; 0 \end{bmatrix}, G_2 = \begin{bmatrix} 0 \; G_c \end{bmatrix}, \; E_1 = \begin{bmatrix} E_a \\ 0 \end{bmatrix}$, and $E_2 = \begin{bmatrix} 0 \\ E_c \end{bmatrix}$. □

5.3.2 H_∞ Performance Analysis of Filtering-error System

In this section, we will give an H_∞ performance analysis result for the filtering-error System (5.67)–(5.68). The following definition is first introduced.

Definition 5.2. System (5.67)–(5.68) is said to be robustly exponentially stable with an H_∞ norm bound γ, if the following hold:

(i) System (5.67)–(5.68) with $\beta(t) \equiv 0$ is robustly exponentially stable; that is, there exist constants $\alpha > 0$ and $\beta > 0$ such that $\|\zeta(t)\| \leq \alpha \sup_{t_0 - \eta \leq s \leq t_0} \|\phi(s)\| e^{-\beta t}$, $t \geq t_0$, for all admissible uncertainties $\Delta\tilde{A}(t)$ and $\Delta\tilde{A}_1(t)$;

(ii) under the assumption of zero initial condition, the controlled output $e(t)$ satisfies $\|e(t)\|_2 \leq \gamma \|\beta(t)\|_2$ for any nonzero $\beta(t) \in \mathcal{L}_2[t_0, \infty)$. □

By using the same method mentioned in Lemma 5.1, we can conclude the following result.

Lemma 5.2. For given scalars τ_m, η, and γ, if there exist matrices P_k ($k = 1, 2, 3$), $W_j > 0$, $T_j > 0$, $R_j > 0$ ($j = 1, 2$), and any matrices N_i, S_i, and M_i ($i = 1, 2, 3, 4$) of appropriate dimensions such that

$$\begin{bmatrix} \Xi_{11} + \mathrm{diag}(W_1, 0, 0, W_2) & * \\ \Xi_{21} & \Xi_{22} \end{bmatrix} < 0, \tag{5.72}$$

$$\begin{bmatrix} P_1 & P_2 \\ P_2^T & P_3 \end{bmatrix} > 0, \tag{5.73}$$

where

$$\Xi_{11} = \begin{bmatrix} \Gamma_{11} & * & * & * \\ \Gamma_{21} & \Gamma_{22} & * & * \\ \Gamma_{31} & \Gamma_{32} & \Gamma_{33} & * \\ \Gamma_{41} & \Gamma_{42} & \Gamma_{43} & \Gamma_{44} \end{bmatrix},$$

$$\Xi_{21} = \begin{bmatrix} \tau_0 P_3 & 0 & -\tau_0 P_3 & \tau_0 P_2^T \\ \eta N_1^T & \eta N_2^T & \eta N_3^T & \tau_M N_4^T \\ \delta S_1^T & \delta S_2^T & \delta S_3^T & \delta S_4^T \\ \tilde{L} & 0 & 0 & 0 \\ -\tilde{B}^T M_1^T & -\tilde{B}^T M_2^T & -\tilde{B}^T M_3^T & -\tilde{B}^T M_4^T \end{bmatrix},$$

$$\Xi_{22} = \mathrm{diag}\left(-\tau_0 T_2, \ -\eta R_1, \ -\delta R_2, \ -I, \ -\gamma^2 I\right),$$

with

$$\Gamma_{11} = P_2 + P_2^T + T_1 + \tau_0 T_2 + N_1 + N_1^T - M_1\left[\tilde{A} + \Delta\tilde{A}(t)\right] - \left[\tilde{A} + \Delta\tilde{A}(t)\right]^T M_1^T,$$

$$\Gamma_{21} = N_2 - N_1^T + S_1^T - M_2\left[\tilde{A} + \Delta\tilde{A}(t)\right] - \left[\tilde{A}_1 + \Delta\tilde{A}_1(t)\right]^T M_1^T,$$

$$\Gamma_{31} = N_3 - P_2^T - S_1^T - M_3\left[\tilde{A} + \Delta\tilde{A}(t)\right],$$

$$\Gamma_{41} = M_1^T + N_4 + P_1 - M_4\left[\tilde{A} + \Delta\tilde{A}(t)\right],$$

$$\Gamma_{22} = -N_2 - N_2^T + S_2 + S_2^T - M_2\left[\tilde{A}_1 + \Delta\tilde{A}_1(t)\right] - \left[\tilde{A}_1 + \Delta\tilde{A}_1(t)\right]^T M_2^T,$$

$$\begin{aligned}
\Gamma_{32} &= -N_3 + S_3 - S_2^T - M_3\left[\tilde{A}_1 + \Delta\tilde{A}_1(t)\right],\\
\Gamma_{42} &= -N_4 + S_4 + M_2^T - M_4\left[\tilde{A}_1 + \Delta\tilde{A}_1(t)\right],\\
\Gamma_{33} &= -T_1 - S_3 - S_3^T,\\
\Gamma_{43} &= -S_4 + M_3^T,\\
\Gamma_{44} &= M_4 + M_4^T + \eta R_1 + 2\delta R_2,
\end{aligned}$$

then, System (5.67)–(5.69) is robustly exponentially stable with an H_∞ norm bound γ. □

The parameter uncertainties $\Delta\tilde{A}(t)$ and $\Delta\tilde{A}_1(t)$ are contained in (5.72). Therefore, Lemma 5.2 cannot directly be used to determine the performance of System (5.67)–(5.68). The following result is given to provide a sufficient condition for guaranteeing the feasibility of (5.72) (its proof can easily be established, and hence is omitted).

Lemma 5.3. For given scalars τ_m, η, and γ, if there exist matrices P_k $(k = 1, 2, 3)$, $T_j > 0$, $R_j > 0$ $(j = 1, 2)$ and any matrices N_i, S_i and M_i $(i = 1, 2, 3, 4)$ of appropriate dimensions and a scalar $\varepsilon > 0$ such that (5.73) and

$$\begin{bmatrix} \Xi'_{11} & * \\ \Xi'_{21} & \Xi'_{22} \end{bmatrix} < 0, \tag{5.74}$$

where

$$\Xi'_{11} = \begin{bmatrix} \Gamma'_{11} & * & * & * \\ \Gamma'_{21} & \Gamma'_{22} & * & * \\ \Gamma'_{31} & \Gamma'_{32} & \Gamma'_{33} & * \\ \Gamma'_{41} & \Gamma'_{42} & \Gamma'_{43} & \Gamma'_{44} \end{bmatrix},$$

$$\Xi'_{21} = \begin{bmatrix} \tau_0 P_3 & 0 & -\tau_0 P_3 & \tau_0 P_2^T \\ \eta N_1^T & \eta N_2^T & \eta N_3^T & \tau_M N_4^T \\ \delta S_1^T & \delta S_2^T & \delta S_3^T & \delta S_4^T \\ \tilde{L} & 0 & 0 & 0 \\ -\tilde{B}^T M_1^T & -\tilde{B}^T M_2^T & -\tilde{B}^T M_3^T & -\tilde{B}^T M_4^T \\ G^T M_1^T & G^T M_2^T & G^T M_3^T & G^T M_4^T \end{bmatrix},$$

$$\begin{aligned}
\Xi'_{22} &= \mathrm{diag}\left(-\tau_0 T_2,\ -\eta R_1,\ -\delta R_2,\ -I,\ -\gamma^2 I,\ -\varepsilon I\right),\\
\Gamma'_{11} &= P_2 + P_2^T + T_1 + \tau_0 T_2 + N_1 + N_1^T - M_1\tilde{A} - \tilde{A}^T M_1^T + \varepsilon\tilde{E}_1^T\tilde{E}_1,\\
\Gamma'_{21} &= N_2 - N_1^T + S_1^T - M_2\tilde{A} - \tilde{A}_1^T M_1^T + \varepsilon\tilde{E}_2^T\tilde{E}_1,\\
\Gamma'_{31} &= N_3 - P_2^T - S_1^T - M_3\tilde{A},\\
\Gamma'_{41} &= M_1^T + N_4 + P_1 - M_4\tilde{A},\\
\Gamma'_{22} &= -N_2 - N_2^T + S_2 + S_2^T - M_2\tilde{A}_1 - \tilde{A}_1^T M_2^T + \varepsilon\tilde{E}_2^T\tilde{E}_2,\\
\Gamma'_{32} &= -N_3 + S_3 - S_2^T - M_3\tilde{A}_1,\\
\Gamma'_{42} &= -N_4 + S_4 + M_2^T - M_4\tilde{A}_1,\\
\Gamma'_{33} &= -T_1 - S_3 - S_3^T,
\end{aligned}$$

$$\Gamma'_{43} = -S_4 + M_3^T,$$
$$\Gamma'_{44} = M_4 + M_4^T + \eta R_1 + 2\delta R_2,$$

then, System (5.67)–(5.69) is robustly exponentially stable with an H_∞ norm bound γ. □

Remark 5.11. Combining Remark 5.10 and Lemma 5.3, we can obtain an H_∞ performance analysis result, referred to as Lemma 5.3′, for System (5.67)′–(5.69)′ just by substituting the matrices $\tilde{A}$, $\tilde{A}_1$, $\tilde{B}$ and $\tilde{L}$ in (5.74) with those in (5.70)–(5.71). □

Remark 5.12. If we substitute the components of Ξ'_{11}, Ξ'_{21}, and Ξ'_{22} into (5.74), we have a 10×10 block matrix on the left-hand side of (5.74). By deleting the eighth and ninth rows and the corresponding eighth and ninth columns from the 10×10 block matrix, we can obtain a delay-dependent stability condition, referred to as (5.74)″, for System (5.67) with $\beta(t) = 0$. It should be pointed out that when we use this stability condition to check the stability of a system with interval time-varying delay, the matrices $\tilde{A}$ and $\tilde{A}_1$ should be replaced with the corresponding system matrices. □

5.3.3 H_∞ Filter Design

Based on Lemma 5.3, we are in a position to derive a criterion for filter design.

Theorem 5.3. For given $\rho_2,\ \rho_3,\ \rho_4 > 0,\ \tau_m,\ \eta$, and γ, if there exist matrices P_{ik} $(i = 1,2,3)$, N_{jk}, S_{jk} $(j = 1,2,3,4; k = 1,2,3,4)$, T_{pq}, R_{pq} $(p = 1,2; q = 1,2,3)$, U_m $(m = 1,2,3)$, X, and Y of appropriate dimensions and a scalar $\varepsilon > 0$ such that

$$\begin{bmatrix} \Omega_{11} & * \\ \Omega_{21} & \Omega_{22} \end{bmatrix} < 0, \tag{5.75}$$

$$X + X^T - Y - Y^T < 0, \tag{5.76}$$

$$\begin{bmatrix} \tilde{P}_1 & \tilde{P}_2 \\ \tilde{P}_2^T & \tilde{P}_3 \end{bmatrix} > 0,\ \tilde{T}_p > 0,\ \tilde{R}_p > 0, \tag{5.77}$$

where

$$\Omega_{11} = \begin{bmatrix} \Pi_{11} & * & * & * \\ \Pi_{21} & \Pi_{22} & * & * \\ \Pi_{31} & \Pi_{32} & \Pi_{33} & * \\ \Pi_{41} & \Pi_{42} & \Pi_{43} & \Pi_{44} \end{bmatrix},$$

$$\Omega_{21} = \begin{bmatrix} \tau_0\tilde{P}_3 & 0 & -\tau_0\tilde{P}_3 & \tau_0\tilde{P}_2 \\ \eta\tilde{N}_1^T & \eta\tilde{N}_2^T & \eta\tilde{N}_3^T & \eta\tilde{N}_4^T \\ \delta\tilde{S}_1^T & \delta\tilde{S}_1^T & \delta\tilde{S}_1^T & \delta\tilde{S}_1^T \\ \mathcal{L} & 0 & 0 & 0 \\ \mathcal{B} & \rho_2\mathcal{B} & \rho_3\mathcal{B} & \rho_4\mathcal{B} \\ \mathcal{G} & \rho_2\mathcal{G} & \rho_3\mathcal{G} & \rho_4\mathcal{G} \end{bmatrix},$$

$$\Omega_{22} = \mathrm{diag}\left(-\tau_0\tilde{T}_2,\ -\eta\tilde{R}_1,\ -\delta\tilde{R}_2,\ -I,\ -\gamma^2 I,\ -\varepsilon I\right),$$

with the definition of Π_{ij} $(i,j=1,2,3,4)$ given in Section 5.4, and

$$\tilde{N}_j=\begin{bmatrix} N_{j1} & N_{j2} \\ N_{j3} & N_{j4}\end{bmatrix},\ \tilde{S}_j=\begin{bmatrix} S_{j1} & S_{j2} \\ S_{j3} & S_{j4}\end{bmatrix}, j=1,2,3,4,$$

$$\tilde{T}_p=\begin{bmatrix} T_{p1} & T_{p2} \\ T_{p2}^T & T_{p3}\end{bmatrix},\ \tilde{R}_p=\begin{bmatrix} R_{p1} & R_{p2} \\ R_{p2}^T & R_{p3}\end{bmatrix}, p=1,2,$$

$$\tilde{P}_i=\begin{bmatrix} P_{i1} & P_{i2} \\ P_{i3} & P_{i4}\end{bmatrix},\ i=1,2,3, \mathcal{L}=\begin{bmatrix} L-U_3 & L\end{bmatrix},$$

$$\mathcal{B}=-\begin{bmatrix} B^TY & B^TX \\ 0 & D^TU_2^T\end{bmatrix},\ \mathcal{G}=\begin{bmatrix} G_1^TY & G_1^TX+G_2^TU_2^T\end{bmatrix},$$

then, the H_∞ filtering problem is solvable. Furthermore, the parameter matrices of the filter (5.64)–(5.65) is given as follows

$$A_f=J^{-1}U_1Y^{-1}W^{-T}, B_f=J^{-1}U_2, L_f=U_3Y^{-1}W^{-T}, \tag{5.78}$$

where J and W are the nonsingular matrix solutions to

$$JW^T=I-XY^{-1}. \tag{5.79}$$

Proof. From (5.75) and noting that $\rho_4>0$, it can be seen that $X+X^T$ and $Y+Y^T$ satisfy

$$X+X^T<0,\ Y+Y^T<0. \tag{5.80}$$

Since (5.80) implies X and Y are nonsingular, from (5.76), we have

$$\left(XY^{-1}-I\right)Y+Y^T(XY^{-1}-I)^T<0, \tag{5.81}$$

which further implies that $XY^{-1}-I$ is nonsingular. Therefore, there exist nonsingular matrices J and W such that (5.79) holds.

Define

$$\Phi_1=\begin{bmatrix} Y^{-1} & I \\ W^T & 0\end{bmatrix} \text{ and } \Phi_2=\begin{bmatrix} I & X \\ 0 & J^T\end{bmatrix}. \tag{5.82}$$

Then,

$$\mathcal{M}^T=\Phi_2\Phi_1^{-1}=\begin{bmatrix} X & J \\ J^T & \Psi\end{bmatrix}, \tag{5.83}$$

where $\Psi=W^{-1}Y^{-T}\left(X^T-Y^T\right)Y^{-1}W^{-T}$.

For (5.74) in Lemma 5.3, letting $M_1=\mathcal{M}$, $M_2=\rho_2\mathcal{M}$, $M_3=\rho_3\mathcal{M}$, and $M_4=\rho_4\mathcal{M}$, and substituting them into (5.74), we can obtain a sufficient condition, referred to as (5.74)′, whose solvability can guarantee that of (5.74). Next, we will show that (5.74) with $\mathcal{M}^T=\begin{bmatrix} X & J \\ J^T & \Psi\end{bmatrix}$ is equivalent to (5.75).

Pre- and post-multiplying (5.74)′ with

$$\text{diag}\left(\tilde{Y}^T\Phi_1^T,\ \tilde{Y}^T\Phi_1^T,\ \tilde{Y}^T\Phi_1^T,\ \tilde{Y}^T\Phi_1^T,\ \tilde{Y}^T\Phi_1^T,\ \tilde{Y}^T\Phi_1^T,\ \tilde{Y}^T\Phi_1^T,\ I,\ I,\ I\right)$$

and its transpose, where $\tilde{Y} = \begin{bmatrix} Y & 0 \\ 0 & I \end{bmatrix}$. By using some routine matrix manipulations and defining $\tilde{N}_i = \tilde{Y}^T \Phi_1^T N_i \Phi_1 \tilde{Y}$, $\tilde{S}_i = \tilde{Y}^T \Phi_1^T S_i \Phi_1 \tilde{Y}$ $(i = 1,2,3,4)$, $\tilde{T}_j = \tilde{Y}^T \Phi_1^T T_j \Phi_1 \tilde{Y}$, $\tilde{R}_j = \tilde{Y}^T \Phi_1^T R_j \Phi_1 \tilde{Y}$ $(j = 1,2)$, $\tilde{P}_k = \tilde{Y}^T \Phi_1^T P_k \Phi_1 \tilde{Y}$ $(k = 1,2,3)$, $U_1 = JA_f W^T Y$, $U_2 = JB_f$, and $U_3 = L_f W^T Y$, it can be shown that (5.74)$'$ is equivalent to (5.75). Pre- and post-multiplying $\begin{bmatrix} P_1 & P_2 \\ P_2^T & P_3 \end{bmatrix} > 0$ with $\text{diag}\left(\tilde{Y}^T \Phi_1^T,\ \tilde{Y}^T \Phi_1^T\right)$ and $\text{diag}\left(\Phi_1 \tilde{Y},\ \Phi_1 \tilde{Y}\right)$, we can show that the equivalence between $\begin{bmatrix} P_1 & P_2 \\ P_2^T & P_3 \end{bmatrix} > 0$ and $\begin{bmatrix} \tilde{P}_1 & \tilde{P}_2 \\ \tilde{P}_2^T & \tilde{P}_3 \end{bmatrix} > 0$. Finally, by (5.75)–(5.77) and Lemma 5.3, we can complete the proof. □

Remark 5.13. From (5.74)$'$ with $\mathcal{M}^T = \begin{bmatrix} X & J \\ J^T & \Psi \end{bmatrix}$, it is easy to see that

$$\Gamma'_{44} = \rho_4 \begin{bmatrix} X + X^T & J + J^T \\ J + J^T & \Psi + \Psi^T \end{bmatrix} + \eta R_1 + 2\delta R_2. \tag{5.84}$$

Obviously, if ρ_4 is chosen as a positive scalar, in order to guarantee (5.74)$'$, it is required that $\Psi + \Psi^T < 0$, which is equivalent to $X + X^T - Y - Y^T < 0$. Therefore, if the tuning scalar parameter ρ_4 is chosen as a positive number, the condition $X + X^T - Y - Y^T < 0$ in (5.76) must be used. On the other hand, when $\rho_4 < 0$ is chosen, it is required to use $X + X^T - Y - Y^T > 0$ in (5.76). □

Remark 5.14. Similar to Theorem 5.3, from Lemma 5.3$'$ and Remark 5.11, we can obtain a network-based filter design result for System (5.50)–(5.52), referred to as Theorem 5.3$'$, which can be established by deleting the terms containing A_1 and C in Π_{ij} $(i, j = 1,2,3,4)$ in Theorem 5.3. □

5.3.4 Numerical Examples

Example 5.3. Consider the following system [7]:

$$\dot{x}(t) = \begin{bmatrix} 0 & 1 \\ -1 & -2 \end{bmatrix} x(t) + \begin{bmatrix} 0 & 0 \\ -1 & 1 \end{bmatrix} x(t-1), \tag{5.85}$$

and

$$\dot{x}(t) = \begin{bmatrix} 0 & 1 \\ -1 & -2 \end{bmatrix} x(t) + \begin{bmatrix} 0 & 0 \\ -1 & 1 \end{bmatrix} x(t - 1 + \eta(t)), \tag{5.86}$$

where $\eta(t)$ satisfies the condition that $|\eta(t)| \le \eta_0$ and $|\dot{\eta}(t)| \le \dot{\eta}_0$, where η_0 and $\dot{\eta}_0$ as defined in [7]. It was found in [7] that System (5.86) remains stable if $\eta_0 < \dfrac{1}{640}\mu_0$ and $\dot{\eta}_0 < 1 - 8\mu_0$, where $\mu_0 \in \left(0, \dfrac{1}{40}\right)$. Now we study the stability of System (5.86) when the restriction on the derivative of $\eta(t)$ is removed. Using Lemma 5.3 and Remark 5.12, it was found that the maximum allowable value of η_0 is 0.27. It is clear to see that for this example the result using the present method is much better than that in [7]. Furthermore, $\eta(t)$ is allowed to be a fast time-varying function. □

Example 5.4. Consider the network-based robust H_∞ filtering problem for the following system:

$$\begin{aligned}
\dot{x}(t) &= [A + \Delta A(t)]\,x(t) + Bw(t),\\
y(t) &= [C_1 + \Delta C_1(t)]\,x(t) + Dw(t),\\
z(t) &= Lx(t),
\end{aligned} \tag{5.87}$$

where $A = \begin{bmatrix} 0 & 3 \\ -4 & -5 \end{bmatrix}$, $B = \begin{bmatrix} -0.5 \\ 0.9 \end{bmatrix}$, $C_1 = \begin{bmatrix} 0 \; 1 \end{bmatrix}$, $D = 1$, $L = \begin{bmatrix} 1 \; 1 \end{bmatrix}$, $\Delta A(t)$ and $\Delta C_1(t)$ can be expressed as $\Delta A(t) = \begin{bmatrix} 0.3 \\ 0.3 \end{bmatrix} F(t) \begin{bmatrix} 1 \; 1 \end{bmatrix}$ and $\Delta C_1(t) = 0.1F(t) \begin{bmatrix} 1 \; 1 \end{bmatrix}$. In terms of Remark 5.10, we choose $G_1 = \begin{bmatrix} 0.3 & 0 \\ 0.3 & 0 \end{bmatrix}$, $E_1 = \begin{bmatrix} 1 & 1 \\ 0 & 0 \end{bmatrix}$, $G_2 = \begin{bmatrix} 0 \; 0.1 \end{bmatrix}$ and $E_2 = \begin{bmatrix} 0 & 0 \\ 1 & 1 \end{bmatrix}$.

The designed filter is of the form (5.53) and (5.54). The measurement $y(t)$ is sampled and transmitted through a common network media. It is assumed that the propagation delay in the network to be monitored is 0.2 s. Therefore, we can choose $\tau_m = 0.2$.

Applying Theorem 5.3′ and Remark 5.14 with $\rho_2 = \rho_3 = 0.2$ and $\rho_4 = 5$ and choosing the H_∞ performance level $\gamma = 1.5$, it is found that the maximum allowable value of η is 0.48, and

$$X = \begin{bmatrix} -1.5150 & -0.5616 \\ -0.6199 & -0.4832 \end{bmatrix}, \quad Y = \begin{bmatrix} -0.4817 & -0.1938 \\ -0.2926 & -0.2196 \end{bmatrix},$$

$$U_1 = \begin{bmatrix} -2.4776 & 1.4792 \\ -2.0235 & -0.7505 \end{bmatrix}, \quad U_2 = \begin{bmatrix} 0.0032 \\ -0.0361 \end{bmatrix},$$

$U_3 = \begin{bmatrix} 1.2796 \; 0.9780 \end{bmatrix}$. To derive the parameter matrices of the filter (5.53) and (5.54), we choose the matrix J as $J = \begin{bmatrix} 2 & 1 \\ 1 & 1 \end{bmatrix}$. Then, from (5.79), we can solve that $W = \begin{bmatrix} -2.5382 & 2.6461 \\ 1.7656 & -3.0617 \end{bmatrix}$. Furthermore, from (5.78), we can finally obtain the parameter matrices of the filter as

$$\begin{aligned}
A_f &= \begin{bmatrix} 5.0775 & 10.6633 \\ -9.0693 & -12.7620 \end{bmatrix}, \quad B_f = \begin{bmatrix} 0.0393 \\ -0.0754 \end{bmatrix},\\
L_f &= \begin{bmatrix} 3.7782 \; 3.6644 \end{bmatrix}.
\end{aligned} \tag{5.88}$$

□

From the solutions of Example 5.4, it can be concluded that, as long as the network condition satisfies

$$(i_{k+1} - i_k)h + \tau_{k+1} \leq 0.48, \quad k = 1, 2, 3, \ldots \tag{5.89}$$

the filter (5.53) and (5.54) with parameter matrices in (5.88) can obtain an estimate $z_f(t)$ of the signal $z(t)$ of the System (5.87) that provides small estimation error $e(t) = z(t) - z_f(t)$ for all $w(t) \in \mathcal{L}_2[t_0, \infty)$ and allowable uncertainties in the System (5.87).

In Example 5.4, for the case of $\tau_m = 0.2$, it was solved that the maximum allowable value of η is 0.48. If setting $\tau_m = 0$, we found that the maximum allowable value of η is 0.45. Obviously, the former is less conservative than the latter.

Example 5.5. Consider the System (5.87) with

$$\begin{aligned} A &= \begin{bmatrix} -0.6 & 4 \\ -4 & -0.6 \end{bmatrix}, \; B = \begin{bmatrix} 0 & 0 \\ 1.5 & 0 \end{bmatrix} \\ C_1 &= \begin{bmatrix} 0 & 1 \end{bmatrix}, \; D = \begin{bmatrix} 0 & 1 \end{bmatrix}, \; L = \begin{bmatrix} 1 & 1 \end{bmatrix}, \end{aligned} \tag{5.90}$$

and the uncertainties $\Delta A(t) = \begin{bmatrix} 0.4 \\ 0 \end{bmatrix} F(t) \begin{bmatrix} 0 & 1 \end{bmatrix}$ and $\Delta C_1(t) = 0$. For this example, a filter with an optimal H_∞ performance of $\gamma_{opt} = 0.7624$ was designed based on the method in [5].

Considering the effect of the network conditions on the system, we will propose the design method of a network-based H_∞ filter for the system (5.90). The assumption on the network conditions is the same as that in Example 5.4.

Applying Theorem 5.3′ and Remark 5.14 with $\rho_2 = \rho_3 = 0.2$, and $\rho_4 = 5$ and choosing the H_∞ performance level $\gamma = 0.7624$, it is found that the maximum allowable value of η is 0.40, and similar to Example 5.4, we can obtain the parameter matrices A_f, B_f and L_f of the filter as

$$\begin{aligned} A_f &= \begin{bmatrix} 15.7425 & 11.4138 \\ -29.2242 & -18.3093 \end{bmatrix}, \; B_f = \begin{bmatrix} -0.0002 \\ -0.0014 \end{bmatrix}, \\ L_f &= \begin{bmatrix} 11.6265 & 16.2950 \end{bmatrix}. \end{aligned} \tag{5.91}$$

From the solutions of this example, it can be shown that for the same H_∞ performance $\gamma = 0.7624$, the designed filter in this chapter can tolerate transmission delay in the measurement $y(t)$ as long as the delay is less than 0.40. Obviously, the filter given in this chapter is more robust to the variation in network conditions. Setting $\tau_m = 0$, we found that the maximum allowable value of η is 0.38. □

5.4 Definition of Π_{ij}

We provide in this section the definition of Π_{ij} for $i, j = 1, 2, 3, 4$.

$$\Pi_{11} = \begin{bmatrix} \Pi_{11}^{(1,1)} & * \\ \Pi_{11}^{(2,1)} & \Pi_{11}^{(2,2)} \end{bmatrix};$$

where

$$\begin{aligned}
\Pi_{11}^{(1,1)} &= P_{21} + P_{21}^T + T_{11} + \tau_0 T_{21} + N_{11} + N_{11}^T \\
&\quad + A^T Y + Y^T A + \varepsilon E_1^T E_1, \\
\Pi_{11}^{(2,1)} &= P_{23} + P_{22}^T + T_{12}^T + \tau_0 T_{22}^T + N_{13} + N_{12}^T \\
&\quad + X^T A + U_2 C + U_1 + Y + \varepsilon E_1^T E_1, \\
\Pi_{11}^{(2,2)} &= P_{24} + P_{24}^T + T_{13} + \tau_0 T_{23} + N_{14} + N_{14}^T \\
&\quad + X^T A + A^T X + U_2 C + C^T U_2^T + \varepsilon E_1^T E_1.
\end{aligned}$$

$$\Pi_{21} = \begin{bmatrix} \Pi_{21}^{(1,1)} & \Pi_{21}^{(1,2)} \\ \Pi_{21}^{(2,1)} & \Pi_{21}^{(2,2)} \end{bmatrix};$$

where

$$\begin{aligned}
\Pi_{21}^{(1,1)} &= N_{21} - N_{11}^T - S_{11}^T - \rho_2 Y^T - A_1^T Y + \varepsilon E_2^T E_1, \\
\Pi_{21}^{(1,2)} &= N_{22} - N_{13}^T - S_{13}^T - \rho_2 Y^T - A_1 X \\
&\quad - C_1^T U_2^T + \varepsilon E_2^T E_1, \\
\Pi_{21}^{(2,1)} &= N_{23} - N_{12}^T - S_{12}^T - \rho_2 X^T A - \rho_2 U_2 C \\
&\quad - \rho_2 U_1 - A_1^T Y + \varepsilon E_2^T E_1, \\
\Pi_{21}^{(2,2)} &= N_{24} - N_{14}^T - S_{14}^T - \rho_2 X^T A - \rho_2 U_2 C \\
&\quad - A_1^T X - C_1^T U_2^T + + \varepsilon E_2^T E_1.
\end{aligned}$$

$$\Pi_{31} = \begin{bmatrix} \Pi_{31}^{(1,1)} & \Pi_{31}^{(1,2)} \\ \Pi_{31}^{(2,1)} & \Pi_{31}^{(2,2)} \end{bmatrix};$$

where

$$\begin{aligned}
\Pi_{31}^{(1,1)} &= N_{31} - P_{21}^T - S_{11}^T - \rho_3 Y^T, \\
\Pi_{31}^{(1,2)} &= N_{32} - P_{23}^T - S_{13}^T - \rho_3 Y^T, \\
\Pi_{31}^{(2,1)} &= N_{33} - P_{22}^T - S_{12}^T - \rho_3 X^T A - \rho_3 U_2 C - \rho_3 U_1, \\
\Pi_{31}^{(2,2)} &= N_{34} - P_{24}^T - S_{14}^T - \rho_3 X^T A - \rho_3 U_2 C.
\end{aligned}$$

$$\Pi_{41} = \begin{bmatrix} \Pi_{41}^{(1,1)} & \Pi_{41}^{(1,2)} \\ \Pi_{41}^{(2,1)} & \Pi_{41}^{(2,2)} \end{bmatrix};$$

where

$$\begin{aligned} \Pi_{41}^{(1,1)} &= N_{41} + P_{11} + Y - \rho_4 Y^T, \\ \Pi_{41}^{(1,2)} &= N_{42} + P_{12} + X - X^T + Y^T - \rho_4 Y^T, \\ \Pi_{41}^{(2,1)} &= N_{43} + P_{13} + Y - \rho_4 X^T A - \rho_4 U_2 C - \rho_4 U_1, \\ \Pi_{41}^{(2,2)} &= N_{44} + P_{14} + X - \rho_4 X^T A - \rho_4 U_2 C. \end{aligned}$$

$$\Pi_{22} = \begin{bmatrix} \Pi_{22}^{(1,1)} & * \\ \Pi_{22}^{(2,1)} & \Pi_{22}^{(2,2)} \end{bmatrix};$$

where

$$\begin{aligned} \Pi_{22}^{(1,1)} &= -N_{21} - N_{21}^T + S_{21} + S_{21}^T - \rho_2 Y^T A_1 \\ &\quad - \rho_2 A_1^T Y + \varepsilon E_2^T E_2, \\ \Pi_{22}^{(2,1)} &= -N_{23} - N_{22}^T + S_{23} + S_{22}^T - \rho_2 X^T A_1 \\ &\quad - \rho_2 U_2 C_1 - \rho_2 A_1^T Y + \varepsilon E_2^T E_2, \\ \Pi_{22}^{(2,2)} &= -N_{24} - N_{24}^T + S_{24} + S_{24}^T - \rho_2 X^T A_1 \\ &\quad - \rho_2 A_1^T X - \rho_2 U_2 C_1 - \rho_2 C_1^T U_2^T + \varepsilon E_2^T E_2. \end{aligned}$$

$$\Pi_{32} = \begin{bmatrix} \Pi_{32}^{(1,1)} & \Pi_{32}^{(1,2)} \\ \Pi_{32}^{(2,1)} & \Pi_{32}^{(2,2)} \end{bmatrix};$$

where

$$\begin{aligned} \Pi_{32}^{(1,1)} &= -N_{31} + S_{31} - S_{21}^T - \rho_3 Y^T A_1, \\ \Pi_{32}^{(1,2)} &= -N_{32} + S_{32} - S_{23}^T - \rho_3 Y^T A_1, \\ \Pi_{32}^{(2,1)} &= -N_{33} + S_{33} - S_{22}^T - \rho_3 X A_1 - \rho_3 U_2 C_1, \\ \Pi_{32}^{(2,2)} &= -N_{34} + S_{34} - S_{24}^T - \rho_3 X^T A_1 - \rho_3 U_2 C_1. \end{aligned}$$

$$\Pi_{42} = \begin{bmatrix} \Pi_{42}^{(1,1)} & \Pi_{42}^{(1,2)} \\ \Pi_{42}^{(2,1)} & \Pi_{42}^{(2,2)} \end{bmatrix};$$

where

$$\begin{aligned}
\Pi_{42}^{(1,1)} &= -N_{41} + S_{41} + \rho_2 Y - \rho_4 Y^T A_1, \\
\Pi_{42}^{(1,2)} &= -N_{42} + S_{42} + \rho_2 X + \rho_2 Y^T \\
&\quad -\rho_2 X^T - \rho_4 Y^T A_1, \\
\Pi_{42}^{(2,1)} &= -N_{43} + S_{43} + \rho_2 Y - \rho_4 X^T A_1 - \rho_4 U_2 C_1, \\
\Pi_{42}^{(2,2)} &= -N_{44} + S_{44} + \rho_2 X - \rho_4 X^T A_1 - \rho_4 U_2 C_1.
\end{aligned}$$

$$\Pi_{33} = \begin{bmatrix} -T_{11} - S_{31} - S_{31}^T & * \\ -T_{12}^T - S_{33} - S_{32}^T & -T_{13} - S_{34} - S_{34}^T \end{bmatrix};$$

$$\Pi_{43} = \begin{bmatrix} -S_{41} + \rho_3 Y & \begin{matrix} -S_{42} + \rho_3 X + \rho_3 Y^T \\ -\rho_3 X^T \end{matrix} \\ -S_{43} + \rho_3 Y & -S_{44} + \rho_3 X \end{bmatrix};$$

$$\Pi_{44} = \begin{bmatrix} \Pi_{44}^{(1,1)} & * \\ \Pi_{44}^{(2,1)} & \Pi_{44}^{(2,2)} \end{bmatrix};$$

where

$$\begin{aligned}
\Pi_{44}^{(1,1)} &= \rho_4 Y + \rho_4 Y^T + \eta R_{11} + 2\delta R_{21}, \\
\Pi_{44}^{(2,1)} &= 2\rho_4 Y + \rho_4 X^T - \rho_4 X + \eta R_{12}^T + 2\delta R_{22}^T, \\
\Pi_{44}^{(2,2)} &= \rho_4 X + \rho_4 X^T + \eta R_{13} + 2\delta R_{23}.
\end{aligned}$$

5.5 Conclusions

The disturbance attenuation problem for NCSs has been investigated based on a Lyapunov–Krasovskii functional method. The criteria for H_∞ performance analysis and H_∞ control synthesis have been derived by introducing some free-weighting matrices and exploiting the information concerning the lower bound of variation of the network-induced delay, which has been shown by the examples to be effective.

In Sections 5.3 and 5.4, we have addressed the problem of robust H_∞ filtering for a class of uncertain systems with interval time-varying delay. A new analysis method for H_∞ performance of the filtering error systems has been proposed based on the lower and upper bounds of the time-varying delay. In terms of the derived criteria, which are expressed as a set of linear matrix inequalities, solvability of the considered filtering problem can be obtained. Then the derived results have been applied to network-based robust H_∞ filtering for uncertain linear systems over a common network connection. Some numerical examples have been given to show the effectiveness and less conservativeness of the proposed method. As a special case, if we do not consider the effect of a network, we can easily obtain the results for sampled systems. It should be pointed out that the methodology in this chapter can be extended to handle the problem of robust H_∞ control for systems with fast time-varying delay.

References

1. Chow M, Tipsuwan Y (2001) Network-based control systems: A tutorial. In: Proc 27th Annual Conference of the IEEE Industrial Electronics Society, Denver, CO, 1593–1602
2. de Souza CE, Palhares RM, Peres PLD (2001) Robust H_∞ filter design for uncertain linear systems with multiple time-varying state delays. IEEE Trans Signal Processing 49:569–576
3. Gao H, Wang C (2003) Delay-dependent robust H_∞ and L_2-L_∞ filtering for a class of uncertain nonlinear time-delay systems. IEEE Trans Automatic Control 48:1661–1666
4. Hu S, Zhu Q (2003) Stochastic optimal control and analysis of stability of networked control systems with long delay. Automatica 39:1877–1884
5. Jin SH, Park JB (2001) Robust H_∞ filtering for polytopic uncertain systems via convex optimisation. IEE Proc – Control Theory Applications 148:55–59
6. Khargonekar PP, Petersen IR, Zhou K (1990) Robust stabilization of uncertain linear systems: quadratic stabilizability and H_∞ control theory. IEEE Trans Automatic Control 35:356–361
7. Kharitonov VL, Niculescu SI (2003) On the stability of linear systems with uncertain delay. IEEE Trans Automatic Control 48:127–132
8. Kim DS, Lee YS, Kwon WH, Park HS (2003) Maximum allowable delay bounds of networked control systems. Control Engineering Practice 11:1301–1313
9. Lin H, Zhai G, Antsaklis P (2003) Robust stability and disturbance attenuation analysis of a class of networked control systems. In: Proc Conference on Decision and Control, Maui, HI, 1182–1187
10. Mao X, Koroleva N, Rodkina A (1998) Robust stability of uncertain stochastic differential delay equations. Systems and Control Letters 35:325–336
11. Montestruque LA, Antsaklis PJ (2003) On the model-based control of networked systems. Automatica 39:1837–1843
12. Nguang S, Shi P (2001) Fuzzy H_∞ output feedback control of nonlinear systems under sampled measurements. In: Proc Conference on Decision and Control, Orlando, Florida, 4370–4375
13. Niculescu SI (1998) H_∞ memoryless control with an α-stability constraint for time-delay systems: an LMI approach. IEEE Transactions on Automatic Control 43:739–743
14. Nilsson J, Bernhardsson B, Wittenmark B (1998) Stochastic analysis and control of real-time systems with random time delays. Automatica 34:57–64
15. Oriov Y, Acho L (2000) Nonlinear H_∞-control via sampled-data measurement feedback: time-scale conversion to continuous measurement case. In: Proc Conference on Decision and Control, Sydney, Australia, 3037–3042
16. Park HS, Kim YH, Kim DS, Kwon WH (2002) A scheduling method for network based control systems. IEEE Trans Control Systems Technology 10:318–330

17. Petersen IR (1987) Disturbance attenuation and H_∞ optimization: a design method based on the algebraic Riccati equation. IEEE Transactions on Automatic Control 32:427–492
18. Sinopoli B, Schenato L, Franceschetti M, Poolla K, Jordan MI, Sastry SS (2004) Kalman filtering with intermittent observations. IEEE Trans Automatic Control 49:1453–1464
19. Toivonen H (1992) Sampled-data control of continuous-time system with an H_∞ optimality criterion. Automatica 28:45–54
20. Walsh G, Ye H, Bushnell L (2002) Stability analysis of networked control systems. IEEE Trans Control Systems Technology 10:438–446
21. Wang Z, Goodall DP, Burnham, KJ (2002) On designing observers for time-delay systems with nonlinear disturbances. International Journal of Control 75:803–811
22. Xie L, de Souza CE (1992) Robust H_∞ control for linear systems with norm-bounded time-varying uncertainty. IEEE Trans Automatic Control 37:1188–1191
23. Xu S, Chen T (2004) An LMI approach to the H_∞ filter design for uncertain systems with distributed delays. IEEE Trans Circuits Systems-II: Express Briefs 51:195–201
24. Yang TC (2004) Networked control system: a brief survey. IEE Proc – Control Theory and Applications 153:403–412
25. Yue D, Han QL (2004) Robust H_∞ filter design of uncertain descriptor systems with discrete and distributed delays. IEEE Trans Signal Processing 52:3200–3212
26. Yue D, Han QL (2006) Network-based roubust H_∞ filtering for uncertain linear systems. IEEE Trans Signal Processing 54:4293–4301
27. Yue D, Han QL, Chen P (2004) State feedback controller design of networked control systems. IEEE Trans Circuits and Systems - II 51:640–644
28. Yue D, Han QL, Lam J (2005) Network-based robust H_∞ control of systems with uncertainty. Automatica 41:999–1007
29. Zhang W, Branicky MS, Phillips SM (2001) Stability of networked control systems. IEEE Control Systems Magazine 21:84–99

6

Switched Feedback Control for Wireless Networked Systems

George Nikolakopoulos, Athanasia Panousopoulou, and Anthony Tzes

University of Patras, Patras, Achaia 26500, Greece
gnikolak@ece.upatras.gr, apanous@ece.upatras.gr, tzes@ece.upatras.gr

Abstract. In this chapter a switched output feedback control scheme for networked systems will be presented. The control scheme is applied on client–server architectures where the feedback control loop is closed over a general purpose wireless communication channel between the plant (server) and the controller (client). The inserted delays from the communication network in general are time-varying and degrade the system dynamic performance, while forcing it to instabilities. To deal with these changes a linear quadratic regulator (LQR)–output feedback control scheme is introduced, whose parameters are tuned accordingly to the variation of the measured round trip latency times. The weights of the LQR controllers are subsequently tuned using the theory of linear matrix inequalities (LMIs) to ensure a prescribed stability margin despite the variable latency time. The overall scheme resembles a gain scheduler controller with the latency times playing the role of scheduling parameter. The proposed control scheme is applied in experimental and simulation studies to a networked control system over different communication channels including: (a) the GPRS, (b) the IEEE 802.11b, and (c) the IEEE 802.11b over a mobile *ad hoc* sensor network (MANET). The underlying mechanisms that generate the time-varying latency times in each case will also be presented and analyzed prior to the control scheme development.

Keywords. Networked control systems, switching control, 802.11b, GPRS, mobile *ad hoc* networks (MANETs), quality of service (QoS).

6.1 Introduction

Remote client–server architectures are becoming dominant due to developments in communication capabilities and improvements in network infrastructures. These networks are susceptible to various issues [22, 27] stemming from the need to exchange information over a common communication link [10, 11, 23, 24].

The utilization of a wired or wireless communication network [6, 12, 18, 25] in control applications where the end user has no control over the provided communication link leads to problems associated with data packet interchange. The main difficulty with the design of such a control loop is the presence of the sensing and actuation delays introduced by the communication networks. Unlike conventional time delay systems, the type of delays introduced by the network are time-varying, since they depend on the traffic currently on the network. For wireless communication channels, the problem is further complicated by the mobility of clients (controllers) and servers (plant) which induces structural changes in the packet routing procedure.

The resulting networked control system (NCS) should be able to adjust the settings of the transmission scheme to account for the possible peculiarities encountered in typical real-time control application problems. The most common problems that can be encountered are related to the need for: (a) maintaining an effective bit-rate [19], (b) synchronizing heterogeneous computers with varying computing power, (c) utilizing a large bandwidth for the sampling applications [30], (d) accounting for loss-of-packets in classical wireless transmissions, (e) using the unreliable UDP rather than the TCP protocol for control related purposes [22], and (f) stabilizing the resulting time-delayed systems (TDS) [28].

The main focus of this chapter is to present a methodology in order to design linear quadratic regulator (LQR)–output feedback controllers for NCS systems that are characterized by time-varying latency times in the utilized transmission channels. The inserted delays will be appropriately embedded in the system representation and the overall system will result in a model of a jump system. The controller proposed to deal with the switching of the controlled system will adapt the weights of the LQR-scheme by the utilization of an algorithm that was initially presented in [27] where the tuning is based on: (a) linear matrix inequalities (LMIs), and (b) a prescribed stability margin approach that will be invariant of the latency times.

The overall scheme resembles a gain scheduler controller with the latency times playing the role of the scheduling parameter. With respect to the utilization of the communication channel for data packet transmission the underlying mechanisms that generate the time-varying latency times at each case will also be presented and analyzed prior to the control scheme development. This presentation extends by providing examples for the network cases of: (a) GPRS [3], (b) IEEE standard 802.11b [6, 12], and (c) the IEEE 802.11b over a mobile *ad hoc* sensor network (MANET) [21]. Relying on the presented approach for the switching rule among the system's transitions, experimental and simulation results will be presented.

This chapter is structured as follows. In Section 6.2, the modeling of an LTI system, with time delays in the feedforward and the feedback control loop, as a switching (jump) system is presented. In Section 6.3, the proposed controller design approach for the LQR–output feedback controller is presented. In Section 6.4, the stability investigation of the proposed scheme and the tuning algorithm of the LQR's gains are covered. Finally, in Section 6.5, experimental and simulation results on NCSs over three communication channels are presented.

6.2 Mathematical Modeling of NCS as a Switched System

In the general case the control process is modeled as a linear time-invariant (LTI) system of the following form:

$$x(k+1) = Ax(k) + Bu(k) \tag{6.1}$$

$$y(k) = Cx(k) \tag{6.2}$$

while the controller is time varying and is represented as:

$$F\tilde{u}(k) = Gy(k-d_2) + Hr(k) \tag{6.3}$$

$$u(k) = \tilde{u}(k-d_1) \tag{6.4}$$

and with d_1, d_2 the time delays that are inserted from the communication network in: (a) the feedforward loop (d_1–delay in the transmission of the control signal), and (b) the feedback loop (d_2–delay in the transmission of the system's output). These delays are assumed to be time varying, or $d_1, d_2 \in Z_0^+$. The existence of these delays, as it is going to be presented in this section, result in the transformation of the LTI–closed loop system into a jump system, with the delay factor (round-trip latency time), to be the switching rule among the various models and the respective controllers.

In Fig. 6.1 the block structure of the system presented in (6.1)–(6.4) is presented, where r is the reference signal, e is the error signal, H is a pre-compensator to the reference, $\tilde{u}$ and u are the non-delayed and the delayed versions of the control signal and y is the system output.

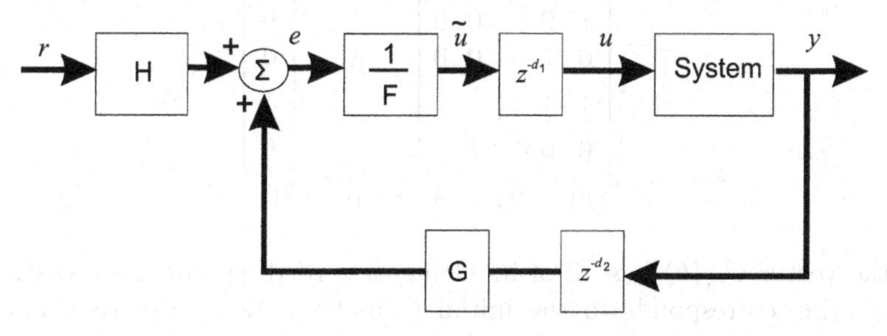

Fig. 6.1. Generic model of a delayed control plant

If we simplify the case presented in Fig. 6.1, by setting $H = G = 1$ the previous model of the system can be represented by the utilization of difference equations from (6.1)–(6.2) where $x(k) \in \mathbb{R}^n$ and $u(k) \in \mathbb{R}$. The block diagram representation of this simplified case is presented in Fig. 6.2 and is the one that will be adopted in this presentation and usually encountered in the literature of NCS. Let $r(k) = 0$ and $r_s(k) = d_1 + d_2$ be the overall delay (measured

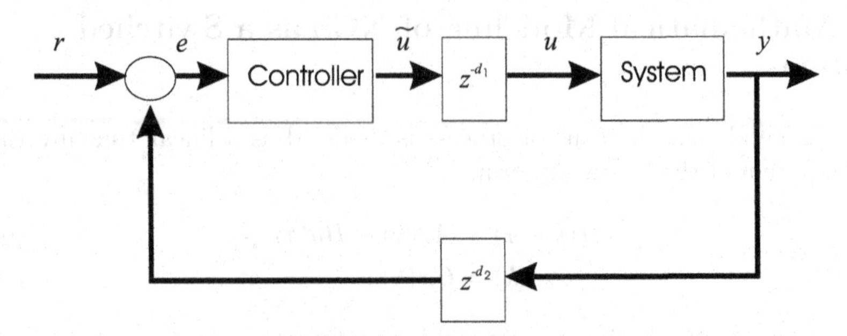

Fig. 6.2. Simplified block diagram of an NCS

round trip latency time) at time instant k. In this case $r_s(k)$ is a random bounded sequence of integers $r_s(k) \in [0, 1, \ldots, D]$ and D is the upper bound of the delay term. In order to embed the time delays in the system model the state vector $\tilde{x}(k)$ is augmented in order to include all the delayed terms as:

$$\tilde{x} = [x(k)^T, x(k-1)^T, \ldots, x(k-D)^T]^T.$$

The dynamics of the open-loop system, at time k, with the augmented state vector take the following form

$$\begin{aligned}\tilde{x}(k+1) &= \tilde{A}\tilde{x}(k) + \tilde{B}u(k),\\ y(k) &= \tilde{C}_{r_s}(k)\tilde{x}(k),\end{aligned}$$

where

$$\tilde{A} = \begin{bmatrix} A & 0 & \cdots & 0 & 0 \\ I & 0 & \cdots & 0 & 0 \\ 0 & I & \cdots & 0 & 0 \\ \vdots & \vdots & \ddots & \vdots & \vdots \\ 0 & 0 & \cdots & I & 0 \end{bmatrix}, \quad \tilde{B} = \begin{bmatrix} B \\ 0 \\ 0 \\ \vdots \\ 0 \end{bmatrix},$$

$$\tilde{C}_{r_s}(k) = \begin{bmatrix} 0 & \cdots & 0 & C & 0 & \cdots & 0 \end{bmatrix},$$

and the vector $\tilde{C}_{r_s}(k)$ has all of its elements zeroed, except the $r_s(k)$th one whose value corresponds to the initial C matrix. As we can see from the structure of $\tilde{A}, \tilde{B}$, and $\tilde{C}$ matrices their dimensions remain constant as the latency time varies and the dimensions are given only as a function of the maximum anticipated time delay D. The transformation to a switching system is caused by the $\tilde{C}$ matrix, where its sub-vector C is translated horizontally, depending on the value of the current delay $r_s(k)$. An alternative switching representation of the $\tilde{C}$ matrix's row elements can be given as:

$$\tilde{C}_{r_s}(1,i) = \begin{cases} 0, & i \neq r_s(k), \\ C, & i = r_s(k), \end{cases} \tag{6.5}$$

for $i = 1, \ldots, D$. As it will be presented in the next section, these switchings of the $\tilde{C}$ matrix causes the overall closed-loop system to make switchings with respect to the $r_s(k)$-switching rule.

6.3 Optimal Output Feedback Control

Let the control objective be the computation of an LQR–output feedback controller, $u(k) = Ky(k)$, that minimizes the following cost [29]:

$$\min_K \sum_{i=0}^{\infty} \left[y^T(i)Ry(i) + u^T(i)Qu(i) \right] e^{\sigma i}, \tag{6.6}$$

with $\sigma \geq 1$. Upon computation of this controller the resulting closed-loop system has its poles $\mathrm{eig}(A + BKC)$ located inside a disk of radius $1/\sigma$.

In an NCS the delays that are inserted in the loop affect the anticipated control command $u(k) = KCx(k)$ that is applied to the plant and is given by:

$$u(k) = KCx(k - r_s(k)). \tag{6.7}$$

The overall closed-loop system is

$$\tilde{x}(k+1) = \left(\tilde{A} + \tilde{B}K\tilde{C}_{r_s}(k)\right)\tilde{x}(k), \tag{6.8}$$

$$y(k) = \tilde{C}_{r_s}(k)\tilde{x}(k). \tag{6.9}$$

Notice that the closed-loop system is switched [9, 17], since the $r_s(k)$ (and thus the feedback term $K\tilde{C}_{r_s}(k)$) is time varying. The closed-loop matrix $\tilde{A} + \tilde{B}K\tilde{C}_{r_s}(k)$ can switch in any of the $(D+1)$-vertices $A_i = \tilde{A} + \tilde{B}K\tilde{C}_i$.

To ensure stability of the closed-loop system, conditions are sought for the stabilization of the switched system

$$\tilde{x}(k+1) = A_i\tilde{x}(k), \; i = 0, \ldots, D.$$

Under the assumption that at every time instance k the bounds of the round-trip latency time $r_s(k)$ can be measured, and therefore the index of the switched-state is known, the system can be described as:

$$x(k+1) = \sum_{i=0}^{D} \xi_i(k) A_i x(k), \tag{6.10}$$

where $\xi_i(k) = [\xi_0(k), \ldots, \xi_D(k)]^T$ and $\xi_i = \begin{cases} 0, & \text{mode} \neq A_i, \\ 1, & \text{mode} = A_i. \end{cases}$

It can be shown [7] that the switched system in (6.10) is stable if $D+1$ positive definite matrices P_i, $i = 0, \ldots, D$, can be found that satisfy the following LMI:

$$\begin{bmatrix} P_i & A_i^T P_j \\ P_j A_i & P_j \end{bmatrix} > 0, \ \forall (i,j) \in I \times I, \tag{6.11}$$

$$P_i > 0, \ \forall i \in I = \{0, 1, \ldots, D\}. \tag{6.12}$$

Based on these P_i-matrices, it is feasible to calculate a positive Lyapunov function of the form:

$$V(k, x(k)) = x(k)^T \left(\sum_{i=0}^{D} \xi_i(k) P_i \right) x(k) \tag{6.13}$$

whose difference $\Delta V(k, x(k)) = V(k+1, x(k+1)) - V(k, x(k)))$ decreases along all $x(k)$ solutions of the switched system, thus ensuring the asymptotic stability of the system. It should be noted that the bounds of the corresponding set can be arbitrary as $I = \{D^{\min}, D^{\max}\}$, while $D^{\min}, D^{\max} \in \mathbb{Z}^+$.

6.3.1 Gain Tuning of Output Feedback Parameter

The computation of the output feedback controller $u(k) = Ky\,(k - r_s(k))$, results in a stable system that can tolerate a communication delay of D-samples ($r_s(k) \in \{0, 1, \ldots, D\}$). It should be noted that the controller design procedure was posed in the following manner: (a) select the cost-weight matrices R and Q and σ-parameter, (b) compute K from the LQR-output minimization problem, and (c) compute the maximum delay D that can be tolerated with this given gain K.

In most cases, the communication factors that produce the communication delay of a typical NCS does not vary rapidly, and remain within certain bounds over large periods of time. With regard to the delay term, we can state that $r_s(k) \in \{D_1, \ldots, D_2\}$ over a large time window, where D_1 and D_2 are predefined lower and upper delay bounds. In this case, the control design problem can be restated as: At sample period k, given r_s, select the weight matrices $Q(k), R(k)$, and compute the largest prescribed stability factor $\sigma(k)$ in order to maintain stability despite the communication delays.

Rather than adjusting in an *ad hoc* manner the weight matrices, we focus on the $\sigma(k)$-quantity. A closed-loop system derived via the usage of a small radius $1/\sigma(k)$ in the optimization step, has a fast system response, since all its poles have small magnitude $|\text{eig}(A + BKC)| \leq 1/\sigma(k)$. However, this system cannot tolerate large delay variations $D_2 - D_1$ and the suggested gain-adjustment relies on this anticipated observation. The $\sigma(k)$-scheduling amounts to computing the largest value, while at the same time justifying the

LMIs of (6.11), (6.12) for a given index set $I = \{D_1, \ldots, D_2\}$. This design philosophy provides the fastest system while tolerating the given delay bounds. The computation of this optimum $\sigma(k)$ and $K(k)$ is based on the following algorithm:

(1) Set $\sigma = 1$.
(2) Compute K from (6.6).
(3) Check whether the LMIs of (6.11) and (6.12) are verified.
(4) If no, then there is no solution with the given Q, R and σ. Go to Step 7.
(5) If yes, then increase σ by $\sigma = \sigma + \Delta\sigma$.
(6) Repeat Step (2), unless satisfied with the obtained bound of prescribed stability.
(7) Algorithm is stopped.

The output of the computation is a pair $(\sigma(k), K(k))$ for each measured time delay $r_s(k)$ that ensures stability of the closed-loop switching system for all the communication delays within the bounds $[D_1, D_2]$. In this presentation, the controller's design is focused on the calculation of the greater set $I = \{I_{\min}, I_{\min} + 1, \ldots, I_{\max}\}$, where $0 \leq I_{\min} < I_{\max} \leq D$, that the controller's gain could stabilize the control plant upon every pattern of switchings contained in the previous set.

Theoretically an upper bound of the calculated D-limit can be found from the solution of the continuous system with time delays. We assume that the state space representation of the control system can be given as:

$$\dot{x}_c(t) = A_c x(t) + B_c u(t), \quad y(t) = C_c x(t). \tag{6.14}$$

For the case of delayed control law $u(t) = Ky(t - d_1 - d_2)$ the closed-loop system takes the following form:

$$\dot{x}_c(t) = A_c x(t) + B_c K C_c x(t - \tau) = A_c x(t) + A_d x(t - \tau),$$

where $\tau = d_1 + d_2$. The maximum allowable time delay $\tau^{\max}$ can be computed from the solution of the following optimization problem, presented here in the form of LMIs:

$$\tau^{\max} = \max \tau, \quad \text{under the constraints:} \tag{6.15}$$
$$\begin{bmatrix} (A_c+A_d)Q_1+Q_1(A_c+A_d)^T+\tau A_d(Q_2+Q_3)A_d^T & \tau A_c Q_1^T & \tau A_d Q_1^T \\ \tau A_c Q_1 & -\tau Q_2 & 0 \\ \tau A_d Q_1 & 0 & -\tau Q_3 \end{bmatrix} < 0,$$
$$Q_i > 0, \ i = 1, 2, 3.$$

Given $\tau^{\max}$ the maximum time delay D could be calculated as $D = \left\lceil \dfrac{\tau^{\max}}{T_s}, 1 \right\rceil$.

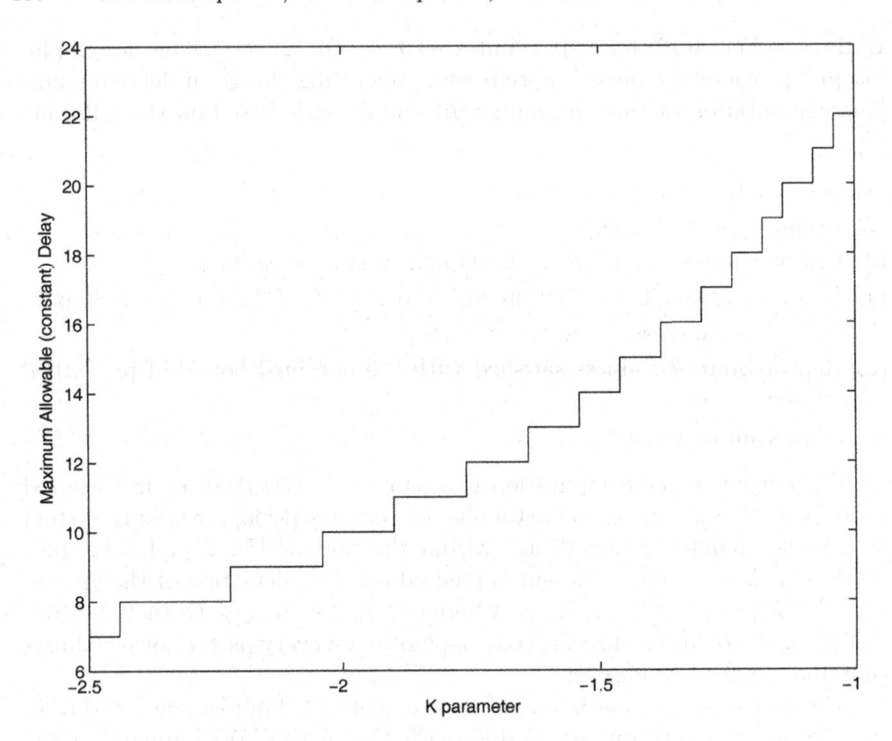

Fig. 6.3. Stability bounds for continuous TDS

6.3.2 Stability Investigation: Numerical Results

In order to investigate the application of the proposed set of LMIs in conjunction with the proposed stability criteria, we consider a prototype SISO-system with a transfer function

$$G(s) = \frac{0.1^3}{(s+0.1)^3}.$$

Application of a controller $u(t) = Ky(t-\tau)$, yields the following TDS:

$$\dot{x}(t) = \begin{bmatrix} -0.3 & -0.03 & -0.001 \\ 1 & 0 & 0 \\ 0 & 1 & 0 \end{bmatrix} x(t) + \begin{bmatrix} 0 & 0 & 0.001K \\ 0 & 0 & 0 \\ 0 & 0 & 0 \end{bmatrix} x(t-\tau). \tag{6.16}$$

In Fig. 6.3 we present the maximum allowable time delay $\tau^{\max}$ that preserves the stability as a function of the controller-gain K for the continuous time case. For example, for the case where $K = 1$ the maximum tolerable constant delay is $\tau^{\max} = 23$ s. It should be noted that this value stems from the solution of the optimization problem (6.15), and by no means is the largest one that can be computed using other relevant theorems (see [32]). Assuming a sampling period of $T_s = 1$ s, the discrete equivalent of the continuous system is (accounting for the ZOH)

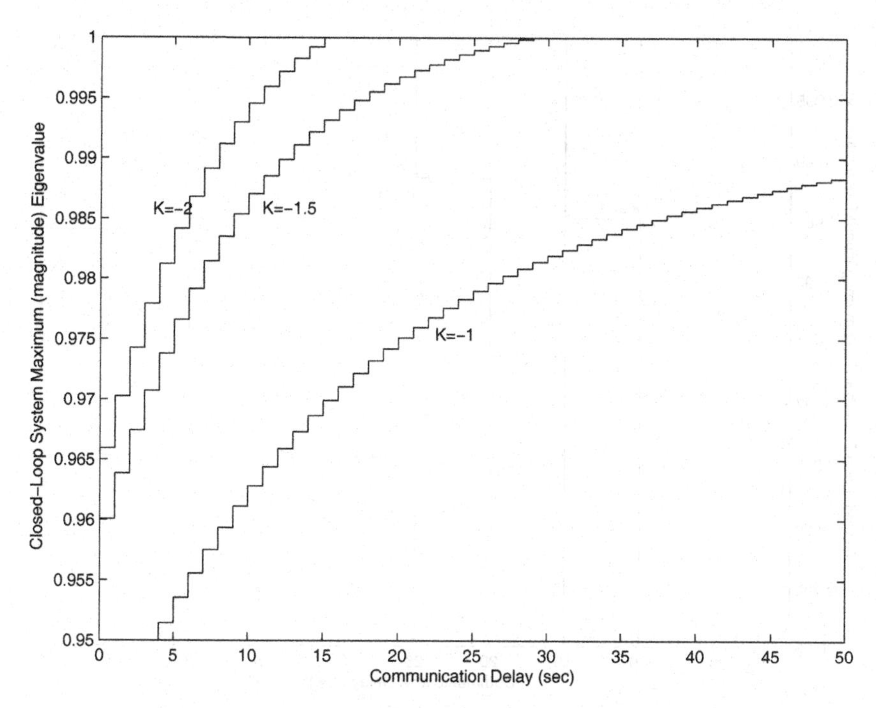

Fig. 6.4. Stability bounds for discrete TDS ($T_s = 1$ s)

$$x((k+1)T_s) = \begin{bmatrix} 3.316 & -3.6640 & 1.35 \\ 1 & 0 & 0 \\ 0 & 1 & 0 \end{bmatrix} x(kT_s) + \begin{bmatrix} 1 \\ 0 \\ 0 \end{bmatrix} u(kT_s)$$

$$y(kT_s) = \begin{bmatrix} 0.1797 & 0.7748 & 0.2088 \end{bmatrix} \times 10^{-3} x(kT_s)$$

where, for brevity in the ensuing text, the notation $x(k)$ will be used instead of $x(kT_s)$. Assume that a discrete controller $u(k) = Ky(k - r_s(k))$ is inserted in the loop; in the sequel the stability bounds (delay vs. controller gain) that the system can sustain are examined.

In Fig. 6.4, we present the amplitude of the maximum eigenvalue of A_{r_s} as a function of the time delay r_sT_s for three different gain values $K = 1,\ 1.5$ and 2.

We should note that, for example, for $K = 2$ (1.5) the system becomes unstable ($|\lambda_{\max} A_{r_s}| \geq 1$) for $r_sT_s \geq 15$ (28) s, while for $K = 1$ the system remains stable for delays smaller than 50 s. A direct comparison with the results from the previous Fig. 6.3 indicates that the results obtained from the discrete domain are not as conservative as the ones from the continuous domain.

A different discretization using $T_s = 5$ s and a similar graph appears in Fig. 6.5, where as expected the results are similar to the aforementioned ones with the only difference set in the size (quantization step) of the sample period

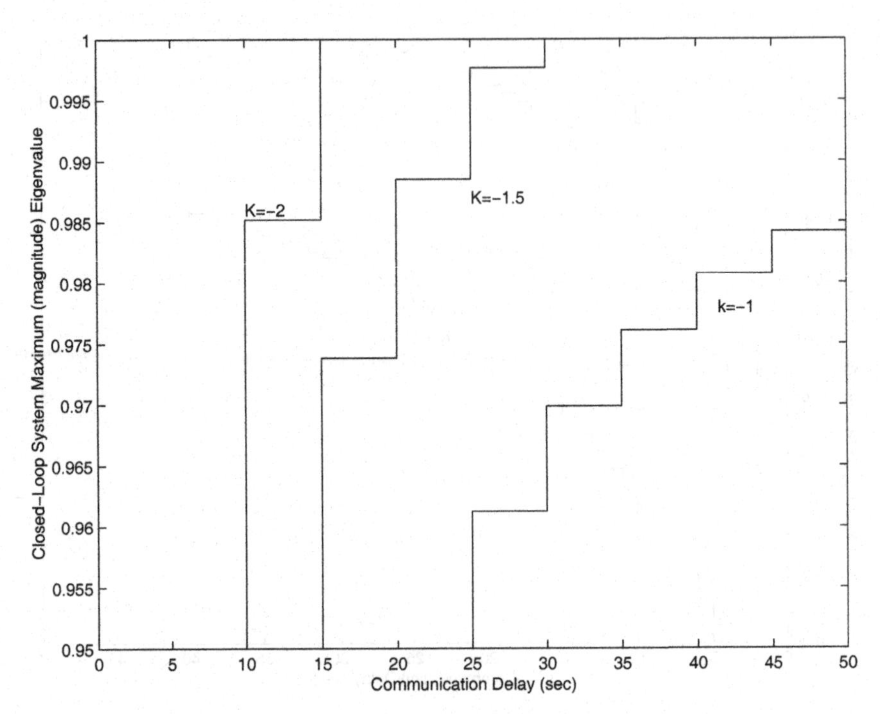

Fig. 6.5. Stability bounds for discrete TDS ($T_s = 5$ s)

T_s. Similarly, the system becomes unstable for $r_s T_s \geq 15$ and 30 s for $K = 1$ and $K = 5$, respectively.

6.4 Experimental and Simulation Results

The presented methodology for the LQR–tuned gain scheduler controller will be applied in experimental and simulation studies to a networked control system over different communication channels including: (a) the GPRS, (b) the IEEE 802.11b, and (c) the IEEE 802.11b over an MANET. These cases will be presented in the following subsections in parallel with the presentation of the underlying mechanisms that generate the time-varying latency times in each case.

6.4.1 Switched Feedback Control Over GPRS

Introduction

The mobile (wireless) telephony (GSM, CDMA) sector embraces a set of characteristics that are detrimental to the stability of closed-loop remote control systems. Its narrowband attribute (GSM 14.4 Kbps, CDMA 19.2 Kbps) along with the communication overhead for data retransmission due to packet losses, when coupled with the time-varying latency times severely affect the performance of the system. These critical issues are caused by the voice rather than data-oriented transmission tuning attributes of the mobile telephony sector. The 2G-mobile network infrastructure was designed to primarily carry voice, while its charging policy depends on the duration of each call. Bi-directional GSM-based data transfer is still considered a relatively expensive scheme, since charging is time-dependent rather than volume-dependent. HSCSD-enabled GSM-phones can multiplex up to three available channels reaching a maximum transmission speed of 43.2 Kbps; however, the time-charging nature prohibits its widespread usage.

In order to overcome the time-dependent pricing structure, 2.5G-mobile phones relying on the GPRS-protocol [3] transmit data at bursty-modes employing a data volume-dependent charging strategy. Although the maximum achievable bandwidth over GPRS is 115.2 Kbps (upload and download) (Class-29), most mobile operators offer up to a 43.2 Kbps-upload (14.4 Kbps-download) (Class-4) available bandwidth to their customers. It should be noted that this bandwidth is not guaranteed but can be offered as long as there are available time-slots at a mobile-cell; this occurs when there is no congestion from voice-calls. Since a higher priority is placed on GSM-based voice-calls, the available bandwidth for GPRS-based data transmission can fluctuate significantly over periods of time (0 to 43.2 Kbps in increments of 14.4 Kbps). Subsequently, this affects the stability and performance of the remote-control system. Situations, where the GPRS-service is interrupted for a limited period of time (a maximum interval of 45 s was observed over a period of six months in our experimental evaluation) can be quite harmful to the control scheme, and proper actions need to be taken. Driven mostly by security reasons, mobile-phone service providers (m-psp) do not offer an extended set of privileges to their customers for data transmission purposes. Classical protocols (i.e., ftp and http) and corresponding actions can be issued from a client with a GPRS-enabled phone to access a remote (server) site with a valid IP-address. However, a server-based initiation of a transmission back to the client (GPRS-phone) is blocked from the firewall of the m-psp. This is necessary to ensure termination of data calls back to the customers of m-psp from untrustworthy sites. Subsequently, there is a need for a "client-centric" remote-control framework, where all actions in the feedback loop (transmitting the command signal $u(k)$ and receiving the system's response $y(k)$) are initiated from the client. Henceforth, the scheme must be "centered" on the client's actions and the supporting software for remote communication should be modified accordingly.

In a mobile networked control system (moNCS) [16], shown in Fig. 6.6, the client computes the control command $u(k)$ and transmits it through a wireless link to a server site. The server receives the data after a certain delay, transfers them to the plant, samples the plant's output $y(k)$ and transmits it back to the client for future processing. The client receives the delayed

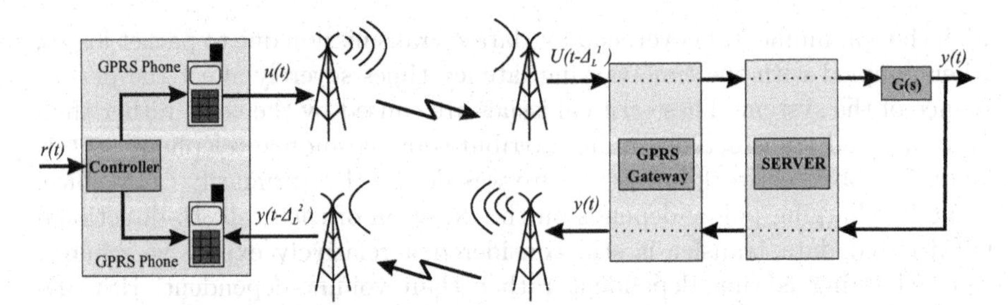

Fig. 6.6. Mobile networked control system architecture

output and repeats the aforementioned process. Due to the inherent delays in the formulation and transmission of signals between the client and server sides [5, 28], there is a need to investigate the stability of this TDS. For this reason, recent theoretical results stemming from LMI theory [4] will be used in this chapter.

Client-centric Mobile NCS Architecture

Within the moNCS architecture presented in Fig. 6.6, the control law is computed remotely at a client computer with the control/response signals transmitted towards/from a server computer located near the plant. The assumed plant's continuous transfer function is $G(s)$, while the latency intervals from the client site to the server and reverse are Δ_L^1 and Δ_L^2, respectively. Assuming a sampling period T_s and an embedded ZOH-device in transferring the discrete signals to the plant, let the discrete control systems' transfer function be

$$G(z^{-1}) = \left(1 - z^{-1}\right) \mathcal{Z}\left\{\frac{G(s)}{s}\right\} \text{ and } d_i = \left\lceil \left(\frac{\Delta_L^i}{T_s}, 1\right) \right\rceil, \; i = 1, 2.$$

Essentially, d_i correspond to the "inserted" delays from the GPRS-network infrastructure during the data-packet exchange.

Within this architecture, the wireless segment poses the most complicated problems to the overall development, since appropriate software drivers must be designed to account for the signaling between the mobile device and the GPRS-network.

Mobile-networked Communication Issues

The client-centric nature of the remote control scheme dictates that the client initiates all data transmissions. Accordingly, the client transmits the control command using the UDP-protocol and records the system's output by issuing an "FTP-get" command. To accommodate the client's requests the server must run locally an FTP-server and have its corresponding UDP-port opened.

In Figs. 6.7 and 6.8 we present procedures governing the data packet exchange between the client and the server. The UDP-latency and FTP-latency times are denoted L^{UDP} and L^{FTP}, respectively.

The highlighted issues in Figs. 6.7 and 6.8 display a set of six cases covering possible problems that can be encountered in the data exchange procedure.

In Fig. 6.7 the top portion (first case) exhibits ideal characteristics: (a) the client uses the UDP-protocol and transmits the control signal, (b) the server receives this packet and converts the digital format of the signal to an analog (voltage) and applies it to the plant, (c) after a certain time, the client initiates the FTP-get command and requests to receive through the server, the digitized value of the system's output, (d) the server samples the output and sends it back to the client through the opened FTP-connection. In the ideal case, this four-step sequence is completed within one sampling period T_s. The second case (middle portion) describes the situation where an instantaneous loss of a UDP-based packet transmission occurs. In this case the server applies to the plant the last (previous) transmitted signal $u(k)$ from the client. In the third case we describe the packet reordering situation, where the UDP-based transmission is delayed and FTP-based reception has already been initiated by the client.

The fourth case (top portion of Fig. 6.8) corresponds to the situation of an instantaneous loss of FTP-based data acquisition. The UDP transmission is performed correctly, but the client's request for the FTP-get command fails. In this case the client computes the next control signal $u(k+1)$ based on the previously recorded (and outdated) $y(k-1)$ output. The fifth case (middle portion) stands for an instantaneous loss of the communication link. During this phase the UDP and FTP data packet are lost. The client computes the next control signal $u(k+1)$ based on the last correctly received, from the FTP-protocol, system output. In the sixth case (bottom portion) we have high latency times due to traffic congestion and the sequence cannot be completed within one sampling period.

In the sequel, we introduce an application of the presented methodology for switched LQR–output feedback controllers for the Cases 1, 5 and 6 where there is either a normal data exchange (Case 1), or a mutual loss of packets on the client and server side (Case 5), or a mutual delay related to the data transmission (Case 6). The cases of (a) packet-reordering, and (b) unidirectional (UDP, or FTP) loss of packets is not covered in this study.

The suggested scheme is applied to an experimental prototype SISO-system with a transfer function

$$G(s) = \frac{0.1^3}{(s+0.1)^3}.$$

Application of a controller $u(t) = Ky(t-\tau)$, yields the same TDS as in (6.16), while the stability analysis of the system has been presented in Figs. 6.4 and 6.5 for fixed delays and for sampling periods of 1 and 5 s, respectively.

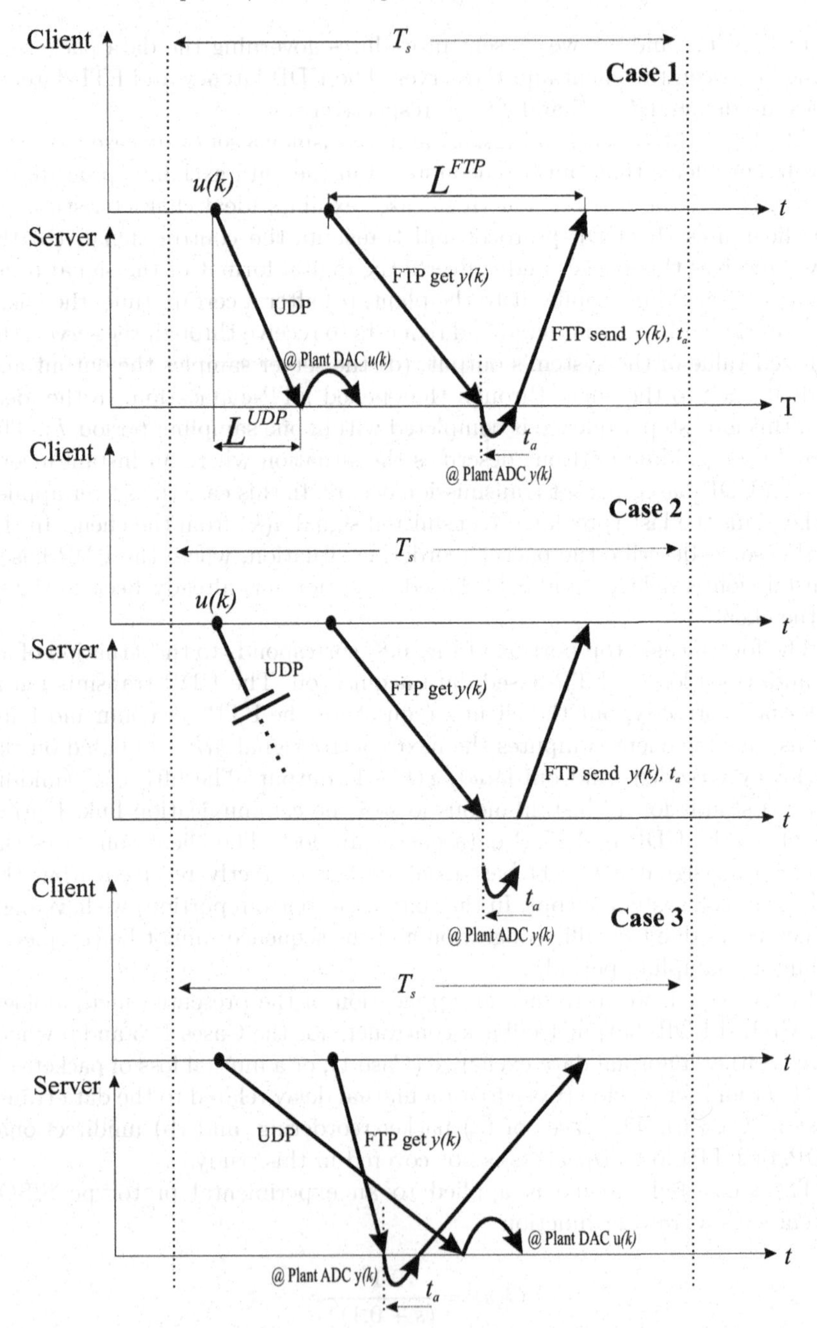

Fig. 6.7. UDP and FTP data packet exchange flowchart (Cases 1–3)

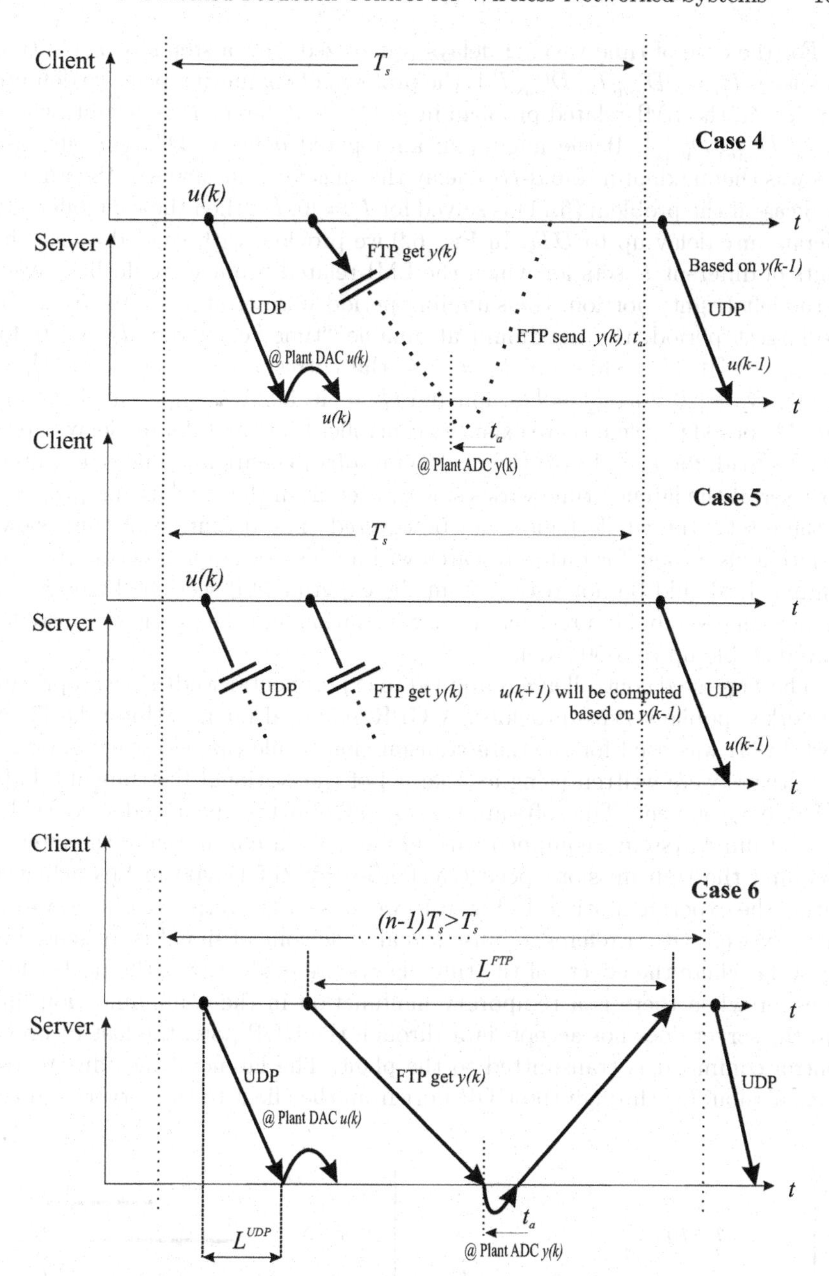

Fig. 6.8. UDP and FTP data packet exchange flowchart (Cases 4–6)

For the case of time-varying delays (quantized with a step size T_s) within subspaces $I_D^i = \left[D_{\text{min}}^i T_s, D_{\text{max}}^i T_s\right]$, the problem of computing positive definite matrices in the LMI-related problem in (6.11) for different I^i is sought, where $I^i = \left[D_{\text{min}}^i, D_{\text{max}}^i\right]$. If the maximum anticipated delay is DT_s (in our case 50 s was the maximum round-trip delay that has been measured), then in the ideal case if the problem (6.11) is solved for $I^i = [0, D]$, then the controller can tolerate any delay up to DT_s. In Fig. 6.9 we provide, with shaded areas, the limits of different I^i sets for which the LMI-related problem could be solved. In the left (right) portion, the sampling period was set at $T_s = 10$ (5) s. For the largest period, the maximum attainable "time sets" were $I_D^1 = [0, 40]$ and $I_D^2 = [10, 50]$, while for $T_s = 5$ s, the corresponding sets were $I_D^1 = [0, 30]$, $I_D^2 = [20, 40]$, $I_D^3 = [25, 45]$, and $I_D^4 = [40, 50]$. It is apparent that from the LMI-posed problem there exists no controller that can tolerate delays up to 50 s. Instead, for $T_s = 10$ (5) the maximum tolerable latency time is 40 (30) s. However, if the latency time varies slowly, then from the overlapping property of these sets, the whole region can be covered. The definition of this "slow-variation" is a topic for future research within this overlapping decomposition context. It should be noted that from the experimental section the observed latency time exhibited a reasonably slow variation and the provided controller proved stable up to a 50 s delay.

The suggested controller was applied to experimental studies over a private network's mobile service provider. A GPRS-enabled phone (Motorola Time-Port T189) was used for the data transmission, while the necessary interface and drivers were written using as a kernel of the National Instruments' Lab-VIEW environment. The software was executed at the client and server sides on Pentium-4 systems, equipped with proper software to measure the latency time and the transmission speed (NetPerSec by Ziff Davis) in bps achieved during the experimentation. In Fig. 6.10 we present the response of the system when excited with a reference pulse signal. The control signal is presented in Fig. 6.11 where the effects of the time delays are evident. For the packet-loss cases, or when there is a temporary malfunction in the communication link and the server does not accept data through the UDP port, the last recorded control command is transmitted to the plant. The latency times due to the transmission: (a) through the UDP-port from the client to the server, and (b)

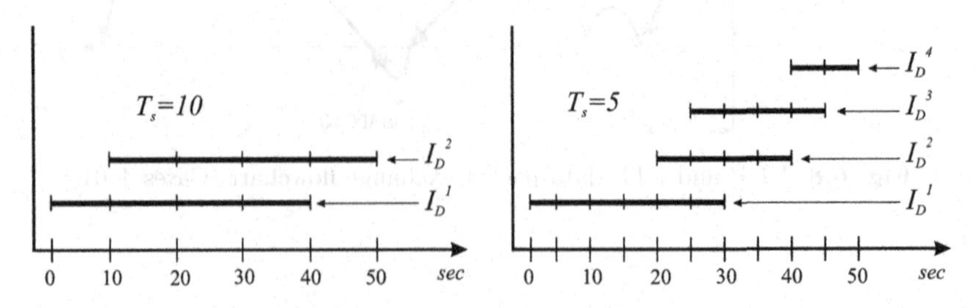

Fig. 6.9. Stability limits of the discrete TDS

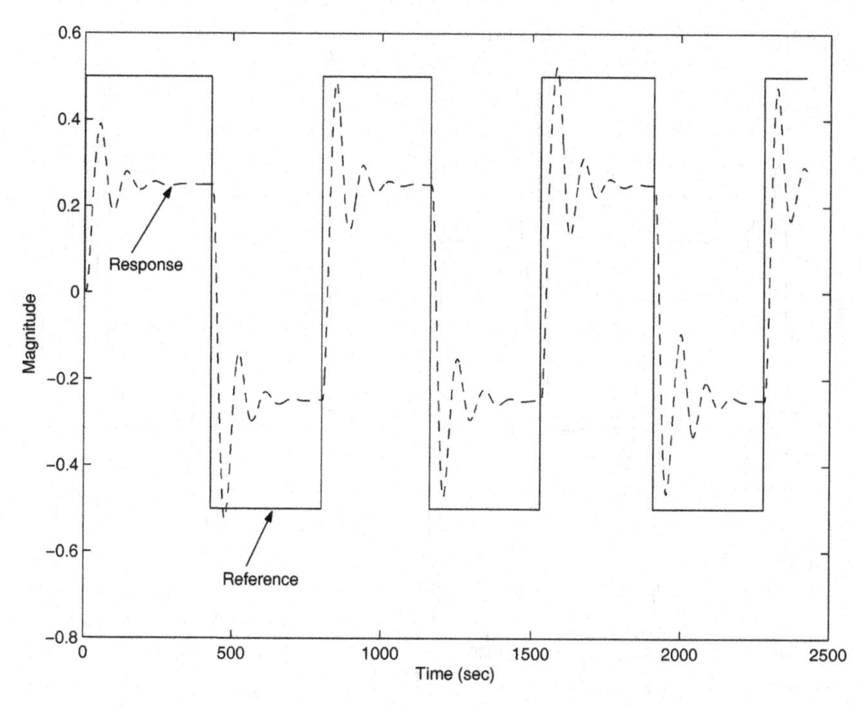

Fig. 6.10. Mobile (GPRS-based) NCS response

through the FTP-agent from the server to the client are shown in Figs. 6.12 and 6.13, respectively. From the recorded data the transmission delays have a mean value of 18 s with a worst case of 35 s.

6.4.2 Switched Feedback Control Over IEEE 802.11b

Introduction

In WiNCS, the communication delays (latency times) cause significant deterioration in the system's performance. Other factors like the reordering and loss of data packets as well as the attainable communication bit-rate impede the controller design process [26].

The last factors are of paramount importance in congested networks, typical in a clustered WNCS, where the transmitter cannot easily find a free time-slot to transmit its data packet. In this case, the transmitter detects a collision and re-attempts its transmission. This increases the communication load and the overhead associated with the transmission. To avoid continuous attempts of packet retransmissions using the UDP-protocol via a wireless link, after a certain time the transmitter quits, and flags that attempt as a failed one (loss-of-packet). However, a large number (percentage) of packet losses

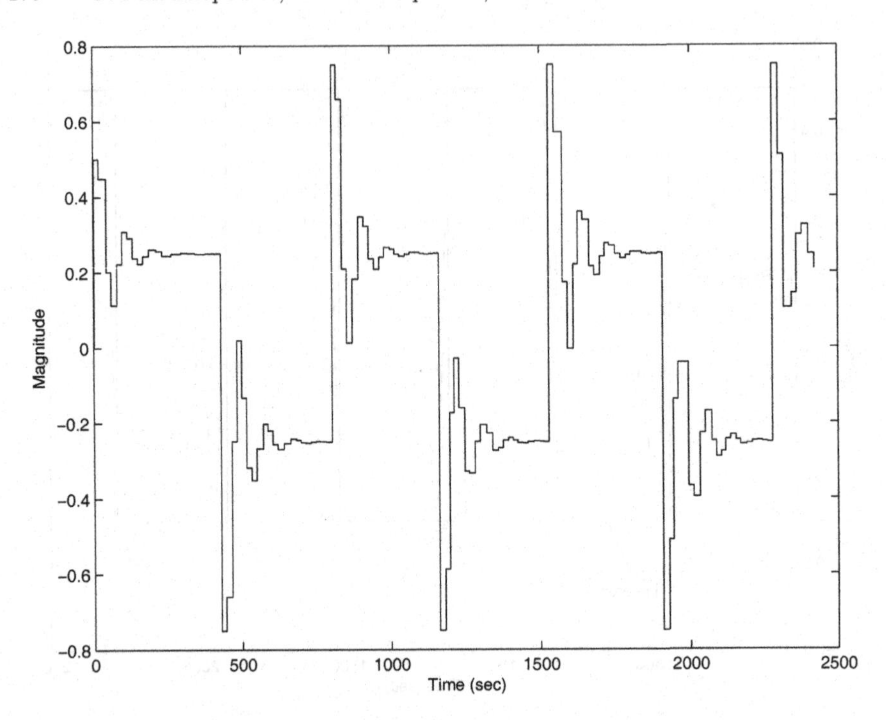

Fig. 6.11. Mobile (GPRS-based) NCS control signal

can significantly affect the performance of the system and reduce stability margins.

The need to retransmit the data packet results in a pure increase of the overall delay $d(k)$ at the kth instant. From a control point of view, the $d(k)$-latency time needs to be as small as possible; at the same time this increases the load (bit-rate) on the communication network due to the fast transmissions of control commands.

A compromise is sought to balance the bit-rate on the wireless link and the need to reduce the latency time for control purposes. Low bit-rates will result in large latency times and the system can be destabilized. Small latency times may cause bottleneck on the queues of transceivers and the link may collapse for short periods of time.

In this subsection, in a prototype WNCS a QoS-module is inserted in each unidirectional communication link to account for the loss-of-packet as shown in Fig. 6.14. This module monitors the recorded lost packets over a sliding window, and assigns a certain time interval prior to the next retransmission. This pre-timed delayed retransmission results in a reduced communication load (traffic) at the expense of applying the control signal to the plant with a large delay. To account for these artificially induced delays (by the QoS-module), rather than maintaining a fixed (time-invariant) controller we utilize a tuning

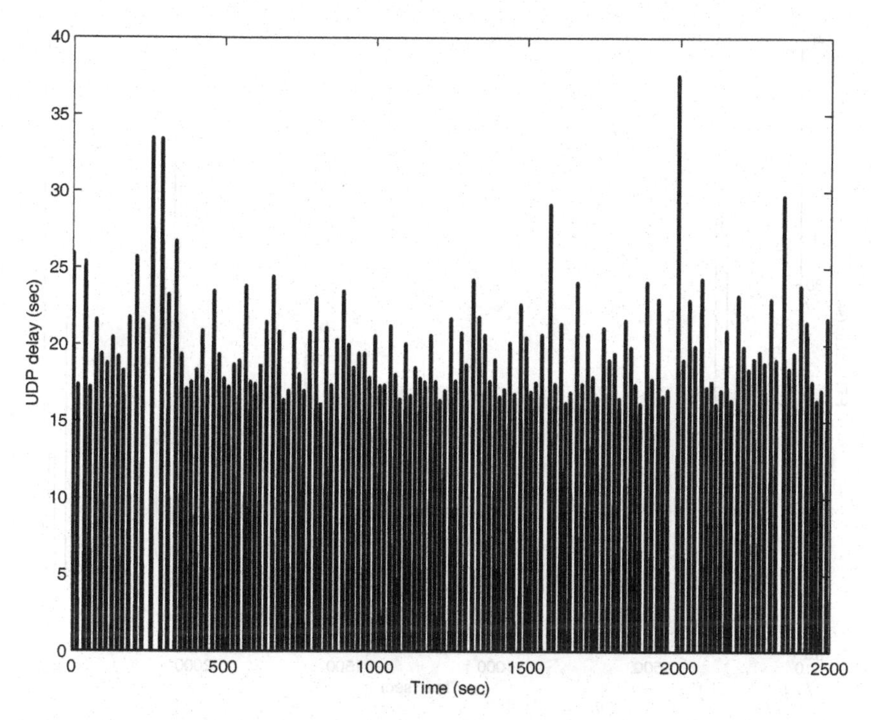

Fig. 6.12. Transmission delay of the mobile UDP-connection

scheme for the controller parameters in order to maintain a certain degree of stability. In this manner, when the latency time is small the controller gains are indirectly adjusted so as to increase the closed-loop system bandwidth. Furthermore, to avoid continuous tuning of the controller parameters and an increase in the overhead this process takes place at certain time intervals (batch-tuning).

Strictly, from an experimental point-of-view, the results presented in the sequel indicate the need for the development and deployment of such a QoS-module and a simplified controller tuning scheme in congested WNCS. An *ad hoc* procedure for tuning the parameters of such a module is examined in the sequel by: (1) observing the intricacies of the communication protocols (802.11b and user datagram protocol (UDP)), and (2) obtaining statistical data regarding the behavior of wireless transmission scheme in lieu of certain parameters (signal strength, distance between receiver/transmitter, line-of-sight variations, etc.). The batch-tuning procedure is examined in the following sections and the experimental results indicate the need to couple this with the developed QoS-module.

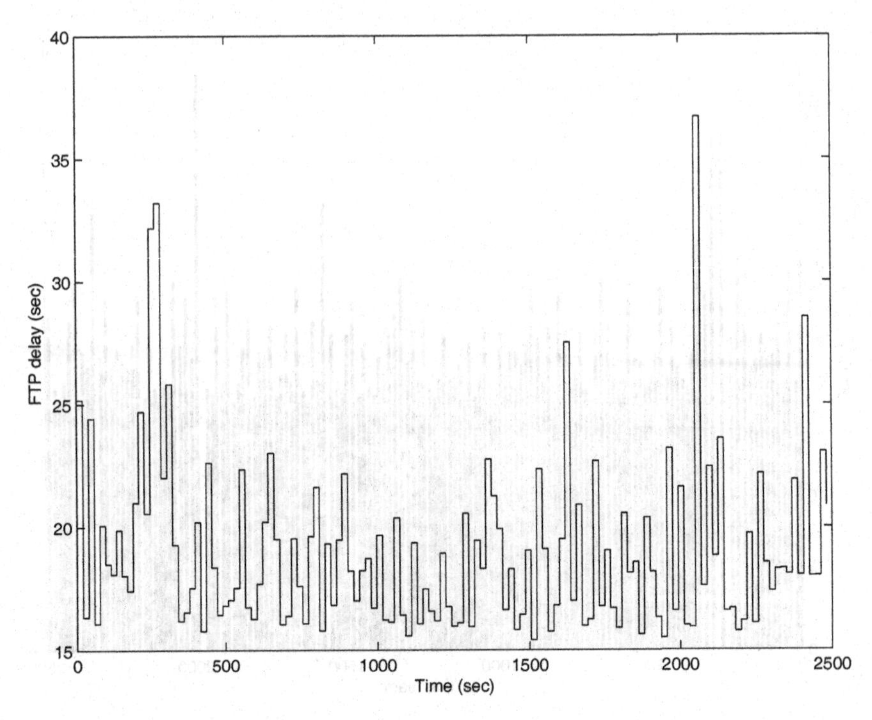

Fig. 6.13. Transmission delay of the mobile FTP-connection

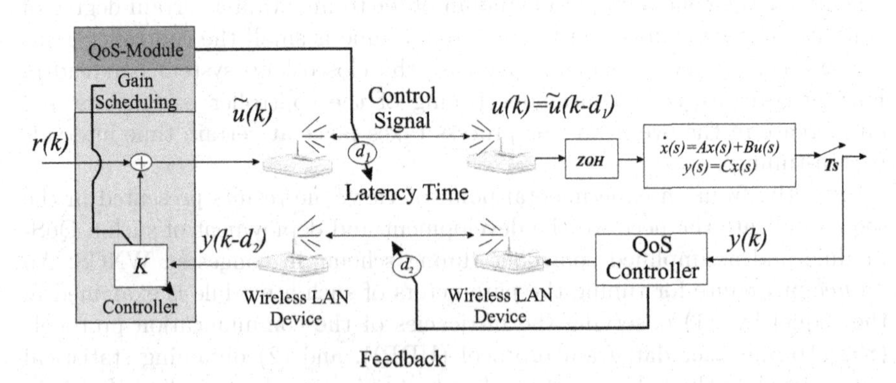

Fig. 6.14. WNCS structural representation

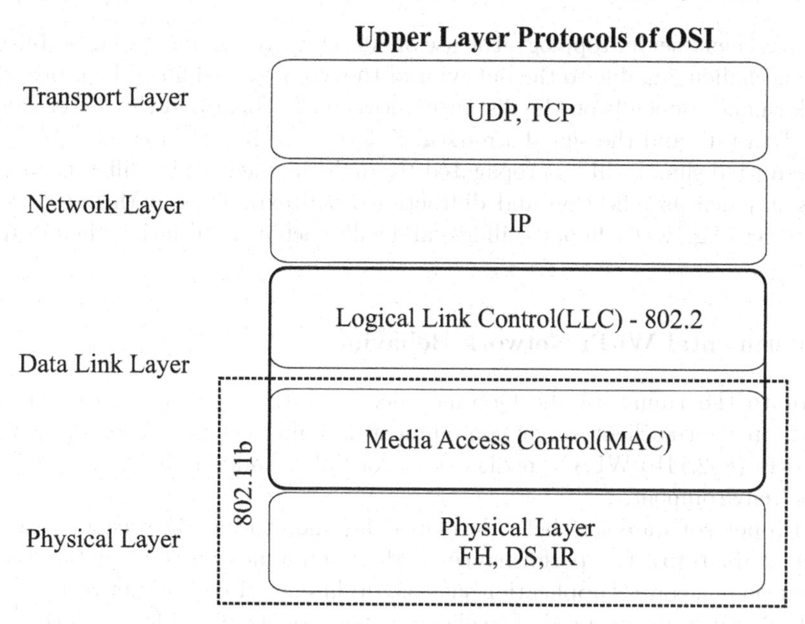

Fig. 6.15. Lower four OSI communication layers (802.11b, UDP/IP protocols)

Communication Protocols

The protocols used in this study are: (a) the IEEE 802.11b for the data link and physical layers, and (b) the IP and UDP for the network and transport layers, respectively, as shown in Fig. 6.15.

The IEEE 802.11b standard [12] is a protocol for RF-communication. The standard covers the lower levels (physical layer and medium access control sub-layer) in the OSI model. By using high rate DSSS for data transmission through the wireless network, 802.11b operates at the 2.4 GHz ISM band frequency, while it offers four transmission data rates: 1 Mbps, 2 Mbps, 5.5 Mbps and 11 Mbps. The MAC layer utilizes the carrier sense multiple access with collision avoidance (CSMA\CA) protocol for accessing the communication medium. Compared with the CSMA with collision detection (CSMA\CD) protocol, which is widely used in local area networks, the CSMA\CA protocol provides a better prediction method in accessing the transmission medium. CSMA\CA's main advantage is the ability to predict the network's response time for data packets transmission, although it cannot provide certain bounds for the delays in the transmission, nor prevent collisions of the data packets. The 802.11b (also known as Wi-Fi) is primarily a data transmission protocol, having its QoS focused on the network's bandwidth and the utilization of high transmission data-rate. The 802.11b is not considered a real-time communication protocol and the inserted communication delays can deteriorate the performance of a WNCS system.

In a client–server topology, the use of 802.11b protocol for system stabilization is challenging due to the behavior of the wireless medium. The protocol's performance depends on the distance between the client and the server sides, the data rate and the signal strength. Referring to indoor environments, the transmitted signal will be propagated through the medium by different mechanisms (such as reflection and diffraction), will experience path loss due to obstacles (e.g., walls, floors, ceilings) and will reach its destination via multiple paths.

Experimental Wi-Fi Network Behavior

Prior to the tuning of the QoS-module, a study is performed to test the attributes of the client–server communication link. Each side is equipped with a Wi-Fi (802.11b) WLAN media access controller and operates in a "peer-to-peer" environment.

Proper software has been developed for monitoring the various aspects that characterize the performance of the communication link, without the overhead of a control application such as: (a) latency times, (b) network traffic, and (c) transmission speed. The client–server model utilized for this primitive quantitative testing of the communication link is presented in Fig. 6.16. In this scheme, the client exchanges data from the server using the UDP-transport protocol. The system operates in an open-loop configuration and there is no handshake to account for loss-of-packets.

Under the assumption of line-of-sight transmission between the client and the server sides, we present in Fig. 6.17 a typical observed channel throughput (in Mbps) over time (four cases). The distinguishing factor in these cases is the distance L between the receiver-transmitter sides, for the following four cases: (1) [top-left] $L = 1$ m, (2) [top-right] $L = 10$ m, (3) [bottom-left] $L = 30$ m, (4) [bottom-right] $L = 30$ m with no line-of-sight. From the presented results, it appears that the distance factor does not play a considerable role in the recorded throughput, whereas the line-of-sight absence affects significantly the observed results ([bottom-right] case). Essentially, the lack of line-of-sight corresponds to a reduction of the received signal strength leading either to: (a) packet-losses, or (b) automatic fallback [20] of the 802.11b transmission

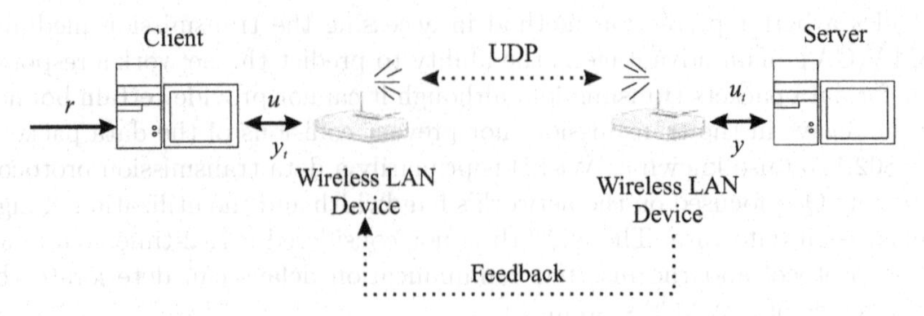

Fig. 6.16. Client–server network model architecture

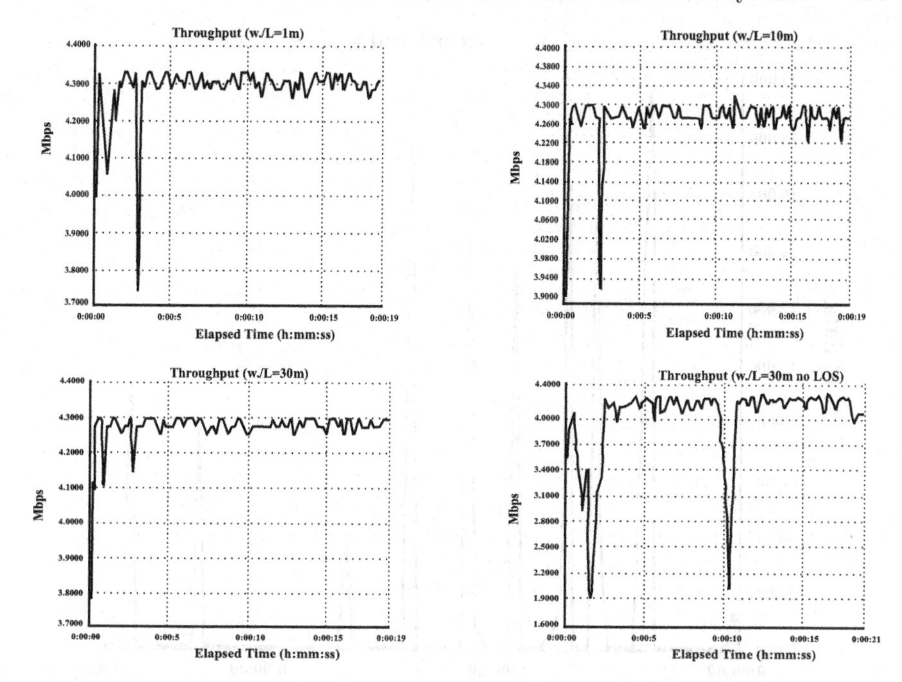

Fig. 6.17. Typical 802.11b throughput parameterized w.r.t. communication link distance

speed at lower data rates $11 \rightarrow 5.5 \rightarrow 2 \rightarrow 1$ Mbps. The percentage of the packet losses appear in Fig. 6.18, where in spurious time instants we obtain percentages as high as 20% in the lost packets over the transmitted packets for the aforementioned third case ($L = 30$ m, line-of-sight transmission). It should be noted that in the previous test-runs, the system was operating in an open-loop manner and the client was essentially streaming data to the server, as shown in Fig. 6.19 (left part) without the need to synchronize both sides typically found in a closed-loop configuration.

Closed-loop Controller Tuning with Wireless-link QoS Optimization

The suggested supervisory scheme, as shown in Fig. 6.14, contains two distinct modules: (a) the QoS-optimization module, and (b) the controller-tuning portion. The need for synchronization between the client and the server sides in a wireless environment poses additional problems compared to the case of unidirectional data-streaming, presented in the previous section. The necessary experimental setup was developed in National Instruments' LabVIEW graphical environment.

From a packet-transversal point of view, consider the case presented in Fig. 6.20. Over a "data packet batch", consisting of a specific amount (F

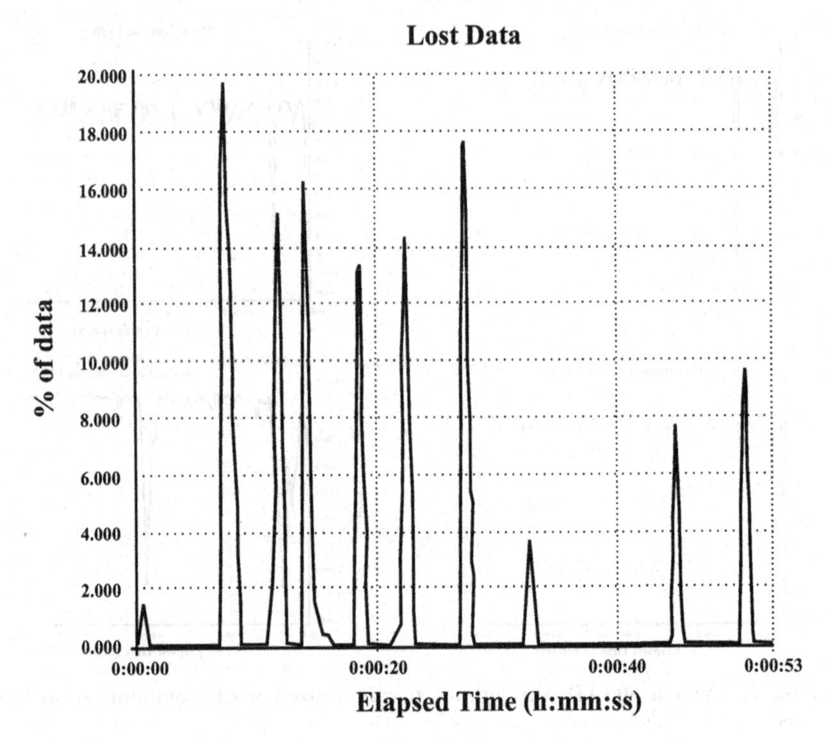

Fig. 6.18. Percentage of packet losses in data-streaming

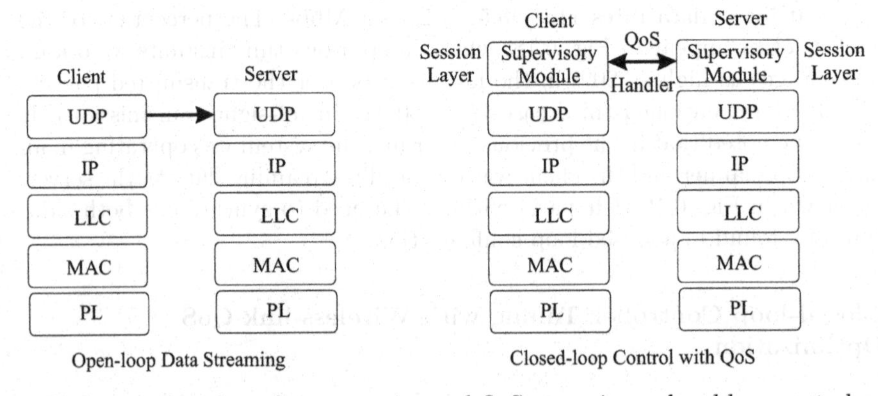

Fig. 6.19. Open-loop data streaming and QoS supervisory closed-loop control

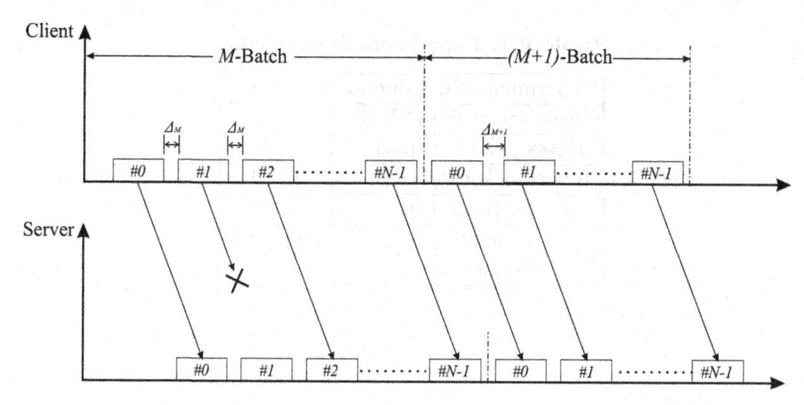

Fig. 6.20. Data packet timing diagram with QoS optimization

bytes) of data segmented in N UDP packets (with P bytes per packet), the client transmits these packets with a fixed inner-packet delay of $d_1 = \Delta_M$. The "batch" period M (s) depends on the value of the inner-packet delay, according to $M = N \cdot (\Delta_0 + \Delta_M)$, where Δ_0 is the packet's transmission time delay. The QoS-module on each side records the arrived packets and keeps statistical data relevant to the channel's quality of service (QoS). The link's QoS is measured in terms of the lost packets recorded in the server's side. Essentially, the latter records the arrived "tagged" packets and maintains in its memory a counter for the percentage of lost packets. When the packet loss exceeds a certain level (on a percentage scale), a supervisory scheme decides on the adjustment of the inner packet delay ($\Delta_M \rightarrow \Delta_{M+1}$) throughout the duration of the next batch-period.

This supervisory scheme is implemented at the OSI's session layer, as shown in Fig. 6.19 (right part), while the experimental parameters are outlined in Table 6.1. This artificially induced delay affects the rate of transmitted packets over the wireless link, since the packet delay-insertion reduces the traffic on the channel (measured in Mbps). Since the network's traffic is reduced, the collision avoidance probability increases (based on the CSMA\CA protocol) during the packet transmission process. Therefore, the percentage of the received packets versus the transmitted packets increases at the expense of reducing the channel's throughput.

For the experimental verification, an $F = 5$ MB file is used, fragmented into $P = 516$ B-UDP packets. We present the percentage of the received packets in open-loop operation versus a constant inserted delay that varies from 0 to 50 ms by the supervisory QoS-handler, as shown in Fig. 6.21. From the reported results, note the small percentage ($< 30\%$) of packet reception when no delay is inserted, compared to the simple network evaluation case (Fig. 6.18) where we had spurious spikes of 20% maximum of lost data. Furthermore, as expected, the recorded percentage increases when the delay is inserted and

Table 6.1. Experimental parameters

Experimental parameters	values
Amount of data (MB)	5
Bytes per UDP packet	516
Inner-packet delay (ms)	0–50
Distance (m)	1–30
Line of sight	supported
Nodal motion	no

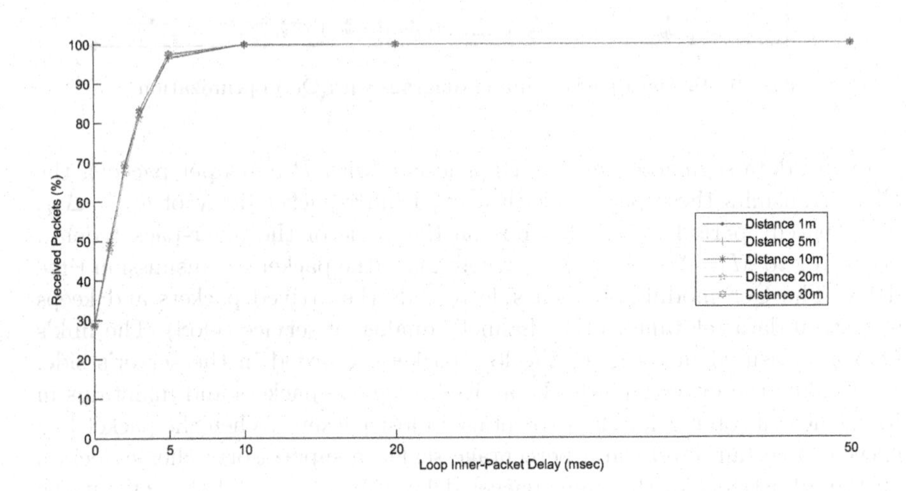

Fig. 6.21. Received packet percentage in open-loop wireless control

reaches a maximum of 99.89% for the case of a QoS-based inserted delay greater than 10 ms.

To emphasize the compromise in the channel throughput when the QoS-module inserts delays, we present in Figs. 6.22 and 6.23 the measured "successful bit rate" (recorded in Mbps) and total transmission delay, respectively. For the best case (i.e., for $\Delta_M = 0$ s) the measured mean value of transmission delay is 4.23 s. For the aforementioned values of F and P, the file's segmentation into UDP packets, corresponds to $\Delta_0 = 436.68$ μs. The value of Δ_0 is too small to affect the overall transmission delay, especially as the loop delay becomes significant (Fig. 6.23). Clearly, there should be a compromise in the insertion of the delay and the bit-rate that the link can provide. Larger QoS-delays lead to higher percentages of received packets (Fig. 6.21) and increased transmission delays (Fig. 6.23) at the expense of reducing the bit rate (Fig. 6.22).

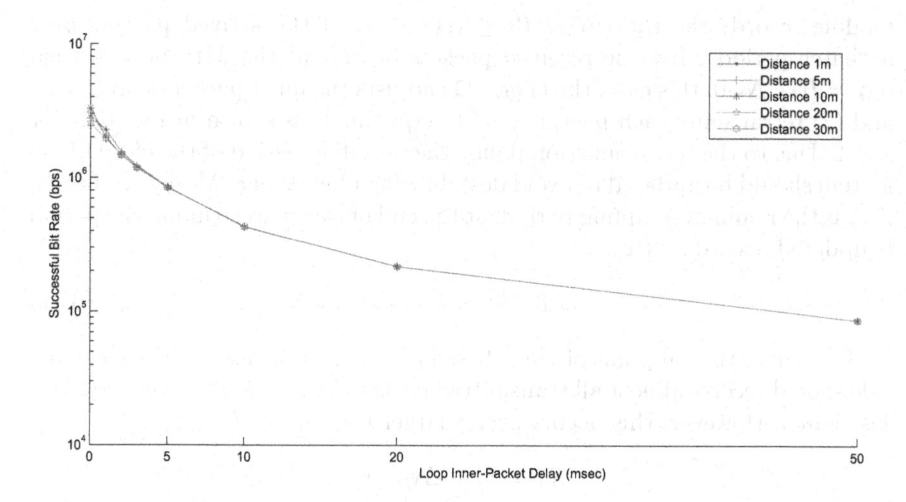

Fig. 6.22. Recorded channel bit-rate in open-loop wireless control

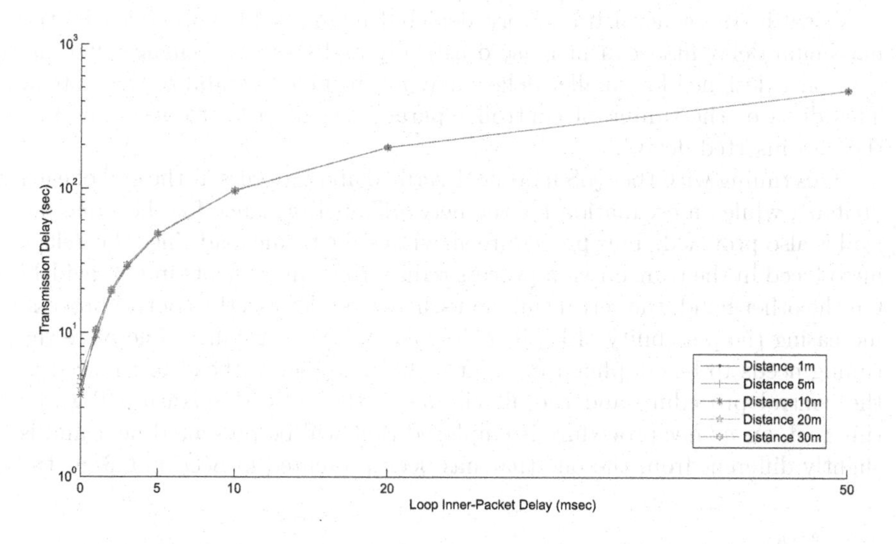

Fig. 6.23. Total transmission delay in open-loop wireless control

Closed-loop Controller Tuning and Transmission Mechanism

For the case where the lost packets exceeds a certain large percentage, the delay-insertion by the QoS-module cannot be used as a stand-alone countermeasure to: (a) ensure the network's stability (bottleneck avoidance), and (b) maintain certain stability margins in the closed-loop system. Here, a retransmission mechanism is proposed coupled to the QoS-module. The QoS-module records the ratio α_M, $(0 \leq \alpha_M \leq 1)$ of the arrived packets over a "batch-period". Let the received packets be H_M at the Mth period, then $\alpha_M = H_M/N$. In this case, the client: (1) adjusts the inner packet delay Δ_{M+1} and (2) retransmits each packet $\lambda = \lceil 1/\alpha_M \rceil$-times, as shown in Fig. 6.24 for $\lambda = 2$. Due to the retransmission policy the sampling period of the closed-loop system should be updated to avoid destabilizing phenomena. More specifically, if T_s is the nominal sampling period, at the end of each batch tuning the period is updated according to:

$$T_s^{M+1} = \lambda \times T_s. \tag{6.17}$$

From a statistical point of view, based on this retransmission the receiving side should receive almost all transmitted packets ($\lambda \times \alpha_M \simeq 1$) over the $(M+1)$st batch. However, this occurs over a larger time span of

$$\lambda \times (\Delta_0 + \Delta_{M+1}).$$

Indirectly, this time-extension that each side receives the control command or the system output is equivalent to prolonging the d_1 and d_2 latency times, as shown in the general framework depicted in Fig. 6.14. Subsequently, the maximum delay $d(k)$ can increase drastically and there is a chance that the controller designed for smaller delays may not be able to stabilize the system. This dictates the tuning of controller parameters in order to accommodate the new inserted delays.

This tuning with the QoS module described above occurs at the end of each "batch", while an estimation for the network's performance for the same period is also provided. This procedure alleviates the traffic load since the delays introduced in the transmission process reduce the requirements in bandwidth. On the other hand, the retransmission scheme decelerates the control process, increasing the possibility of leading the system into instability. Therefore the tuning needs to be simplified in order to balance the overhead introduced in the control procedure and the alleviation of the network's status. Towards this goal, an *ad hoc* procedure is employed that will be presented next and is slightly different from the one that has been presented in Section 6.3; at the

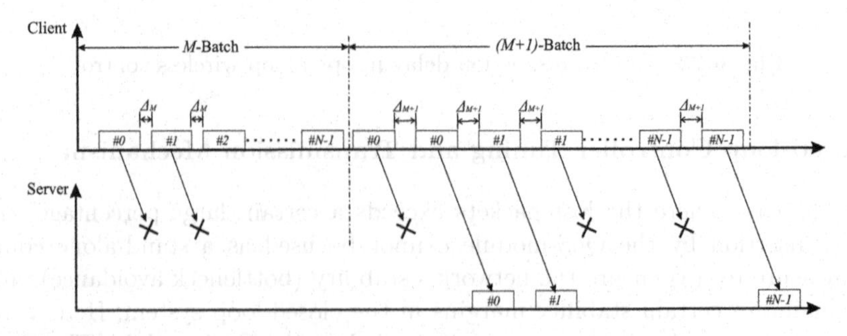

Fig. 6.24. Data packet timing diagram with QoS optimization and retransmission

Table 6.2. Simplified feasibility search algorithm

1	$K_{M+1}(j+1) = K_{M+1}(j) - \Delta(K_{M+1})$
2	If (6.11)–(6.12) are satisfied with $K_{M+1}(j+1)$, go to 1
3	$\Delta(K_{M+1}) = -\dfrac{\Delta(K_{M+1})}{2}$
4	If $\Delta(K_{M+1}) \leq \epsilon \simeq 0$ stop, else go to 1

end of a "batch" tuning period, the controller's gain K_M is adapted according to the network performance metrics at the Mth batch, in order to guarantee a certain degree of stability at the $(M+1)$st batch. More specifically, under the assumption of a T_s^{M+1}-sampling interval, this stability should be ensured over

$$I = \left\{ \frac{d + \lambda \times (\Delta_0 + \Delta_{M+1})}{T_s} - \gamma, \ldots, \frac{d + \lambda \times (\Delta_0 + \Delta_{M+1})}{T_s} + \gamma \right\}, \quad (6.18)$$

where γ is a parameter related to the width of time-delay uncertainty that the controller can tolerate. In this scheme, the gain K_M is tuned according to the inserted time delay Δ_M. The objective is to find an optimal value for the controller's gain that will satisfy the LMI posed in (6.11) for a nominal gain K in the delay-set I stated in (6.18). For a given plant and an output feedback controller with an initial gain K, a gain scheduling step $K_{\mathbf{step}}$ with maximum precission threshold, the problem is transformed to:

Compute K_{M+1} such that,

$$K_{M+1}: \max_{K_{M+1}<0} C(sI - A)^{-1} B K_{M+1}, \quad (6.19)$$

subject to the LMI-constraints of (6.11)–(6.12). Since the cost in (6.19) can be made arbitrarily large by decreasing K_{M+1}, the only restriction to the selection of K_{M+1} is the satisfaction of (6.11)–(6.12). Henceforth, the problem is similarly restated as:

Select the minimum K_{M+1} that satisfies the set of LMIs in (6.11)–(6.12).

Because of K is a scalar, a simplified feasibility search algorithm can be employed for this problem. A typical pseudocode of this algorithm appears in Table 6.2. Let $K_{M+1}(j)$ be the value of K_{M+1} at the jth iteration of the algorithm, assuming that this value satisfies the LMIs (6.11)–(6.12).

Experimental Closed-loop Studies

The aforementioned controller is applied in a prototype WNCS. The plant that is controlled corresponds to a sampled ($T_s = 150$ ms) continuous system with a transfer function of

$$\frac{10^3}{(s+10)^3}.$$

However, due to the QoS module and the retransmission mechanism adopted, the actual value of the closed-loop sampling period T_s^M is updated at the end of each batch tuning period according to Equation (6.20)

$$T_s^{M+1} = \lambda \times T_s. \tag{6.20}$$

Our research goal focuses on the study of the proposed optimization scheme in near-congestion conditions. The experiment consists of two nodes only, that under normal operation would exchange small, periodical amount of control data and consequently would have limited bandwidth requirements. Therefore, in order to examine the behavior of the control process in a highly competitive environment, exogenous traffic of size $F = 5$ MB is inserted artificially on the Wi-Fi link. For the specified amount of data F the mean total transmission delay, as shown in Fig. 6.23, equals 4.2314 s for the best case scenario, which corresponds to:

$$\left(\frac{5\ (\text{MB})}{4.2314\ (\text{s})}\right) \times 8\ (\text{bits}) = 9.45\ \text{Mbps},$$

a value near the nominal transmission rate of the technology utilized (i.e, 11 Mbps). The control-related information is randomly mixed up with useless packets. The message created is sent to the plant within a closed-loop period. The plant separates the useful information from the rest of the data received, which are discarded. The observed latency time appears in Fig. 6.25, where two delay-sets are revealed; $I_1 = \{1, 2, 3, 4\}$ for $\lambda = 1$, i.e., $T_s^M = T_s$ and $I_2 = \{2, 4, 6, 8\}$ for $\lambda = 2$, i.e., $T_s^M = 2 \times T_s$.

The QoS-gain scheduling optimization algorithm is applied in order to examine the stability of the plant in delay-sets I_1 and I_2. In order to expand the delay-sets into a wider region, the delay set I_3, where $I_3 = \{3, 6, 9, 12\}$ is also examined. The original controller's gain is set to $K = -0.5$, while the scheduling step K_{step} varies from 0.002 up to 0.125. Solutions of the algorithm are shown in Fig. 6.26.

The optimal values derived are:

$$\begin{aligned} I_1 &= \{1, 2, 3, 4\}\,,\ K^1 = -1.0234, \\ I_2 &= \{2, 4, 6, 8\}\,,\ K^2 = -0.76172, \\ I_3 &= \{3, 6, 9, 12\}\,,\ K^3 = -0.65625. \end{aligned}$$

The optimal values K^1, K^2 and K^3 are applied to the WNCS prototype, where two cases are examined; (1) the datagram containing the control command is uniformly distributed among the overall exogenously generated traffic (F bytes) and (2) the useful information is uniformly distributed among the second half of the artificial traffic F. The appended system's response appears in Fig. 6.27. The effect of the controller tuning is evident on the system's response. The sudden deviations observed in the system's response are a result of the controller's gain adaptation, according to the network's performance.

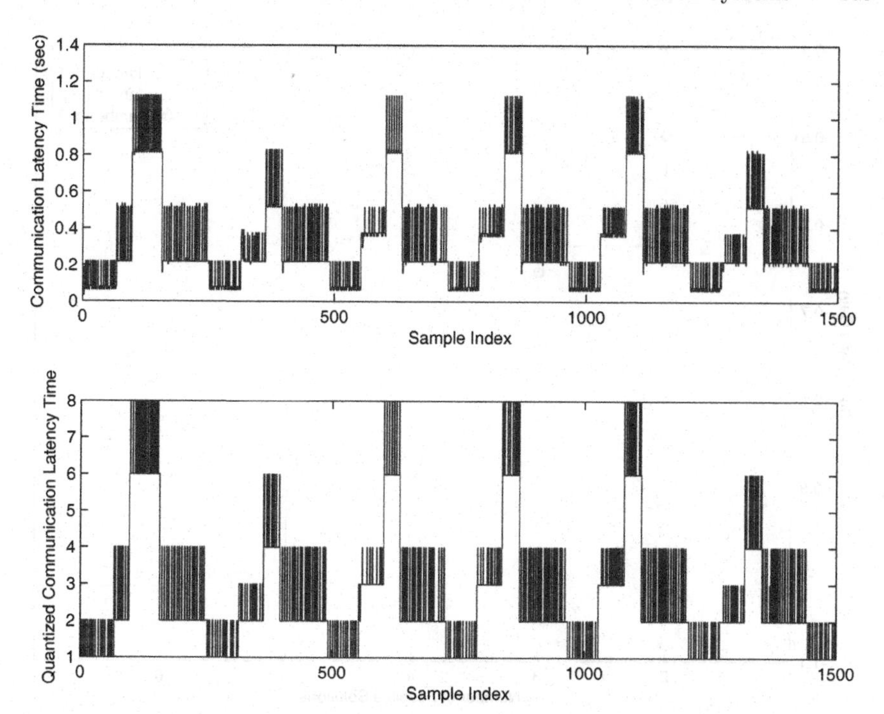

Fig. 6.25. Experimental communication latency time (a) in seconds (top) and (b) quantized (bottom)

In Fig. 6.28 the observed latency time is outlined for each experimental case. The dependence of system's response on the network performance for the second case is dominant since as the reception conditions worsen the system has to switch its operation from I_2 to I_3, whereas in the first case examined, the system typically operates into the delay set I_1.

For comparison purposes, a similar system response is provided in Fig. 6.29 with fixed gain $K = -1$, where there are no provisions for handling the QoS-issue. It should be noted the destabilization effect that takes place after the 500th sample, where there is considerable loss of packets. Since there are no retransmissions nor any artificial inserted delays the two sides (client and server) struggle to obtain a free time-slot to exchange their data packets. This causes additional traffic on the link, leading to the noted destabilization effect. A direct comparison of these results with those presented in Fig. 6.29 indicates the advantages enjoyed by our suggested approach.

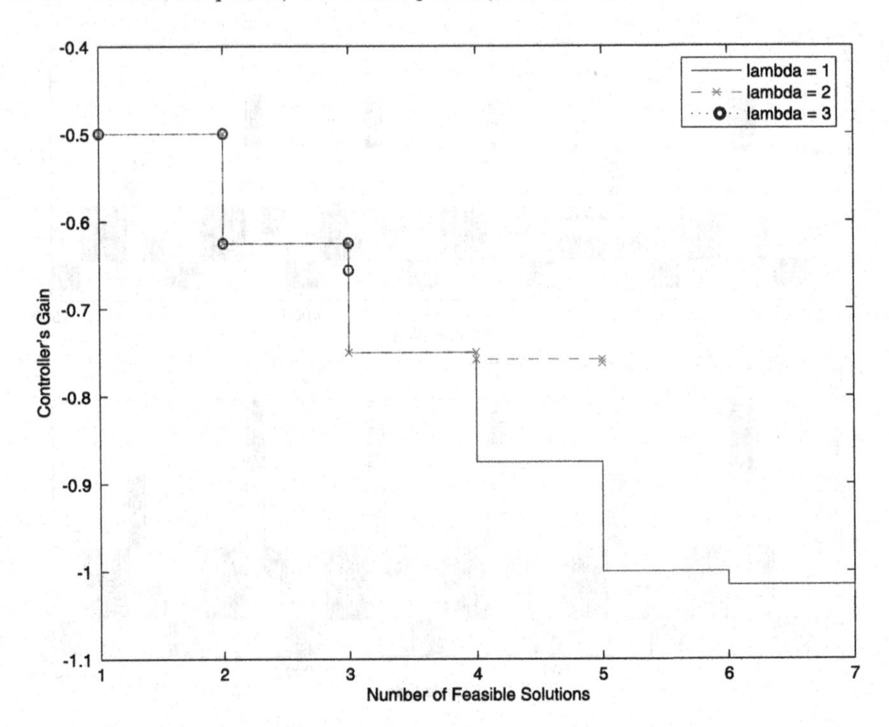

Fig. 6.26. Solutions of the QoS optimization algorithm for delay-sets I_1, I_2, and I_3

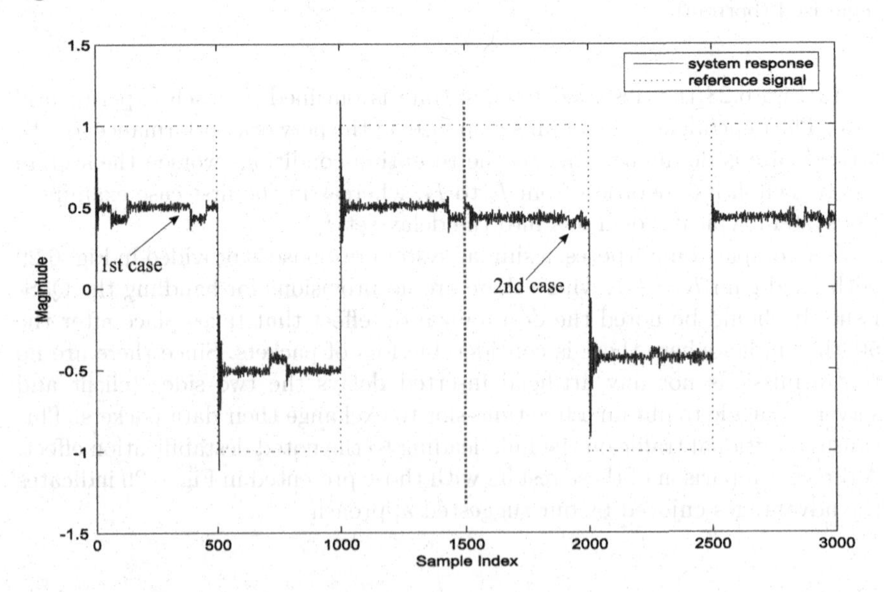

Fig. 6.27. Experimental WNCS appended optimal response for the two cases examined

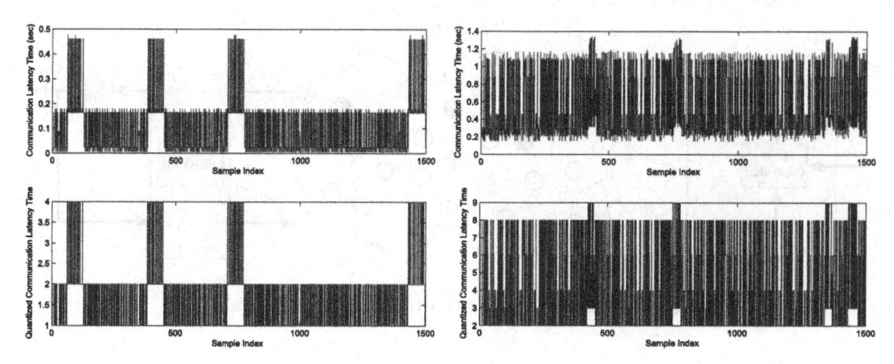

Fig. 6.28. The observed latency time for the first (left) and second (right) experimental cases examined

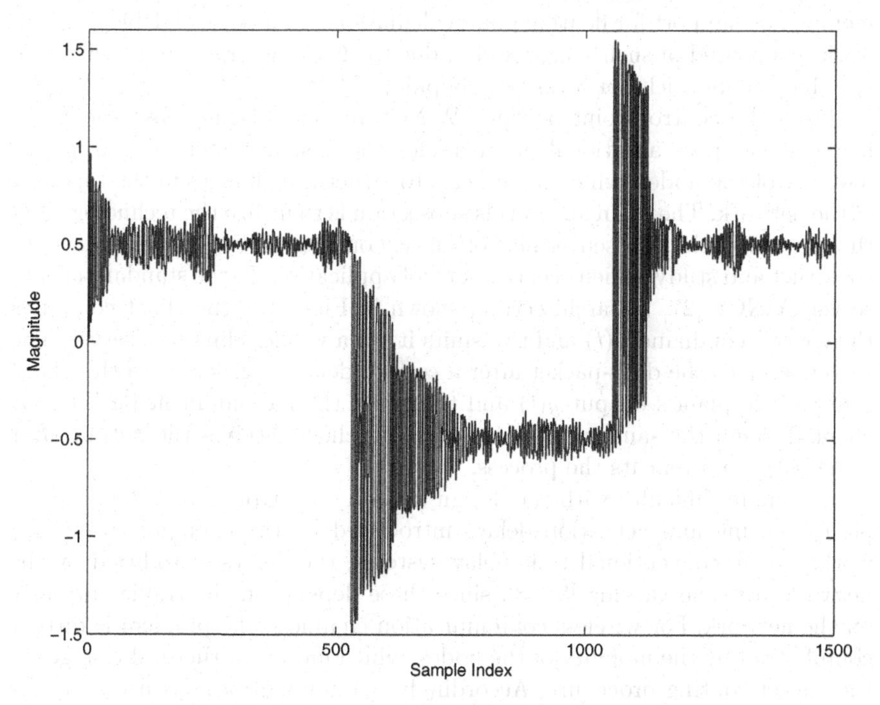

Fig. 6.29. Experimental fixed WNCS response

6.4.3 Switched Optimal Feedback Control Over IEEE 802.11b in MANETs

Introduction

Among NCS, wireless NCS (WiNCS), and in particular WiNCS built around wireless sensor networks (WSN) [1, 2] is a rapidly evolving area at the moment. Recent technological advances have resulted in small, integrated sens-

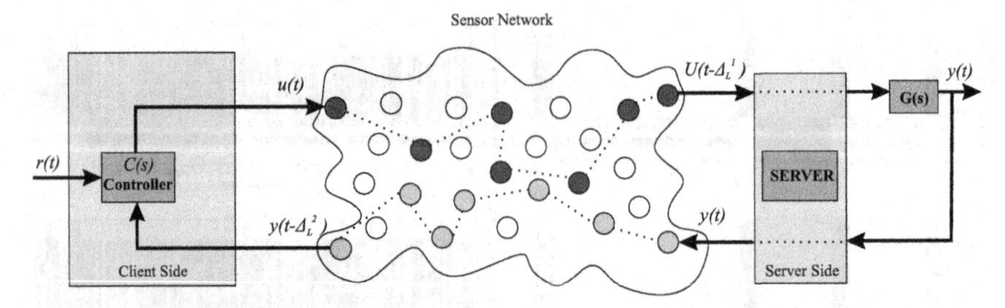

Fig. 6.30. Client–server architecture based on a WSN

ing devices, capable of running a complete protocol stack. These devices have been optimized for communication with limited resources (transmission power, memory, no support for floating point calculations). The ease of deploying networks comprised of such sensor nodes, due to their low prices and their small size, have made such networks very popular.

From the control point of view, WiNCS in general and sensor networks in particular pose additional problems for the designer, stemming from the mobility of the nodes, which often leads to structural changes in the topology of the network. The main aim of this subsection is to utilize the technology and the characteristics of a sensor network based on the IEEE 802.11b protocol to construct and study a client-centric control application. In the standard client–server WiNCS [27, 16] architecture, shown in Fig. 6.30, the client computes the control command $u(t)$ and transmits it via a wireless link to a server. The server receives the data-packet after a certain delay, transfers it to the plant, samples the plant's output $y(t)$ and transmits the measurement back to the client through the same sensor network. The client receives the output after some delay and repeats the process.

The main difficulty with the design of such a control loop is the presence of the sensing and actuation delays introduced by the communication network. Unlike conventional time delay systems, the delays introduced by the network are time varying [5, 28], since these depend on the traffic currently on the network. For wireless communication channels, the problem is further complicated by the mobility of the nodes, which induces structural changes in the packet routing procedure. Accordingly, the number of hops necessary for a packet to reach from the client to the server (and reverse) varies in a step manner introducing an additional source of varying delays by the network.

As was presented in the previous sections, we will utilize the LMIs to select the parameters of the LQR problem to ensure that poles of the closed loop system maintains a prescribed stability margin despite the variability of network delays. In this case we will extend this approach to investigate the effect of multi-hopping in the process. Based on results of the maximum delay that the control system can tolerate, obtained for the peer-to-peer system, a gain scheduling controller, which uses the number of hops to select appropriate LQR gains is developed and tested in simulation.

Properties of Wireless Sensor Networks

Although WSN maintain several characteristics of conventional networks, they also have key differences [2]. WSNs combine three important components: sensing, data processing and communication [1]. The nodes that comprise a sensor network are spatially distributed, energy-constrained, self configuring and self-aware. WSNs can provide quite effective performance in noisy environments, since they allow sensors to be placed close to signal sources, therefore yielding high signal to noise ratios. Moreover, the scalability of sensor networks permits the monitoring of phenomena widely distributed across space and time, and their versatility makes them an ideal infrastructure for robust, reliable and self-repairing systems.

An important issue related to scalability [15] is the fact that after some point, communication becomes more expensive than computation. The requirements for collaboration and adaptation to "stochastic networking" features (usually due to exogenous factors) impose the need for the development of novel protocols dedicated to sensor networks, such as [33]. Another major concern is energy consumption, which requires a compromise between node collaboration and energy constraints [8] and affects the maximum active communication area. These features that affect the routing of communication packets sent over a WSN which requires multiple hops to complete the origin-destination travel. The dynamic nature of the network further implies that the number of hops may be variable. In our case we investigate how these features affect the design of controllers that attempt to close the loop over such a network. Towards this goal, an MANET in a noisy, crowded environment is simulated, as the communication medium that transfers the data packets between a remote controller and the plant.

WiNCS Simulation

The network scenarios are tested with the NS-2 [13] simulator for the physical, MAC and network layers of each node. Providing a variety of networking protocols, several scenarios can be simulated, based on the cases examined. The parameters that have been utilized for our test case are outlined in Table 6.3. The simulated WSN consists of n-mobile nodes, communicating over a wireless link based on the IEEE 802.11b standard. For the examined case we assume that all nodes "wake up" at the same instant. While every node in the network may potentially exchange information with other nodes, two nodes are of particular interest: the node attached to the plant (server) and the node attached to the controller (client). The routing protocol assumed is the dynamic source routing protocol (DSR) [14]. In the transport layer information exchange is based on the user datagram protocol (UDP).

Due to node mobility, routing is not fixed [31]. Therefore, the number of hops during the transmission of a packet changes as the node moves from

Table 6.3. Simulation parameters

Network characteristics	values
Simulation time (s)	500
Number of nodes	20
Number of connections	20
Maximum number of packets transmitted per connection N_o	10000
UDP transmission interval per connection (s)	0.2
Maximum speed per node (m/s)	20
Coverage area	670×670
Agent type	UDP
Routing protocol	DSR

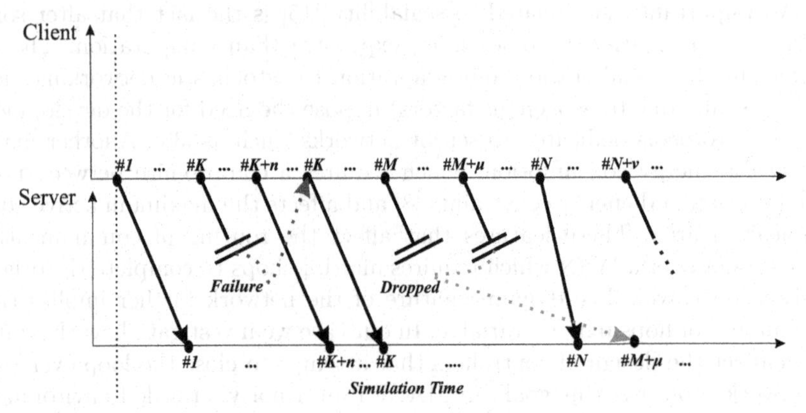

Fig. 6.31. Phenomena observed during a UDP data transmission from client to server

one position to another. Moreover, due to the connectionless services provided by the transport layer, other interesting phenomena are also observed. For example a packet that fails to reach its destination or an intermediate node, may be dropped, or sent back to its source node. The retroactivity phenomenon observed, as a consequence of the unreliable services that UDP provides, is becoming even more dominant as a delay factor in heavy network traffic conditions. Some of the observed events are described in the cases that are presented in Fig. 6.31.

Simulation Studies

The suggested scheme was applied in the same simulated prototype SISO-system as before, with the following transfer function:

$$G(s) = \frac{0.1^3}{(s+0.1)^3}.$$

Assuming a sampling period of $T_s = 5$ s, the discrete equivalent of the continuous system is (accounting for the ZOH)

$$x(k+1) = \begin{bmatrix} 1.82 & -1.104 & 0.2231 \\ 1 & 0 & 0 \\ 0 & 1 & 0 \end{bmatrix} x(k) + \begin{bmatrix} 1 \\ 0 \\ 0 \end{bmatrix} u(k),$$

$$y(k) = [0.001439 \;\; 0.003973 \;\; 0.006794]\, x(k).$$

Assume that a discrete controller $u(k) = Ky(k - r_s(k))$ is inserted in the loop.

Theoretical Results

In this case the controller's gain values are $K \in \{-1.1927, -1.4439, -1.7225\}$ and were computed from the suggested algorithm minimization of (6.6), where $Q = R = 1$ and $\sigma^{\max} = 1/0.85$. As an example we should note for $K = -1.7225$ (-1.4439) the system is stable ($|\lambda_{\max}(A_{r_s})| \leq 1$) for $r_s T_s \leq 20$ (30) sec.

Based on exhaustive simulation of data packets traffic in the sensor network, the dependence among the number of hops and bounds in the communication latency times that are introduced is presented in Fig. 6.32 for the MANET-parameters presented in Table 6.1.

Based on these worst case bounds on the communication latency times with respect to the number of hops, L, the switched controller's gains are determined as:

$$I_1 = 0, 1, 2, 3, \quad K^1 = -1.7225, \quad 0 \leq L \leq 2,$$
$$I_2 = 1, 2, 3, 4, 5, \quad K^2 = -1.4439, \quad 3 \leq L \leq 9,$$
$$I_3 = 4, 5, 6, \quad K^3 = -1.1927, \quad 10 \leq L \leq 15.$$

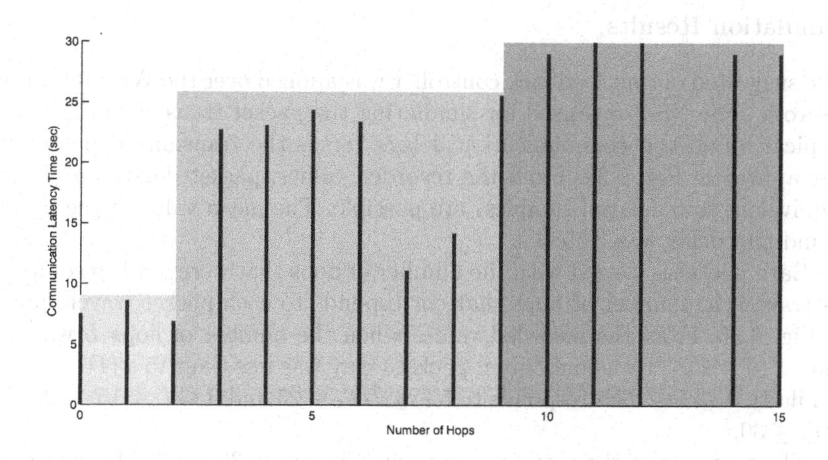

Fig. 6.32. Communication latency time dependence on the number of hops

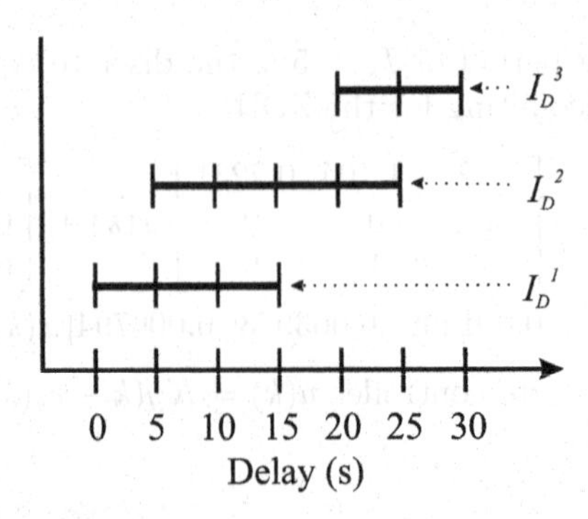

Fig. 6.33. Stability limits of a discrete control TDS, $T_s = 5$ s

It can be shown that the LMIs in (6.11) and (6.12) are satisfied with these gains.

Next, we try to maximize the range of delays $I^i = \left[D_i^{\min}, D_i^{\max}\right]$ for which solutions to the LMI problem (6.11) exist for each gain. In Fig. 6.33 the shaded areas show the ranges I^i sets for which the LMI problem could be solved for each K^i. The corresponding sets were $I_D^1 = [0, 15]$, $I_D^2 = [5, 25]$, and $I_D^3 = [20, 30]$.

It is apparent that from the LMI problem there exists no controller that can tolerate delays up to 30 s. However, if the latency time varies slowly, then from the overlapping property of these sets, the whole region can be covered and stability can be guaranteed by minimum dwell time arguments. The details of this argument are a topic for future research. It should be noted, that from the experimental section the observed latency time exhibited a reasonably slow variation and the switching controller provided proved stable up to a 30 s delay.

Simulation Results

The suggested output feedback controller was applied over the WiFi (802.11b) network. The NS-2 was used for simulating the packet transmission process. Typical round-trip communication delays versus the transmitted packet index appear in Fig. 6.34. From the recorded values, packet delays up to 30 s (equivalent to 6 delayed samples) are possible. The mean value of the packet round-trip delay was 2.5586 s.

Each packet is tagged with the number of hops that were used to complete its travel; the number of hops that corresponds to each packet-travel appear in Fig. 6.35. From the recorded values when the number of hops L was less than 3 ($0 \leq L \leq 3$) the maximum packet delay was less than 15 s ($DT_s \leq 15$). Similarly $3 \leq L \leq 9$ corresponds to $5+ \leq DT_s \leq 25$ and $9 \leq L \leq 15 \rightarrow 20+ \leq DT_s \leq 30$.

The response of the system is presented in Fig. 6.36, while the switching of the LQR-output feedback controller's gains are presented in Fig. 6.37. It

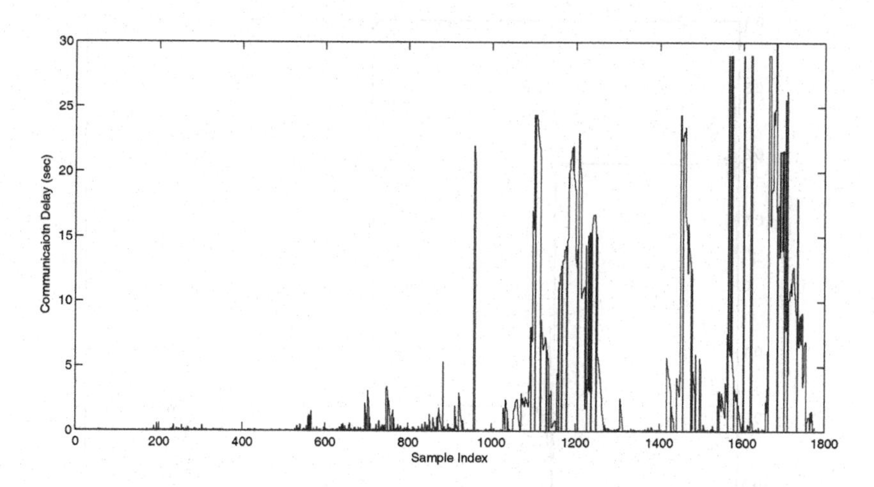

Fig. 6.34. Round-trip communication delay versus the sample index

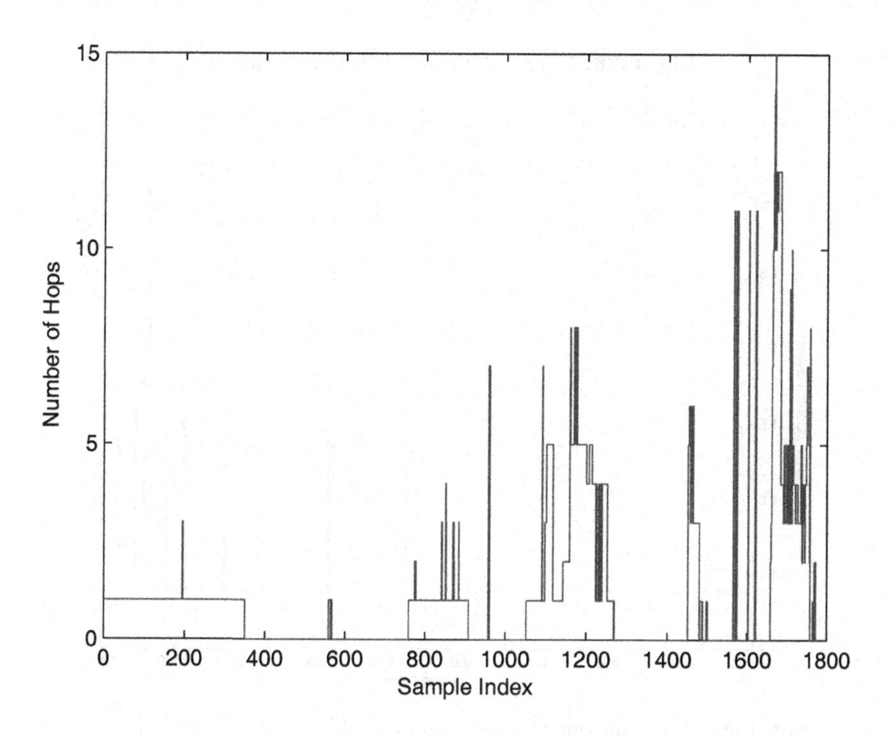

Fig. 6.35. Number of hops versus the sample index

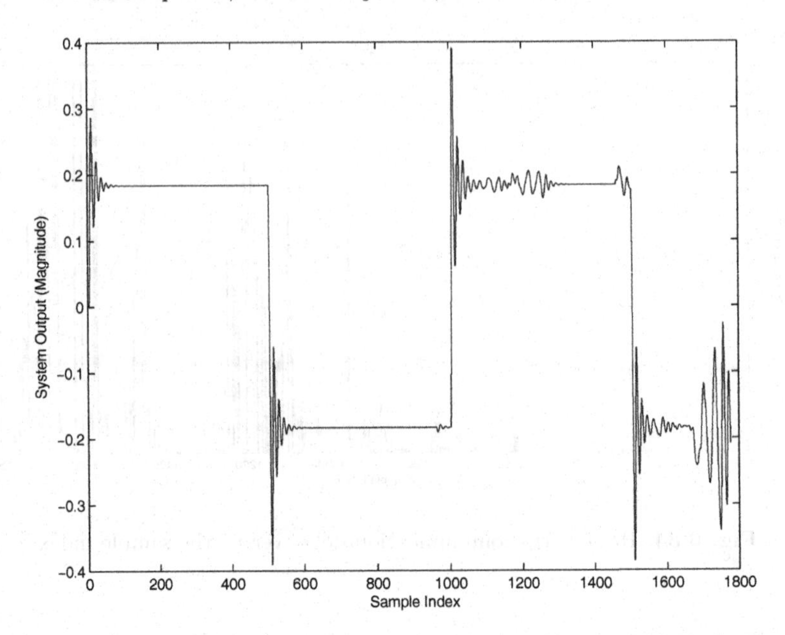

Fig. 6.36. Networked control system response

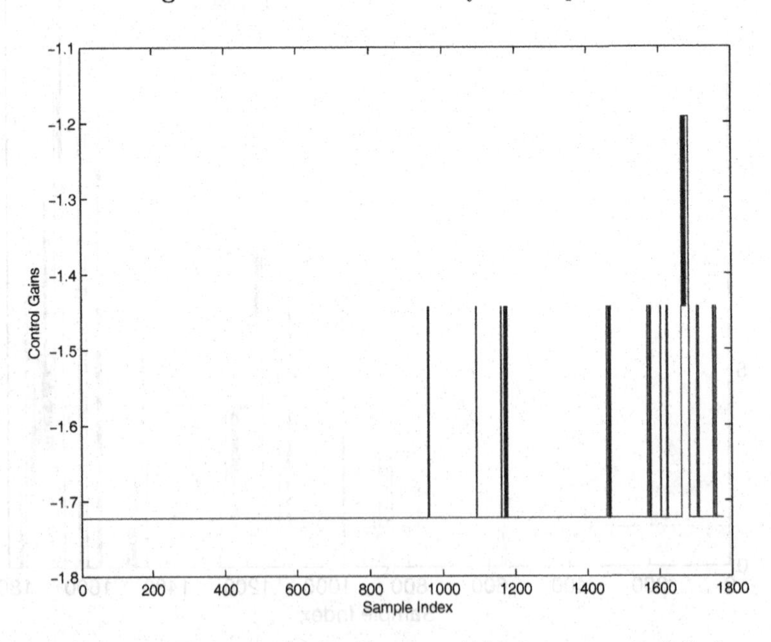

Fig. 6.37. LQR–output feedback gain switching versus sample index

should be noted that the delay-classification into these three distinct packet-hop classes $L \in \{0,\dots,3\}$ or $\{3,\dots,9\}$ or $\{9,\dots,15\}$ is arbitrary and more efficient clustering techniques can be used.

With regard to controller design, the parameters used in optimization cost were $Q = R = 1$ while the initial value of the prescribed stability radius σ was set to one ($\sigma = 1$). For each one of the three classes $I_1 = \{0,\dots,3\}$, $I_2 = \{1,\dots,5\}$ and $I_3 = \{4,\dots,6\}$ the output feedback gain was computed according to the aforementioned scheduling procedure. In this scheme, $\Delta\sigma = 0.01$ and this process was terminated when $\sigma \leq 0.95$.

6.5 Conclusions

The development of controllers for networked systems requires particular attention due to the inherent delays encountered in the network. Subsequently, conventional control schemes should be modified in order to be robust against these time-varying delays. In this chapter, theoretical and experimental results are presented for networked control systems over: (a) GPRS, (b) 802.11b, and (c) 802.11b in MANETs. It is shown that the user should focus on the QoS, while at the same time ensuring the scheme's robustness. In summary, control development of networked systems, necessitate the design of customized solutions with a careful selection of controller's parameters.

References

1. Akyildiz I, Su W, Sankarasubramamian Y, Cayirci E (2002) Wireless sensor networks: a survey. Computer Networks 38:393–422
2. Akyildiz I, Wang X, Wang W (2005) Wireless mesh networks: a survey. Computer Networks 47:445–487
3. Bettstetter C, Vogel H, Eberspecher J (1999) GSM phase 2+ general packet radio service (GPRS): architecture, protocols, and air interface. IEEE Communications Surveys 2, 3rd Quarter
4. Boyd S, Balakrishnan V, Feron E, Ghaoui L (1993) Control system analysis and synthesis via linear matrix inequalities. In: Proc American Control Conference, San Francisco, CA, 2:2147–2154
5. Branicky M, Phillips S, Zhang W (2000) Stability of networked control systems: explicit analysis of delay. In: Proc American Control Conference, Urbana, IL, 2352–2357
6. Crow B, Widjaja I, Kim J, Sakai P (1997) IEEE 802.11 wireless local area networks. IEEE Communications Magazine 35:116–126
7. Daafouz J, Riedinger P, Iung C (2002) Stability analysis and control synthesis for switched systems: a switched lyapunov function approach. IEEE Trans Automatic Control 47:1883–1887
8. Duarte-Melo EJ, Liu M (2003) Data gathering wireless sensor networks: organization and capacity. Computer Networks 43:393–422

9. Ge S, Sun Z, Lee T (2001) Reachability and controllability of switched linear discrete-time systems. IEEE Trans Automatic Control 46:1437–1441
10. Goodwin GC, Haimovich H, Quevendo DE, Welsh JS (2004) A moving horizon approach to networked controlled systems design. IEEE Trans Automatic Control 49:1427–1445
11. Halevi Y, Ray A (1988) Integrated communication and control systems: part I-analysis. J Dynamic Systems, Measurement and Control 110:367–373
12. IEEE Std 802.11 (2001) IEEE standard for local and metropolitan area networks: wireless lan medium access control (MAC) and physical layer (PHY) specification
13. Information science institute network simulator: http://www.isi.edu/nsnam/ns/
14. Jonhson DB, Maltz DA, Broch J (2001) Ad-hoc networking. Addison-Wesley, Reading, MA
15. Langendoen K, Reijers N (2003) Distributed localization in wireless sensor networks: a quantitive comparison. Computer Networks 43:499–518
16. Lian F, Moyne J, Tilbury D (2002) Network design consideration for distributed control systems. IEEE Trans Control Systems Technology 10:297–307
17. Liou L, Ray A (1990) Integrated communication and control systems: part III-nonidentical sensor and controller sampling. J Dynamic Systems, Measurement and Control 112:357–364
18. Moller N, Johansson KH, Jalmarsson H (2004) Making retransmission delays in wireless links friendlier to TCP. In: Proc 43rd IEEE Conference on Decision and Control, Paradise Island, Bahamas, 5134–5139
19. Nair G, Evans R, Mareels I, Moran W (2003) Feedback data rates for nonlinear systems. In: Proc European Control Conference, Cambridge, UK, 731–736
20. Natkaniec M, Pach AR (2000) An analysis of the backoff mechanism used in IEEE 802.11 networks. In: Proc IEEE Symposium on Computer and Communications, Antibes-Juan les Pins, France, 444–449
21. Nikolakopoulos G, Panousopoulou A, Tzes A, Lygeros J (2007) Multihopping induced gain scheduling for wireless networked controlled systems. Asian J Control, to appear; a shorter version appears in Proc 44th IEEE Conf Decision and Control, Seville, Spain, 470–475
22. Overstreet J, Tzes A (1999) An internet-based real time control engineering laboratory. IEEE Control Systems Magazine 99:19–34
23. Ray A, Halevi Y (1988) Integrated communication and control systems: part II-design considerations. J Dynamic Systems, Measurement and Control 110:374–381
24. Recht B, Andrea R (2004) Distributed control of systems over discrete groups. IEEE Trans Automatic Control 49:1446–1542
25. Sweet C, Sidhu D (1999) Perfomance Analysis of the IEEE 802.11 wireless standard. In: Proc IEEE Global Telecommunications Conference, Rio de Janeiro, Brazil
26. Tipsuwan Y, Chow M (2003) Control methodologies in networked control systems. Control Engineering Practice 11:1099–1111
27. Tzes A, Nikolakopoulos G, Koutroulis I (2005) Development and experimental verification of a mobile client-centric networked controlled system. European J Control 11:229–241
28. Walsh GC, Ye H, Bushnell L (2002) Stability analysis of networked control systems. IEEE Trans Control Systems Technology 10:438–446

29. Weinmann A (1991) Uncertain models and robust control. Springer Verlag New York
30. Wong WS, Brockett R (1999) Systems with finite communication bandwidth constraints-II: stabilization with limited information feedback. IEEE Trans Automatic Control 44:1049–1053
31. Yoon J, Liu M, Noble B (2003) Random waypoint model considered harmful. In: Proc International Conference on Mobile Computing and Networking, San Diego, CA, 1312–1321
32. Zhang J, Knopse C, Tsiotras P (2001) Stability of time-delay systems: equivalence between Lyapunov and scaled small gain conditions. IEEE Trans Automatic Control 46:482–486
33. Zheng J, Lee M (2004) A comprehensive study of IEEE 802.15.4. IEEE Press, New York

7

Networked Control for T–S Fuzzy Systems with Time Delay

Dedong Yang and Huaguang Zhang

Northeastern University, Shenyang 110004, P. R. China
ydd12677@163.com, zhanghuaguang@ise.neu.edu.cn

Abstract. In this chapter, we first develop a guaranteed cost networked control (GCNC) method for Takagi–Sugeno (T–S) fuzzy systems with time delay. State feedback controller is designed via the networked control system (NCS) theory. The stability of the overall fuzzy system using GCNC is established. Some deductions are also extended to uncertain systems. Simulation results show the validity of this control scheme. Second, a robust H_∞ networked control method for T–S fuzzy systems with uncertainty and time delay is presented. Sufficient conditions for robust stability with H_∞ performance are obtained. An example is included to illustrate the effectiveness of the proposed method.

Keywords. Guaranteed cost control, networked control systems, robust H_∞ control, T–S fuzzy systems.

7.1 Introduction

In the real world, nonlinear dynamic systems with time delays exist extensively in industrial control fields. Because of nonlinear and time-delay characteristics, the design of a closed-loop controller is difficult. Aimed at these problems, many intelligent control methods are developed for complex industrial plants. In these methods, fuzzy control is a well-known way to solve the control problems of nonlinear systems. In the last few years, the stability analysis and synthesis problem of fuzzy system as an important issue was studied by many researchers [3, 4, 6, 7, 9, 10, 17, 19, 24, 25]. The T–S fuzzy system proposed in [15] is widely applied to industrial control fields because of its simple structure with local dynamics. In the T–S fuzzy model, local dynamics in different state-space regions is represented by many linear models such that the linear system theory can conveniently be employed to analyze the stability of the overall closed-loop system and to design the feedback controller. The typical design approach is carried out based on a fuzzy model via the so-called parallel distributed compensation (PDC) method [18]. In recent years, some controller design methods based on the linear matrix inequality (LMI) technique are also used for the stability analysis and controller design of the T–S fuzzy system.

Considering the time-delay characteristic of nonlinear systems, a straightforward idea is that the traditional T–S fuzzy model can be extended to a fuzzy model with time delay term. Recently, many results for the T–S fuzzy model with time delay term have been obtained to deal with stability and stabilization problems of nonlinear systems with time delays [3, 4, 6, 7, 9, 10, 17, 19]. The analysis and synthesis problem for continuous and discrete-time nonlinear systems via the PDC approach was considered in [3, 4]. Based on the Lyapunov criterion and the Razumikhin theorem, some sufficient conditions were derived under which parallel-distributed fuzzy control can stabilize the whole uncertain fuzzy time-delay system asymptotically [17]. The guaranteed cost control for a T–S fuzzy system with time delays was also presented in [6] and [9]. In [9], a delay-dependent guaranteed cost control method was introduced and the proposed generalized output feedback controller via the so-called generalized PDC (GPDC) technique was developed within the framework of the dual indexed controller. The time-varying delay cases were also generalized from general time delay cases with state feedback, and an observer-based output feedback controller was obtained in [6]. A delay-dependent robust fuzzy H_∞ controller was designed via state feedback in [7]. Robust H_∞ control via an output feedback controller was designed for a class of uncertain discrete-time fuzzy systems with time delays in [19]. The output feedback robust H_∞ control problem for a class of uncertain fuzzy dynamic systems with time-varying delayed state was presented in [10].

It is well known that a traditional isolated control system may be replaced by a networked control system (NCS) as computer network technology is being developed rapidly. Much attention to stability analysis and controller design of NCS has been paid in [1, 11, 12, 13, 14, 16, 21, 22, 23]. The most popular method is to model the NCS as a system with time-varying delays. So the stability of an NCS is equivalent to the stability of a system with time-varying delays. Moreover the sampling behavior also has an important impact on the design of NCS controller because the states of feedback controller are not continuous as a result of the existence of zeroth-order-hold (ZOH). In [23], a detailed summary was given of a review of existing works, and the relationship between the sampling rate and the network-induced delay was captured using a stability region plot. In [21], a model of NCS was provided considering the network-induced delay and the packet dropout in the transmission. Using Lyapunov theory, the designed parameters of the feedback controller and the maximum allowable value of the network-induced delay can be derived by solving a set of LMIs. Robust controllers for uncertain NCS were also obtained in [22]. How to analyze the stability of nonlinear NCS is a challenging and interesting topic. Some results about the stability of nonlinear NCS were obtained in [13, 14, 16]. Input-to-state stability (ISS) and input-to-output stability (IOS) were analyzed in [13, 14]. The nonlinear NCS can be stabilized under certain assumptions [16]. However, these methods often require some strict assumptions for the system model so practical applications are difficult to achieve.

In the following sections, we propose a novel control scheme which will be called guaranteed cost networked control (GCNC) method for the T–S fuzzy system with time delays in a network situation. Utilizing a fuzzy control method and considering quality of service (QoS) in network systems, the corresponding state feedback control law is obtained. Comparing with [6] and [9], we consider the stabilization problem of the T–S fuzzy system with time delays in the network situation. Some results are obtained for the T–S fuzzy model with multiple time delays and both network-induced delay and packet dropout in network transmission are addressed in our analysis. Further, some sufficient stability conditions of GCNC law are proposed by solving a set of LMIs. We also propose another control scheme called the robust H_∞ networked control method for T–S fuzzy systems with uncertainty and time delay under network conditions. The robust H_∞ performance index of the controlled model after considering network action is satisfied. We consider the robust stability problem of the T–S fuzzy system with uncertainty and time delay under network conditions. Further, some sufficient stability conditions of this control scheme are proposed by solving a set of LMIs.

In this chapter, matrices are assumed to have compatible dimensions. The identity matrix and zero matrix are denoted by I and 0, respectively. The notation $*$ always denotes the symmetric block in a symmetric matrix. The standard notation $>$ ($<$) is used to denote the positive (negative)-definite ordering of matrices. Inequality $X > Y$ shows that the matrix $X - Y$ is positive definite.

7.2 Guaranteed Cost Networked Control for T–S Fuzzy Systems with Time Delay

In general, a nonlinear time-delay system can be represented by the T–S fuzzy system with time delays, which expresses the nonlinear system as a weighted sum of linear systems. The ith rule of such a fuzzy system is of the following format:

$$
\begin{aligned}
&\textbf{Rule } i: \\
&\text{IF } \theta_1(t) \text{ is } F_{i1}, \cdots, \text{ and } \theta_{\bar{n}}(t) \text{ is } F_{i\bar{n}} \\
&\text{THEN} \\
&\dot{x}(t) = A_i x(t) + A_{d_i} x(t-d) + B_i u(t), \\
&y(t) = C_i x(t), \\
&x(t) = \varphi(t), \ -\bar{\tau} \le t < 0,
\end{aligned}
$$

where $i = 1, 2, \cdots, r$ is the index number of the fuzzy rule, $x(t) \in \mathbb{R}^n$ and $y(t) \in \mathbb{R}^s$ denote the state vector and the output vector, respectively, $u(t) \in \mathbb{R}^m$ is the control input, A_i and $A_{d_i} \in \mathbb{R}^{n\times n}$ are known system matrices, $B_i \in \mathbb{R}^{n\times m}$ is the input matrix, $C_i \in \mathbb{R}^{s\times n}$ is the output matrix, d is the constant bounded time delay in the state and is assumed to be $0 < d \le \bar{\tau}$, $\theta_1(t), \theta_2(t), \cdots, \theta_{\bar{n}}(t)$ are premise variables, the initial condition $\varphi(t)$ is a differentiable function or constant vector, F_{ig} is a fuzzy set $(g = 1, 2, \cdots, \bar{n})$. The inferred system is described by

$$\dot{x}(t) = \sum_{i=1}^{r} h_i(\theta(t))[A_i x(t) + A_{d_i} x(t-d) + B_i u(t)],$$

where

$$\mu_i(\theta(t)) = \prod_{g=1}^{\bar{n}} F_{ig}(\theta_g(t)),$$

$$h_i(\theta(t)) = \frac{\mu_i(\theta(t))}{\sum_{i=1}^{r} \mu_i(\theta(t))},$$

and $F_{ig}(\theta_g(t))$ is the grade of membership of $\theta_g(t)$ in the fuzzy set F_{ig}. Notice the following facts:

$$\mu_i(\theta(t)) \ge 0 \text{ and } \sum_{i=1}^{r} \mu_i(\theta(t)) > 0,$$

for all t . Then, we can see that

$$h_i(\theta(t)) \ge 0, \text{ for } i = 1, 2, \cdots, r,$$

and

$$\sum_{i=1}^{r} h_i(\theta(t)) = 1.$$

While considering network action, the state feedback controller is

$$u(t) = \sum_{i=1}^{r} h_i(\theta(t)) K_i x(t_k), \text{ for } t \in [t_k + \tau_k, \ t_{k+1} + \tau_{k+1}),$$

where t_k is the sampling instant, and $x(t_k)$ is the state vector of plant at the instant t_k, which is a piecewise constant function obtained using a ZOH. The inferred fuzzy system is reconstructed using the following form:

$$\dot{x}(t) = \sum_{i=1}^{r}\sum_{j=1}^{r} h_i(\theta(t))h_j(\theta(t))[A_i x(t) + A_{d_i} x(t-d) \\ + B_i K_j x(t_k)], \text{ for } t \in [t_k + \tau_k,\ t_{k+1} + \tau_{k+1}), \tag{7.1}$$

where τ_k denotes the network-induced delay, $k = 0, 1, 2, \cdots$.

Remark 7.1. The packet is transmitted at the instant t_k, which contains the measured value of the plant state vector, $x(t_k)$. Note that $x(t_k)$ remains constant in the interval $t \in [t_k + \tau_k, t_{k+1} + \tau_{k+1})$ until the next update. It is assumed that no controller-to-actuator delay exists, so $u(t)$ can be sent to the plant as control input instantaneously. Obviously,

$$\lim_{N\to\infty} \bigcup_{k=0}^{N} [t_k + \tau_k,\ t_{k+1} + \tau_{k+1}) = [t_0 + \tau_0, \infty),\ t_0 \geq 0. \quad \square$$

In this section, the GCNC via state feedback will be designed according to (7.1). Before giving controller design method, we make the following assumptions.

Assumption 7.1. The sensor is time-driven. The controller and actuator are event-driven. The clocks among them are synchronized. □

Assumption 7.2. The signal transmission is with a single packet. Also the computational delay is negligible. □

Assumption 7.3. The local dynamics is controllable. □

Assumption 7.4. We introduce the notion of a maximum allowable transfer interval $\delta > 0$. The maximum allowable transfer interval is a deadline; if a transmission of packet takes place at time t_k and the control signal will reach the plant at the instant $t_k + \tau_k$, then the next control signal must arrive within the time interval $(t_k,\ t_k + \delta]$. It is explicit that the next control signal will arrive at the instant $t_{k+p} + \tau_{k+p}$ if the packet dropout in network transmission is considered. The following condition is assumed

$$t_{k+p} - t_k + \tau_{k+p} \leq \delta,\ k = 0, 1, 2, \cdots,\ p = 1, 2, \cdots, p_{\max}, \tag{7.2}$$

where p, $p_{\max}$ are positive integers, which denote the sampling index number and the maximum sampling number within δ, and in fact δ has an upper bound in order to guarantee the stability of the closed-loop system. We assume that the upper bound is smaller than $\bar{\tau}$. □

Remark 7.2. We should notice that the network-induced delay is different from the system delays, because it is time varying and unknown. When the transmission time of a packet exceeds a threshold designed by the common network protocols, the packet is to be regarded as a data dropout. For example, $t_{k+2} + \tau_{k+2} < t_{k+1} + \tau_{k+1}$ means that the new data packet may reach the plant before the old one. In fact, we first suppose that δ exists. From (7.2), i.e., $p = 2$, it is required that $t_{k+2} - t_k + \tau_{k+2} \leq \delta$. Thus, the old data packet containing $x(t_{k+1})$ will be discarded. Therefore, when $p > 1$, some packets may be discarded while the whole closed-loop system is still stable under the condition (7.2). From the point of view of the QoS, the network resource is saved by decreasing the network-induced delay or discarding the old packet, based on δ, which can be realized by a suitable network scheduling method. It is explicit that if $p = 1$, (7.2) becomes

$$t_{k+1} - t_k + \tau_{k+1} \leq \delta,\ k = 0, 1, 2, \dots. \tag{7.3}$$

It means that packet dropout is not allowed in the transmission. □

For simplicity, we assume $p = 1$, $t_0 = 0$ and $\tau_0 = 0$ in the following discussion.

Lemma 7.1 ([8]). For any constant symmetric matrix $M \in \mathbb{R}^{n\times n}$, $M = M^T > 0$, scalar $\alpha > 0$, vector function ξ: $[0, \alpha] \to \mathbb{R}^n$, such that the integrations in the following are well defined, we have

$$\alpha \int_0^\alpha \xi^T(\beta) M \xi(\beta) d\beta \geq \left(\int_0^\alpha \xi(\beta) d\beta\right)^T M \left(\int_0^\alpha \xi(\beta) d\beta\right). \tag{7.4}$$

□

Theorem 7.1. If there exist matrices $P_{1,1} = P_{1,1}^T > 0$, $P_{1,2} = P_{1,2}^T > 0$, $T_1 = T_1^T > 0$, matrices K_i and matrices $Y_{1,1}$, $Y_{1,2}$, $Y_{1,3}$ of appropriate dimensions, and for given constant matrices $Y_{1,4}$, $Y_{1,5}$, $Y_{1,6}$ of appropriate dimensions, matrices $X_1 = X_1^T > 0$, $X_2 = X_2^T > 0$, $X_3 = X_3^T > 0$, and a scalar $\delta > 0$, where δ satisfies the condition (7.3), the following LMIs hold

$$\begin{aligned} &\Xi_{ii} < 0,\ 1 \leq i \leq r, \\ &\Theta_{ij} < 0,\ 1 \leq i < j \leq r, \end{aligned} \tag{7.5}$$

then the closed-loop system (7.1) is asymptotically stable with guaranteed cost bound W_1, where

$$\Xi_{ii} = \begin{bmatrix} \Gamma_{ii} & \Upsilon_1 \\ * & \Upsilon_2 \end{bmatrix},\quad \Gamma_{ij} = \begin{bmatrix} \Phi_{1,1} & \Phi_{1,2} & \Phi_{1,3} & \Phi_{1,4} & \Phi_{1,5} \\ * & \Phi_{2,2} & \Phi_{2,3} & \Phi_{2,4} & \Phi_{2,5} \\ * & * & \Phi_{3,3} & \Phi_{3,4} & 0 \\ * & * & * & \Phi_{4,4} & \Phi_{4,5} \\ * & * & * & * & \Phi_{5,5} \end{bmatrix},$$

$$\begin{aligned}
&\Upsilon_1 = \begin{bmatrix} 0\ 0\ 0\ K_i\ 0 \end{bmatrix}^T,\ \Upsilon_2 = -X_3, \\
&\Phi_{1,1} = P_{1,2} + Y_{1,1} + Y_{1,1}^T - Y_{1,4}A_i - A_i^T Y_{1,4}^T + X_1, \\
&\Phi_{1,2} = P_{1,1} + Y_{1,3}^T + Y_{1,4} - A_i^T Y_{1,6}^T, \\
&\Phi_{1,3} = -Y_{1,4}A_{d_i}, \\
&\Phi_{1,4} = -Y_{1,1} + Y_{1,2}^T - Y_{1,4}B_i K_j - A_i^T Y_{1,5}^T,
\end{aligned}$$

$\Phi_{1,5} = \delta Y_{1,1}$,
$\Phi_{2,2} = \delta T_1 + Y_{1,6} + Y_{1,6}^T$,
$\Phi_{2,3} = -Y_{1,6} A_{d_i}$,
$\Phi_{2,4} = -Y_{1,3} + Y_{1,5}^T - Y_{1,6} B_i K_j$,
$\Phi_{2,5} = \delta Y_{1,3}$,
$\Phi_{3,3} = -P_{1,2} + X_2$,
$\Phi_{3,4} = -A_{d_i}^T Y_{1,5}^T$,
$\Phi_{4,4} = -Y_{1,2} - Y_{1,2}^T - Y_{1,5} B_i K_j - K_j^T B_i^T Y_{1,5}^T$,
$\Phi_{4,5} = \delta Y_{1,2}$,
$\Phi_{5,5} = -\delta T_1$,
$\Theta_{ij} = \begin{bmatrix} \Gamma_{ij} + \Gamma_{ji} & \bar{\Upsilon}_1 \\ * & \bar{\Upsilon}_2 \end{bmatrix}$, $\bar{\Upsilon}_1 = \begin{bmatrix} 0\,0\,0\,K_i\,0 \\ 0\,0\,0\,K_j\,0 \end{bmatrix}^T$, $\bar{\Upsilon}_2 = \begin{bmatrix} -X_3 & 0 \\ 0 & -X_3 \end{bmatrix}$,
$X_1 = Q_1, \ X_2 = Q_2, \ X_3 = Q_3^{-1}$.

The guaranteed cost bound is described by:

$$W_1 = \varphi^T(0) P_{1,1} \varphi(0) + \int_{-\bar{\tau}}^{0} \varphi^T(s) P_{1,2} \varphi(s) ds + \int_{-\delta}^{0} \int_{s}^{0} \dot{\varphi}^T(v) T_1 \dot{\varphi}(v) dv ds.$$

□

We require the following definition in our proof for Theorem 7.1.

Definition 7.1. For (7.1), if there exists a fuzzy control law $u(t)$ and a scalar ψ_0 such that the closed-loop system is asymptotically stable and the closed-loop value of the cost function (7.14) satisfies $J \le \psi_0$, then ψ_0 is said to be a guaranteed cost and the control law $u(t)$ is said to be a guaranteed cost control law. □

Proof of Theorem 7.1. Consider a Lyapunov functional given by

$$V(t) = x^T(t) P_{1,1} x(t) + \int_{t-d}^{t} x^T(s) P_{1,2} x(s) ds + \int_{t-\delta}^{t} \int_{s}^{t} \dot{x}^T(v) T_1 \dot{x}(v) dv ds,$$

where $P_{1,1} = P_{1,1}^T > 0$, $P_{1,2} = P_{1,2}^T > 0$ and $T_1 = T_1^T > 0$. Because the following equations hold for matrices $Y_{1,l}$ $(l = 1,2,3,4,5,6)$ of appropriate dimensions,

$$[x^T(t) Y_{1,1} + x^T(t_k) Y_{1,2} + \dot{x}^T(t) Y_{1,3}] \times \left[x(t) - x(t_k) - \int_{t_k}^{t} \dot{x}(s) ds \right] = 0, \tag{7.6}$$

$$[x^T(t) Y_{1,4} + x^T(t_k) Y_{1,5} + \dot{x}^T(t) Y_{1,6}] \times \left[-\sum_{i=1}^{r} \sum_{j=1}^{r} h_i(\theta(t)) h_j(\theta(t)) [A_i x(t) \right.$$
$$\left. + A_{d_i} x(t-d) + B_i K_j x(t_k)] + \dot{x}(t) \right] = 0. \tag{7.7}$$

Combining (7.1), (7.6) and (7.7), the corresponding time derivative of $V(t)$, for $t \in [t_k + \tau_k,\ t_{k+1} + \tau_{k+1})$ is given by

$$
\begin{aligned}
\dot{V}(t) = {} & 2\dot{x}^T(t)P_{1,1}x(t) + x^T(t)P_{1,2}x(t) - x^T(t-d)P_{1,2}x(t-d) + 2(x^T(t)Y_{1,1} \\
& + x^T(t_k)Y_{1,2} + \dot{x}^T(t)Y_{1,3})\left[x(t) - x(t_k) - \int_{t_k}^{t}\dot{x}(s)ds\right] + 2(x^T(t)Y_{1,4} \\
& + x^T(t_k)Y_{1,5} + \dot{x}^T(t)Y_{1,6}) \times \Bigg(- \sum_{i=1}^{r}\sum_{j=1}^{r} h_i(\theta(t))h_j(\theta(t))[A_i x(t) \\
& + A_{d_i}x(t-d) + B_i K_j x(t_k)] + \dot{x}(t)\Bigg) + \delta\dot{x}^T(t)T_1\dot{x}(t) \\
& - \int_{t-\delta}^{t}\dot{x}^T(s)T_1\dot{x}(s)ds,
\end{aligned}
\tag{7.8}
$$

where $Y_{1,l}$ $(l = 1,2,3,4,5,6)$ are matrices of appropriate dimensions.

We can obtain the following fact similar to the proof in [20]. Let $\mathbb{Q}$ be any $\bar{l} \times n$ matrix, we have for any constant $\epsilon > 0$ and any positive-definite symmetric matrix $\mathbb{T}$ that

$$
2\zeta^T\mathbb{Q}\eta \le \epsilon\zeta^T\mathbb{Q}\mathbb{T}^{-1}\mathbb{Q}^T\zeta + \frac{1}{\epsilon}\eta^T\mathbb{T}\eta
\tag{7.9}
$$

for all $\zeta \in \mathbb{R}^{\bar{l}}$, $\eta \in \mathbb{R}^n$ and $\mathbb{T} \in \mathbb{R}^{n\times n}$.

We notice that the first integration term in (7.8) will satisfy the following inequality according to (7.9),

$$
\begin{aligned}
& -2(x^T(t)Y_{1,1} + x^T(t_k)Y_{1,2} + \dot{x}^T(t)Y_{1,3})\int_{t_k}^{t}\dot{x}(s)ds \le \\
& \delta\Lambda^T(t)\bar{Y}T_1^{-1}\bar{Y}^T\Lambda(t) + \frac{1}{\delta}\left[\int_{t_k}^{t}\dot{x}(s)ds\right]^T T_1\left[\int_{t_k}^{t}\dot{x}(s)ds\right],
\end{aligned}
$$

where $\bar{Y} = [Y_{1,1}^T\ \ Y_{1,3}^T\ \ 0\ \ Y_{1,2}^T]^T$ and $\Lambda(t) = [x^T(t)\ \ \dot{x}^T(t)\ \ x^T(t-d)\ \ x^T(t_k)]^T$.

Utilizing (7.4), the following result will be obtained:

$$
\begin{aligned}
& \delta\Lambda^T(t)\bar{Y}T_1^{-1}\bar{Y}^T\Lambda(t) + \frac{1}{\delta}\left[\int_{t_k}^{t}\dot{x}(s)ds\right]^T T_1\left[\int_{t_k}^{t}\dot{x}(s)ds\right] \le \\
& \delta\Lambda^T(t)\bar{Y}T_1^{-1}\bar{Y}^T\Lambda(t) + \frac{t-t_k}{\delta}\int_{t_k}^{t}\dot{x}^T(s)T_1\dot{x}(s)ds.
\end{aligned}
\tag{7.10}
$$

From (7.3), we obtain, for $t \in [t_k + \tau_k,\ t_{k+1} + \tau_{k+1})$,

$$
\int_{t_k}^{t}\dot{x}^T(s)T_1\dot{x}(s)ds \le \int_{t-\delta}^{t}\dot{x}^T(s)T_1\dot{x}(s)ds.
\tag{7.11}
$$

According to (7.3) and (7.11), (7.10) can be expressed as

$$\delta\Lambda^T(t)\bar{Y}T_1^{-1}\bar{Y}^T\Lambda(t)+\frac{t-t_k}{\delta}\int_{t_k}^{t}\dot{x}^T(s)T_1\dot{x}(s)ds\leq$$

$$\delta\Lambda^T(t)\bar{Y}T_1^{-1}\bar{Y}^T\Lambda(t)+\int_{t-\delta}^{t}\dot{x}^T(s)T_1\dot{x}(s)ds. \tag{7.12}$$

Now given positive-definite symmetric matrices Q_1, Q_2 and Q_3, we consider the guaranteed cost function related to global dynamic fuzzy system in (7.1):

$$\begin{aligned}J_k=&\int_{t_k+\tau_k}^{t_{k+1}+\tau_{k+1}}[x^T(t)Q_1x(t)+x^T(t-d)Q_2x(t-d)\\&+u^T(t)Q_3u(t)]dt,\quad t\in[t_k+\tau_k,\ t_{k+1}+\tau_{k+1}),\end{aligned} \tag{7.13}$$

and the sum of guaranteed cost function is described by

$$J=\lim_{N\to\infty}\sum_{k=0}^{N}J_k. \tag{7.14}$$

After the second integration term in (7.8) is counteracted through (7.12), and utilizing the inequality $K_i^TQ_3K_j+K_j^TQ_3K_i\leq K_i^TQ_3K_i+K_j^TQ_3K_j$, (7.8) can be described as follows:

$$\begin{aligned}\dot{V}(t)\leq&\sum_{i=1}^{r}\sum_{j=1}^{r}h_i(\theta(t))h_j(\theta(t))[\Lambda^T(t)\Psi_{ij}\Lambda(t)]\\&-\sum_{i=1}^{r}\sum_{j=1}^{r}h_i(\theta(t))h_j(\theta(t))[x^T(t)Q_1x(t)\\&+x^T(t-d)Q_2x(t-d)+x^T(t_k)K_i^TQ_3K_jx(t_k)]\\\leq&\sum_{i=1}^{r}h_i(\theta(t))h_i(\theta(t))[\Lambda^T(t)\Psi_{ii}\Lambda(t)]\\&+\sum_{i=1}^{r-1}\sum_{j>i}^{r}h_i(\theta(t))h_j(\theta(t))[\Lambda^T(t)\Omega_{ij}\Lambda(t)]\\&-\sum_{i=1}^{r}\sum_{j=1}^{r}h_i(\theta(t))h_j(\theta(t))[x^T(t)Q_1x(t)\\&+x^T(t-d)Q_2x(t-d)+x^T(t_k)K_i^TQ_3K_jx(t_k)]\end{aligned} \tag{7.15}$$

where

$$\Psi_{ij}=\begin{bmatrix}\Pi_{1,1}&\Pi_{1,2}&\Pi_{1,3}&\Pi_{1,4}\\ *&\Pi_{2,2}&\Pi_{2,3}&\Pi_{2,4}\\ *&*&\Pi_{3,3}&\Pi_{3,4}\\ *&*&*&\Pi_{4,4}\end{bmatrix}+\delta\bar{Y}T_1^{-1}\bar{Y}^T,$$

$$\Omega_{ij} = \begin{bmatrix} \bar{\Pi}_{1,1} & \bar{\Pi}_{1,2} & \bar{\Pi}_{1,3} & \bar{\Pi}_{1,4} \\ * & \bar{\Pi}_{2,2} & \bar{\Pi}_{2,3} & \bar{\Pi}_{2,4} \\ * & * & \bar{\Pi}_{3,3} & \bar{\Pi}_{3,4} \\ * & * & * & \bar{\Pi}_{4,4} \end{bmatrix} + 2\delta \bar{Y} T_1^{-1} \bar{Y}^T,$$

$$\begin{aligned}
\Pi_{1,1} &= P_{1,2} + Y_{1,1} + Y_{1,1}^T - Y_{1,4}A_i - A_i^T Y_{1,4}^T + Q_1, \\
\Pi_{1,2} &= P_{1,1} + Y_{1,3}^T + Y_{1,4} - A_i^T Y_{1,6}^T, \\
\Pi_{1,3} &= -Y_{1,4}A_{d_i}, \\
\Pi_{1,4} &= -Y_{1,1} + Y_{1,2}^T - Y_{1,4}B_iK_j - A_i^T Y_{1,5}^T, \\
\Pi_{2,2} &= \delta T_1 + Y_{1,6} + Y_{1,6}^T, \\
\Pi_{2,3} &= -Y_{1,6}A_{d_i}, \\
\Pi_{2,4} &= -Y_{1,3} + Y_{1,5}^T - Y_{1,6}B_iK_j, \\
\Pi_{3,3} &= -P_{1,2} + Q_2, \\
\Pi_{3,4} &= -A_{d_i}^T Y_{1,5}^T, \\
\Pi_{4,4} &= -Y_{1,2} - Y_{1,2}^T - Y_{1,5}B_iK_j - K_j^T B_i^T Y_{1,5}^T + K_i^T Q_3 K_j, \\
\bar{\Pi}_{1,1} &= 2P_{1,2} + 2Y_{1,1} + 2Y_{1,1}^T - Y_{1,4}A_i - Y_{1,4}A_j - A_i^T Y_{1,4}^T - A_j^T Y_{1,4}^T + 2Q_1, \\
\bar{\Pi}_{1,2} &= 2P_{1,1} + 2Y_{1,3}^T + 2Y_{1,4} - A_i^T Y_{1,6}^T - A_j^T Y_{1,6}^T, \\
\bar{\Pi}_{1,3} &= -Y_{1,4}A_{d_i} - Y_{1,4}A_{d_j}, \\
\bar{\Pi}_{1,4} &= -2Y_{1,1} + 2Y_{1,2}^T - Y_{1,4}B_iK_j - Y_{1,4}B_jK_i - A_i^T Y_{1,5}^T - A_j^T Y_{1,5}^T, \\
\bar{\Pi}_{2,2} &= 2\delta T_1 + 2Y_{1,6} + 2Y_{1,6}^T, \\
\bar{\Pi}_{2,3} &= -Y_{1,6}A_{d_i} - Y_{1,6}A_{d_j}, \\
\bar{\Pi}_{2,4} &= -2Y_{1,3} + 2Y_{1,5}^T - Y_{1,6}B_iK_j - Y_{1,6}B_jK_i, \\
\bar{\Pi}_{3,3} &= -2P_{1,2} + 2Q_2, \\
\bar{\Pi}_{3,4} &= -A_{d_i}^T Y_{1,5}^T - A_{d_j}^T Y_{1,5}^T, \\
\bar{\Pi}_{4,4} &= -2Y_{1,2} - 2Y_{1,2}^T - Y_{1,5}B_iK_j - Y_{1,5}B_jK_i - K_j^T B_i^T Y_{1,5}^T - K_i^T B_j^T Y_{1,5}^T \\
&\quad + K_i^T Q_3 K_i + K_j^T Q_3 K_j.
\end{aligned}$$

From (7.13)–(7.15), we can obtain that if $\Psi_{ii} < 0$ and $\Omega_{ij} < 0$ hold for any $1 \le i < j \le r$, then $\dot{V}(t) < 0$ for any nonzero $\Lambda(t)$ and $V_{k+1} - V_k \le -J_k$. As $t \to \infty$, $V(\infty) - V(0) \le -J$ can be derived. Moreover $J \le V(0)$ can be inferred as the closed-loop system to be asymptotically stable. It is easy to see that the guaranteed cost function satisfies

$$J \le V(0) \le W_1 = \varphi^T(0)P_{1,1}\varphi(0) + \int_{-\bar{\tau}}^{0} \varphi^T(s)P_{1,2}\varphi(s)ds + \int_{-\delta}^{0}\int_{s}^{0} \dot{\varphi}^T(v)T_1\dot{\varphi}(v)dvds.$$

Notice that there exist nonlinear terms $\delta \bar{Y} T_1^{-1} \bar{Y}^T$ and $K_i^T Q_3 K_i$ in Ψ_{ii}. And there also exist nonlinear terms $2\delta \bar{Y} T_1^{-1} \bar{Y}^T$, $K_i^T Q_3 K_i$ and $K_j^T Q_3 K_j$ in Ω_{ij}. Utilizing the Schur complement [2], the conditions in Theorem 7.1 can be obtained and the proof is completed. □

Multiple time delays often exist in practical plants. Now, we consider the following form of T–S fuzzy model with multiple time delays:

$$\dot{x}(t) = \sum_{i=1}^{r}\sum_{j=1}^{r} h_i(\theta(t))h_j(\theta(t))\Big[A_i x(t) + \sum_{\tilde{m}=1}^{q} A_{di\tilde{m}} x(t-d_{\tilde{m}}) + B_i K_j x(t_k)\Big], \text{ for } t \in [t_k+\tau_k,\ t_{k+1}+\tau_{k+1}), \tag{7.16}$$

where $d_1, d_2, \ldots, d_q$ $(0 < d_{\tilde{m}} \le \bar{\tau}, \tilde{m} = 1, \ldots, q)$ are assumed to be known bounded time delays of the state, and q denotes the number of terms with time delays.

Further, we establish the following stability results for T–S fuzzy systems with multiple time delays.

Theorem 7.2. If there exist matrices $P_{2,1} = P_{2,1}^T > 0$, $P_{2,2} = P_{2,2}^T > 0$, $T_2 = T_2^T > 0$, matrices K_i and matrices $Y_{2,1}$, $Y_{2,2}$, $Y_{2,3}$ of appropriate dimensions such that for given constant matrices $Y_{2,4}$, $Y_{2,5}$, $Y_{2,6}$ of appropriate dimensions, matrices $X_1' = X_1'^T > 0$, $X_2' = X_2'^T > 0$, $X_3' = X_3'^T > 0$, and a scalar $\delta > 0$, where δ satisfies the condition (7.3), the following LMIs hold,

$$\begin{aligned} &\Xi_{ii}^M < 0, 1 \le i \le r, \\ &\Theta_{ij}^M < 0, 1 \le i < j \le r, \end{aligned} \tag{7.17}$$

then the closed-loop system (7.16) is asymptotically stable with guaranteed cost bound W_1', where

$$\Xi_{ii}^M = \begin{bmatrix} \Gamma_{ii}^M & \Upsilon_1^M \\ * & \Upsilon_2^M \end{bmatrix},$$

$$\Gamma_{ij}^M = \begin{bmatrix} \Phi_{1,1}^M & \Phi_{1,2}^M & \Phi_{1,3}^M & \cdots & \Phi_{1,\tilde{m}}^M & \cdots & \Phi_{1,q+2}^M & \Phi_{1,q+3}^M & \Phi_{1,q+4}^M \\ * & \Phi_{2,2}^M & \Phi_{2,3}^M & \cdots & \Phi_{2,\tilde{m}}^M & \cdots & \Phi_{2,q+2}^M & \Phi_{2,q+3}^M & \Phi_{2,q+4}^M \\ * & * & \Phi_{3,3}^M & \cdots & 0 & \cdots & 0 & \Phi_{3,q+3}^M & 0 \\ * & * & * & \cdots & \cdots & \cdots & \cdots & \cdots & \cdots \\ * & * & * & * & \Phi_{\tilde{m},\tilde{m}}^M & \cdots & 0 & \Phi_{\tilde{m},q+3}^M & 0 \\ * & * & * & * & * & \cdots & \cdots & \cdots & \cdots \\ * & * & * & * & * & * & \Phi_{q+2,q+2}^M & \Phi_{q+2,q+3}^M & 0 \\ * & * & * & * & * & * & * & \Phi_{q+3,q+3}^M & \Phi_{q+3,q+4}^M \\ * & * & * & * & * & * & * & * & \Phi_{q+4,q+4}^M \end{bmatrix},$$

$\Upsilon_1^M = \begin{bmatrix} 0\ 0\ 0 \cdots 0 \cdots 0\ K_i\ 0 \end{bmatrix}^T$, $\Upsilon_2^M = -X_3'$,

$\Phi_{1,1}^M = qP_{2,2} + Y_{2,1} + Y_{2,1}^T - Y_{2,4}A_i - A_i^T Y_{2,4}^T + X_1'$,

$\Phi_{1,2}^M = P_{2,1} + Y_{2,3}^T + Y_{2,4} - A_i^T Y_{2,6}^T$,

$\Phi_{1,3}^M = -Y_{2,4}A_{di1}$,

$\vdots$

$\Phi_{1,\tilde{m}}^M = -Y_{2,4}A_{di(\tilde{m}-2)}$ $(3 < \tilde{m} < q+2)$,

$\vdots$

$\Phi_{1,q+2}^M = -Y_{2,4}A_{diq}$,

$\Phi_{1,q+3}^M = -Y_{2,1} + Y_{2,2}^T - Y_{2,4}B_i K_j - A_i^T Y_{2,5}^T$,

$$\begin{aligned}
&\Phi^M_{1,q+4} = \delta Y_{2,1},\\
&\Phi^M_{2,2} = \delta T_2 + Y_{2,6} + Y^T_{2,6},\\
&\Phi^M_{2,3} = -Y_{2,6}A_{d_{i1}},\\
&\quad\vdots\\
&\Phi^M_{2,\bar{m}} = -Y_{2,6}A_{d_{i(\bar{m}-2)}} \ (3 < \bar{m} < q+2),\\
&\quad\vdots\\
&\Phi^M_{2,q+2} = -Y_{2,6}A_{d_{iq}},\\
&\Phi^M_{2,q+3} = -Y_{2,3} + Y^T_{2,5} - Y_{2,6}B_iK_j,\\
&\Phi^M_{2,q+4} = \delta Y_{2,3},\\
&\Phi^M_{3,3} = \Phi^M_{4,4} = \cdots = \Phi^M_{q+2,q+2} = -P_{2,2} + X'_2,\\
&\Phi^M_{3,q+3} = -A^T_{d_{i1}} Y^T_{2,5},\\
&\quad\vdots\\
&\Phi^M_{\bar{m},q+3} = -A^T_{d_{i(\bar{m}-2)}} Y^T_{2,5} \ (3 < \bar{m} < q+2),\\
&\quad\vdots\\
&\Phi^M_{q+2,q+3} = -A^T_{d_{iq}} Y^T_{2,5},\\
&\Phi^M_{q+3,q+3} = -Y_{2,2} - Y^T_{2,2} - Y_{2,5}B_iK_j - K^T_j B^T_i Y^T_{2,5},\\
&\Phi^M_{q+3,q+4} = \delta Y_{2,2},\\
&\Phi^M_{q+4,q+4} = -\delta T_2,\\
&\Theta^M_{ij} = \begin{bmatrix} \Gamma^M_{ij} + \Gamma^M_{ji} & \bar{\Upsilon}^M_1 \\ * & \bar{\Upsilon}^M_2 \end{bmatrix},\\
&\bar{\Upsilon}^M_1 = \begin{bmatrix} 0\,0\,0 \cdots 0 \cdots 0\ K_i\ 0 \\ 0\,0\,0 \cdots 0 \cdots 0\ K_j\ 0 \end{bmatrix}^T,\ \bar{\Upsilon}^M_2 = \begin{bmatrix} -X'_3 & 0 \\ 0 & -X'_3 \end{bmatrix},\\
&X'_1 = Q'_1,\ X'_2 = Q'_2,\ X'_3 = Q'^{-1}_3.
\end{aligned}$$

The guaranteed cost bound is described by:

$$W'_1 = \varphi^T(0)P_{2,1}\varphi(0) + q\int_{-\bar{\tau}}^{0} \varphi^T(s)P_{2,2}\varphi(s)ds + \int_{-\delta}^{0}\int_{s}^{0} \dot{\varphi}^T(v)T_2\dot{\varphi}(v)dvds.$$

Proof. Consider a Lyapunov functional given by

$$V(t) = x^T(t)P_{2,1}x(t) + \sum_{\bar{m}=1}^{q}\int_{t-d_{\bar{m}}}^{t} x^T(s)P_{2,2}x(s)ds + \int_{t-\delta}^{t}\int_{s}^{t} \dot{x}^T(v)T_2\dot{x}(v)dvds,$$

where $P_{2,1} = P^T_{2,1} > 0$, $P_{2,2} = P^T_{2,2} > 0$ and $T_2 = T^T_2 > 0$.

The guaranteed cost function related to the global dynamic fuzzy system in (7.16) is described by the following form:

$$J'_k = \int_{t_k+\tau_k}^{t_{k+1}+\tau_{k+1}} \left[x^T(t)Q'_1x(t) + \sum_{\tilde{m}=1}^{q} x^T(t-d_{\tilde{m}})Q'_2x(t-d_{\tilde{m}}) + u^T(t)Q'_3u(t) \right] dt,$$

$$t \in [t_k + \tau_k,\ t_{k+1} + \tau_{k+1}).$$

And the sum of guaranteed cost functions is described by

$$J' = \lim_{N\to\infty} \sum_{k=0}^{N} J'_k. \tag{7.18}$$

Following similar lines to those in the proof of Theorem 7.1, the results in Theorem 7.2 can be obtained. □

In addition, some sufficient conditions can be derived with applications to uncertain systems. After considering the uncertain property, the fuzzy system model (7.1) takes the following form:

$$\begin{aligned}\dot{x}(t) =& \sum_{i=1}^{r}\sum_{j=1}^{r} h_i(\theta(t))h_j(\theta(t))[(A_i + \Delta A_i)x(t) + (A_{d_i} + \Delta A_{d_i})x(t-d) \\ &+ (B_i + \Delta B_i)K_j x(t_k)],\end{aligned} \tag{7.19}$$

where ΔA_i, ΔA_{d_i} and ΔB_i denote the uncertainties in the system.

Assumption 7.5. We assume that the admissible uncertainties satisfy $\Delta A_i = M_{1i}F(t)N_{1i}$, $\Delta A_{d_i} = M_{2i}F(t)N_{2i}$, $\Delta B_i = M_{3i}F(t)N_{3i}$, where $M_{k_1 i}$ ($k_1 = 1,2,3$), $N_{k_2 i}$ ($k_2 = 1,2,3$) and $F^T(t)$ are real matrices with appropriate dimensions, and satisfy $F^T(t)F(t) \le I$. □

Lemma 7.2 ([17]). For any two matrices X and Y, we have

$$X^TY + Y^TX \le \varepsilon X^TX + \varepsilon^{-1}Y^TY, \tag{7.20}$$

where $X \in \mathbb{R}^{\tilde{l}\times n}$ and $Y \in \mathbb{R}^{\tilde{l}\times n}$, and ε is any positive constant. □

Theorem 7.3. If there exist matrices $P_{3,1} = P_{3,1}^T > 0$, $P_{3,2} = P_{3,2}^T > 0$, $T_3 = T_3^T > 0$, matrices K_i, and matrices $Y_{3,1}$, $Y_{3,2}$, $Y_{3,3}$ of appropriate dimensions such that the following LMIs

$$\begin{aligned}&\Xi'_{ii} < 0,\ 1 \le i \le r,\\ &\Theta'_{ij} < 0,\ 1 \le i < j \le r,\end{aligned} \tag{7.21}$$

hold for given constant matrices $Y_{3,4}$, $Y_{3,5}$, $Y_{3,6}$ of appropriate dimensions, matrices $X_1 = X_1^T > 0$, $X_2 = X_2^T > 0$, $X_3 = X_3^T > 0$, scalars $\varepsilon > 0$ and $\delta > 0$, where δ satisfies the condition (7.3), then the uncertain system (7.19) is asymptotically stable with guaranteed cost bound W_2, where

$$\Xi'_{ii} = \begin{bmatrix} \Gamma'_{ii} & \Upsilon'_1 \\ * & \Upsilon'_2 \end{bmatrix},\quad \Gamma'_{ij} = \begin{bmatrix} \Phi'_{1,1} & \Phi'_{1,2} & \Phi'_{1,3} & \Phi'_{1,4} & \Phi'_{1,5} \\ * & \Phi'_{2,2} & \Phi'_{2,3} & \Phi'_{2,4} & \Phi'_{2,5} \\ * & * & \Phi'_{3,3} & \Phi'_{3,4} & 0 \\ * & * & * & \Phi'_{4,4} & \Phi'_{4,5} \\ * & * & * & * & \Phi'_{5,5} \end{bmatrix},$$

$$\Upsilon_1' = \begin{bmatrix} 0\,0\,0 & K_i & 0 \\ 0\,0\,0 & \sqrt{3\varepsilon^{-1}}N_{3i}K_i & 0 \end{bmatrix}^T, \quad \Upsilon_2' = \begin{bmatrix} -X_3 & 0 \\ 0 & -I \end{bmatrix},$$

$$\begin{aligned}
\Phi'_{1,1} &= P_{3,2} + Y_{3,1} + Y_{3,1}^T - Y_{3,4}A_i - A_i^T Y_{3,4}^T + X_1 + \varepsilon Y_{3,4}(M_{1i}M_{1i}^T \\
&\quad + M_{2i}M_{2i}^T + M_{3i}M_{3i}^T)Y_{3,4}^T + (\varepsilon^{-1} + 2\varepsilon)N_{1i}^T N_{1i}, \\
\Phi'_{1,2} &= P_{3,1} + Y_{3,3}^T + Y_{3,4} - A_i^T Y_{3,6}^T, \\
\Phi'_{1,3} &= -Y_{3,4}A_{d_i}, \\
\Phi'_{1,4} &= -Y_{3,1} + Y_{3,2}^T - Y_{3,4}B_iK_j - A_i^T Y_{3,5}^T, \\
\Phi'_{1,5} &= \delta Y_{3,1}, \\
\Phi'_{2,2} &= \delta T_3 + Y_{3,6} + Y_{3,6}^T + \varepsilon^{-1}Y_{3,6}M_{1i}M_{1i}^T Y_{3,6}^T + \varepsilon Y_{3,6}(M_{2i}M_{2i}^T + M_{3i}M_{3i}^T)Y_{3,6}^T, \\
\Phi'_{2,3} &= -Y_{3,6}A_{d_i}, \\
\Phi'_{2,4} &= -Y_{3,3} + Y_{3,5}^T - Y_{3,6}B_iK_j, \\
\Phi'_{2,5} &= \delta Y_{3,3}, \\
\Phi'_{3,3} &= -P_{3,2} + X_2 + (2\varepsilon^{-1} + \varepsilon)N_{2i}^T N_{2i}, \\
\Phi'_{3,4} &= -A_{d_i}^T Y_{3,5}^T, \\
\Phi'_{4,4} &= -Y_{3,2} - Y_{3,2}^T - Y_{3,5}B_iK_j - K_j^T B_i^T Y_{3,5}^T \\
&\quad + \varepsilon^{-1}Y_{3,5}(M_{1i}M_{1i}^T + M_{2i}M_{2i}^T)Y_{3,5}^T + \varepsilon Y_{3,5}M_{3i}M_{3i}^T Y_{3,5}^T, \\
\Phi'_{4,5} &= \delta Y_{3,2}, \\
\Phi'_{5,5} &= -\delta T_3, \\
\Theta'_{ij} &= \begin{bmatrix} \Gamma'_{ij} + \Gamma'_{ji} & \bar{\Upsilon}_1' \\ * & \bar{\Upsilon}_2' \end{bmatrix},
\end{aligned}$$

$$\bar{\Upsilon}_1' = \begin{bmatrix} 0\,0\,0 & K_i & 0 \\ 0\,0\,0 & \sqrt{3\varepsilon^{-1}}N_{3i}K_i & 0 \\ 0\,0\,0 & K_j & 0 \\ 0\,0\,0 & \sqrt{3\varepsilon^{-1}}N_{3j}K_j & 0 \end{bmatrix}^T, \quad \bar{\Upsilon}_2' = \begin{bmatrix} -X_3 & 0 & 0 & 0 \\ 0 & -I & 0 & 0 \\ 0 & 0 & -X_3 & 0 \\ 0 & 0 & 0 & -I \end{bmatrix},$$

$$X_1 = Q_1, \ X_2 = Q_2, \ X_3 = Q_3^{-1}.$$

The guaranteed cost bound is described by:

$$W_2 = \varphi^T(0)P_{3,1}\varphi(0) + \int_{-\bar{\tau}}^{0} \varphi^T(s)P_{3,2}\varphi(s)ds + \int_{-\delta}^{0}\int_{s}^{0} \dot{\varphi}^T(v)T_3\dot{\varphi}(v)dvds.$$

Proof. Similar to the proof of Theorem 7.1, selecting guaranteed cost function (7.14) and the Lyapunov functional

$$V(t) = x^T(t)P_{3,1}x(t) + \int_{t-d}^{t} x^T(s)P_{3,2}x(s)ds + \int_{t-\delta}^{t}\int_{s}^{t} \dot{x}^T(v)T_3\dot{x}(v)dvds,$$

where $P_{3,1} = P_{3,1}^T > 0$, $P_{3,2} = P_{3,2}^T > 0$ and $T_3 = T_3^T > 0$. We can obtain the following results:

$$\Psi'_{ii} < 0, \ 1 \le i \le r,$$
$$\Omega'_{ij} < 0, \ 1 \le i < j \le r,$$

where

$$\Psi'_{ii} = \begin{bmatrix} \Pi'_{1,1} & \Pi'_{1,2} & \Pi'_{1,3} & \Pi'_{1,4} \\ * & \Pi'_{2,2} & \Pi'_{2,3} & \Pi'_{2,4} \\ * & * & \Pi'_{3,3} & \Pi'_{3,4} \\ * & * & * & \Pi'_{4,4} \end{bmatrix} + \delta\bar{Y}T_3^{-1}\bar{Y}^T,$$

$$\begin{aligned}
\Pi'_{1,1} &= P_{3,2} + Y_{3,1} + Y_{3,1}^T - Y_{3,4}(A_i + \Delta A_i) - (A_i + \Delta A_i)^T Y_{3,4}^T + Q_1, \\
\Pi'_{1,2} &= P_{3,1} + Y_{3,3}^T + Y_{3,4} - (A_i + \Delta A_i)^T Y_{3,6}^T, \\
\Pi'_{1,3} &= -Y_{3,4}(A_{d_i} + \Delta A_{d_i}), \\
\Pi'_{1,4} &= -Y_{3,1} + Y_{3,2}^T - Y_{3,4}(B_i + \Delta B_i)K_i - (A_i + \Delta A_i)^T Y_{3,5}^T, \\
\Pi'_{2,2} &= \delta T_3 + Y_{3,6} + Y_{3,6}^T, \\
\Pi'_{2,3} &= -Y_{3,6}(A_{d_i} + \Delta A_{d_i}), \\
\Pi'_{2,4} &= -Y_{3,3} + Y_{3,5}^T - Y_{3,6}(B_i + \Delta B_i)K_i, \\
\Pi'_{3,3} &= -P_{3,2} + Q_2, \\
\Pi'_{3,4} &= -(A_{d_i} + \Delta A_{d_i})^T Y_{3,5}^T, \\
\Pi'_{4,4} &= -Y_{3,2} - Y_{3,2}^T - Y_{3,5}(B_i + \Delta B_i)K_i - K_i^T(B_i + \Delta B_i)^T Y_{3,5}^T + K_i^T Q_3 K_i,
\end{aligned}$$

$$\Omega'_{ij} = \begin{bmatrix} \bar{\Pi}'_{1,1} & \bar{\Pi}'_{1,2} & \bar{\Pi}'_{1,3} & \bar{\Pi}'_{1,4} \\ * & \bar{\Pi}'_{2,2} & \bar{\Pi}'_{2,3} & \bar{\Pi}'_{2,4} \\ * & * & \bar{\Pi}'_{3,3} & \bar{\Pi}'_{3,4} \\ * & * & * & \bar{\Pi}'_{4,4} \end{bmatrix} + 2\delta\bar{Y}T_3^{-1}\bar{Y}^T,$$

$$\begin{aligned}
\bar{\Pi}'_{1,1} &= 2P_{3,2} + 2Y_{3,1} + 2Y_{3,1}^T - Y_{3,4}(A_i + \Delta A_i) - Y_{3,4}(A_j + \Delta A_j) \\
&\quad -(A_i + \Delta A_i)^T Y_{3,4}^T - (A_j + \Delta abcde A_j)^T Y_{3,4}^T + 2Q_1, \\
\bar{\Pi}'_{1,2} &= 2P_{3,1} + 2Y_{3,3}^T + 2Y_{3,4} - (A_i + \Delta A_i)^T Y_{3,6}^T - (A_j + \Delta A_j)^T Y_{3,6}^T, \\
\bar{\Pi}'_{1,3} &= -Y_{3,4}(A_{d_i} + \Delta A_{d_i}) - Y_{3,4}(A_{d_j} + \Delta A_{d_j}), \\
\bar{\Pi}'_{1,4} &= -2Y_{3,1} + 2Y_{3,2}^T - Y_{3,4}(B_i + \Delta B_i)K_j - Y_{3,4}(B_j + \Delta B_j)K_i \\
&\quad -(A_i + \Delta A_i)^T Y_{3,5} - (A_j + \Delta A_j)^T Y_{3,5}^T, \\
\bar{\Pi}'_{2,2} &= 2\delta T_3 + 2Y_{3,6} + 2Y_{3,6}^T, \\
\bar{\Pi}'_{2,3} &= -Y_{3,6}(A_{d_i} + \Delta A_{d_i}) - Y_{3,6}(A_{d_j} + \Delta A_{d_j}), \\
\bar{\Pi}'_{2,4} &= -2Y_{3,3} + 2Y_{3,5}^T - Y_{3,6}(B_i + \Delta B_i)K_j - Y_{3,6}(B_j + \Delta B_j)K_i, \\
\bar{\Pi}'_{3,3} &= -2P_{3,2} + 2Q_2, \\
\bar{\Pi}'_{3,4} &= -(A_{d_i} + \Delta A_{d_i})^T Y_{3,5}^T - (A_{d_j} + \Delta A_{d_j})^T Y_{3,5}^T, \\
\bar{\Pi}'_{4,4} &= -2Y_{3,2} - 2Y_{3,2}^T - Y_{3,5}(B_i + \Delta B_i)K_j - Y_{3,5}(B_j + \Delta B_j)K_i \\
&\quad -K_j^T(B_i + \Delta B_i)^T Y_{3,5}^T - K_i^T(B_j + \Delta B_j)^T Y_{3,5}^T + K_i^T Q_3 K_i \\
&\quad +K_j^T Q_3 K_j.
\end{aligned}$$

Utilizing (7.20), Assumption 7.5 and the Schur complement, the conditions in Theorem 7.3 can be obtained and the proof is completed. □

Next, some results are given for the uncertain T–S fuzzy system with multiple time delays. We consider the following form for the uncertain T–S fuzzy model with multiple time delays,

$$\dot{x}(t) = \sum_{i=1}^{r}\sum_{j=1}^{r} h_i(\theta(t))h_j(\theta(t)) \Bigg[(A_i + \Delta A_i)x(t) + \sum_{\tilde{m}=1}^{q} (A_{di\tilde{m}} + \Delta A_{di\tilde{m}}) \times x(t - d_{\tilde{m}}) + (B_i + \Delta B_i)K_j x(t_k) \Bigg], \text{ for } t \in [t_k + \tau_k,\ t_{k+1} + \tau_{k+1}). \tag{7.22}$$

Theorem 7.4. If there exist matrices $P_{4,1} = P_{4,1}^T > 0$, $P_{4,2} = P_{4,2}^T > 0$, $T_4 = T_4^T > 0$, matrices K_i, and matrices $Y_{4,1}$, $Y_{4,2}$, $Y_{4,3}$ of appropriate dimensions such that the following LMIs

$$\begin{aligned} &\Xi_{ii}^{M'} < 0,\ 1 \le i \le r, \\ &\Theta_{ij}^{M'} < 0,\ 1 \le i < j \le r, \end{aligned} \tag{7.23}$$

hold for given matrices $X_1' = X_1'^T > 0$, $X_2' = X_2'^T > 0$, $X_3' = X_3'^T > 0$, constant matrices $Y_{4,4}$, $Y_{4,5}$, $Y_{4,6}$ of appropriate dimensions, scalars $\varepsilon > 0$ and $\delta > 0$, where δ satisfies the condition (7.3), then the uncertain system (7.22) is asymptotically stable with guaranteed cost bound W_2', where

$$\Xi_{ii}^{M'} = \begin{bmatrix} \Gamma_{ii}^{M'} & \Upsilon_1^{M'} \\ * & \Upsilon_2^{M'} \end{bmatrix},$$

$$\Gamma_{ij}^{M'} = \begin{bmatrix} \Phi_{1,1}^{M'} & \Phi_{1,2}^{M'} & \Phi_{1,3}^{M'} & \cdots & \Phi_{1,\tilde{m}}^{M'} & \cdots & \Phi_{1,q+2}^{M'} & \Phi_{1,q+3}^{M'} & \Phi_{1,q+4}^{M'} \\ * & \Phi_{2,2}^{M'} & \Phi_{2,3}^{M'} & \cdots & \Phi_{2,\tilde{m}}^{M'} & \cdots & \Phi_{2,q+2}^{M'} & \Phi_{2,q+3}^{M'} & \Phi_{2,q+4}^{M'} \\ * & * & \Phi_{3,3}^{M'} & \cdots & 0 & \cdots & 0 & \Phi_{3,q+3}^{M'} & 0 \\ * & * & * & \cdots & \cdots & \cdots & \cdots & \cdots & \cdots \\ * & * & * & * & \Phi_{\tilde{m},\tilde{m}}^{M'} & \cdots & 0 & \Phi_{\tilde{m},q+3}^{M'} & 0 \\ * & * & * & * & * & \cdots & \cdots & \cdots & \cdots \\ * & * & * & * & * & * & \Phi_{q+2,q+2}^{M'} & \Phi_{q+2,q+3}^{M'} & 0 \\ * & * & * & * & * & * & * & \Phi_{q+3,q+3}^{M'} & \Phi_{q+3,q+4}^{M'} \\ * & * & * & * & * & * & * & * & \Phi_{q+4,q+4}^{M'} \end{bmatrix},$$

$$\Upsilon_1^{M'} = \begin{bmatrix} 0 & 0 & \cdots & 0 & \cdots & 0 & K_i & 0 \\ 0 & 0 & \cdots & 0 & \cdots & 0 & \sqrt{3\varepsilon^{-1}}N_{3i}K_i & 0 \end{bmatrix}^T, \quad \Upsilon_2^{M'} = \begin{bmatrix} -X_3' & 0 \\ 0 & -I \end{bmatrix},$$

$$\Phi_{1,1}^{M'} = qP_{4,2} + Y_{4,1} + Y_{4,1}^T - Y_{4,4}A_i - A_i^T Y_{4,4}^T + X_1' + \varepsilon Y_{4,4}(M_{1i}M_{1i}^T + qM_{2i}M_{2i}^T + M_{3i}M_{3i}^T)Y_{4,4}^T + (\varepsilon^{-1} + 2\varepsilon)N_{1i}^T N_{1i},$$

$$\Phi_{1,2}^{M'} = P_{4,1} + Y_{4,3}^T + Y_{4,4} - A_i^T Y_{4,6}^T,$$

$$\Phi_{1,3}^{M'} = -Y_{4,4}A_{di1},$$

$$\vdots$$

$$\Phi_{1,\tilde{m}}^{M'} = -Y_{4,4}A_{di(\tilde{m}-2)} \quad (3 < \tilde{m} < q+2),$$

$$\vdots$$

$$\Phi_{1,q+2}^{M'} = -Y_{4,4}A_{diq},$$

$$\Phi_{1,q+3}^{M'} = -Y_{4,1} + Y_{4,2}^T - Y_{4,4}B_iK_j - A_i^T Y_{4,5}^T,$$

$$\Phi_{1,q+4}^{M'} = \delta Y_{4,1},$$

$$\Phi^{M'}_{2,2} = \delta T_4 + Y_{4,6} + Y^T_{4,6} + \varepsilon^{-1}Y_{4,6}M_{1i}M^T_{1i}Y^T_{4,6} \\ +\varepsilon Y_{4,6}(qM_{2i}M^T_{2i} + M_{3i}M^T_{3i})Y^T_{4,6},$$
$$\Phi^{M'}_{2,3} = -Y_{4,6}A_{d_{i1}},$$
$$\vdots$$
$$\Phi^{M'}_{2,\bar{m}} = -Y_{4,6}A_{d_{i(\bar{m}-2)}} \ (3 < \bar{m} < q+2),$$
$$\vdots$$
$$\Phi^{M'}_{2,q+2} = -Y_{4,6}A_{d_{iq}},$$
$$\Phi^{M'}_{2,q+3} = -Y_{4,3} + Y^T_{4,5} - Y_{4,6}B_iK_j,$$
$$\Phi^{M'}_{2,q+4} = \delta Y_{4,3},$$
$$\Phi^{M'}_{3,3} = \cdots = \Phi^{M'}_{q+2,q+2} = -P_{4,2} + X'_2 + (2\varepsilon^{-1} + \varepsilon)N^T_{2i}N_{2i},$$
$$\Phi^{M'}_{3,q+3} = -A^T_{d_{i1}}Y^T_{4,5},$$
$$\vdots$$
$$\Phi^{M'}_{\bar{m},q+3} = -A^T_{d_{i(\bar{m}-2)}}Y^T_{4,5} \ (3 < \bar{m} < q+2),$$
$$\vdots$$
$$\Phi^{M'}_{q+2,q+3} = -A^T_{d_{iq}}Y^T_{4,5},$$
$$\Phi^{M'}_{q+3,q+3} = -Y_{4,2} - Y^T_{4,2} - Y_{4,5}B_iK_j - K^T_jB^T_iY^T_{4,5} + \varepsilon^{-1}Y_{4,5}(M_{1i}M^T_{1i} \\ +qM_{2i}M^T_{2i})Y^T_{4,5} + \varepsilon Y_{4,5}M_{3i}M^T_{3i}Y^T_{4,5},$$
$$\Phi^{M'}_{q+3,q+4} = \delta Y_{4,2},$$
$$\Phi^{M'}_{q+4,q+4} = -\delta T_4,$$
$$\Theta^{M'}_{ij} = \begin{bmatrix} \Gamma^{M'}_{ij} + \Gamma^{M'}_{ji} & \Upsilon^{M'}_1 \\ * & \Upsilon^{M'}_2 \end{bmatrix},$$
$$\Upsilon^{M'}_1 = \begin{bmatrix} 0\,0\cdots0\cdots0 & K_i & 0 \\ 0\,0\cdots0\cdots0 & \sqrt{3\varepsilon^{-1}}N_{3i}K_i & 0 \\ 0\,0\cdots0\cdots0 & K_j & 0 \\ 0\,0\cdots0\cdots0 & \sqrt{3\varepsilon^{-1}}N_{3j}K_j & 0 \end{bmatrix}^T, \quad \Upsilon^{M'}_2 = \begin{bmatrix} -X'_3 & 0 & 0 & 0 \\ 0 & -I & 0 & 0 \\ 0 & 0 & -X'_3 & 0 \\ 0 & 0 & 0 & -I \end{bmatrix},$$
$$X'_1 = Q'_1, \ X'_2 = Q'_2, \ X'_3 = Q'^{-1}_3.$$

The guaranteed cost bound is described by

$$W'_2 = \varphi^T(0)P_{4,1}\varphi(0) + q\int_{-\bar{\tau}}^{0}\varphi^T(s)P_{4,2}\varphi(s)ds + \int_{-\delta}^{0}\int_{s}^{0}\dot{\varphi}^T(v)T_4\dot{\varphi}(v)dvds.$$

Proof. We assume that the admissible uncertainties in multiple time delays satisfy $\Delta A_{d_{i\tilde{m}}} = M_{2i}F(t)N_{2i}$ ($\tilde{m} = 1, \ldots, q$). Similar to the proof of Theorem 7.2, selecting guaranteed cost function (7.18) and Lyapunov functional

$$V(t) = x^T(t)P_{4,1}x(t) + \sum_{\tilde{m}=1}^{q}\int_{t-d_{\tilde{m}}}^{t}x^T(s)P_{4,2}x(s)ds \\ + \int_{t-\delta}^{t}\int_{s}^{t}\dot{x}^T(v)T_4\dot{x}(v)dvds,$$

where $P_{4,1} = P_{4,1}^T > 0$, $P_{4,2} = P_{4,2}^T > 0$ and $T_4 = T_4^T > 0$, the results of Theorem 7.4 can be obtained. □

If each $\Delta A_{d_{i\tilde{m}}}$ in time delay terms is different, a similar proof can also be obtained.

7.3 Simulation Results

In this section, some examples are presented to show the validity of our control scheme.

Example 7.1. Consider the unstable nonlinear system with the following differential equation [9]:

$$\ddot{s}(t) + f(s(t), \dot{s}(t)) - 0.1s(t) = u(t)$$

where

$$f(s(t), \dot{s}(t)) = 0.5s(t) + 0.75 \sin \frac{\dot{s}(t)}{0.5}.$$

Choose the state variable $x(t) = [s(t) \;\; \dot{s}(t)]^T$. The delay state matrix is

$$A_d = \begin{bmatrix} 0.1 & 0 \\ 0.1 & -0.2 \end{bmatrix}.$$

It can be represented by the following fuzzy model consisting of two rules [9]:

$$\begin{aligned}
\textbf{Rule (1)} \;\; & \text{IF } (x_2(t)/0.5) \text{ is about } 0, \text{ THEN} \\
& \dot{x}(t) = A_1 x(t) + A_{d_1} x(t-d) + B_1 u(t); \\
\textbf{Rule (2)} \;\; & \text{IF } (x_2(t)/0.5) \text{ is about } \pi \text{ or } -\pi, \text{ THEN} \\
& \dot{x}(t) = A_2 x(t) + A_{d_2} x(t-d) + B_2 u(t);
\end{aligned}$$

where

$$A_1 = \begin{bmatrix} 0 & 1 \\ 0.1 & -2 \end{bmatrix}, \; A_2 = \begin{bmatrix} 0 & 1 \\ 0.1 & -0.5 - 1.5\beta \end{bmatrix}, \; B_1 = B_2 = \begin{bmatrix} 0 \\ 1 \end{bmatrix},$$

$A_{d_1} = A_{d_2} = A_d$, $\beta = 0.01/\pi$, $d = 0.5$ and β is used to avoid system matrices being singular. The membership functions of "about 0" and "about π or $-\pi$" are selected as

$$F_1(t) = \left[1 - \frac{1}{1 + \exp\left\{-3\left(\frac{x_2}{0.5} - \frac{\pi}{2}\right)\right\}} \right] \times \frac{1}{1 + \exp\left\{-3\left(\frac{x_2}{0.5} + \frac{\pi}{2}\right)\right\}},$$

$$F_2(t) = 1 - F_1(t),$$

respectively.

Next, we select

$Y_{1,4} = \begin{bmatrix} -19.65 & -8.52 \\ -4.95 & -11.26 \end{bmatrix}$, $Y_{1,5} = \begin{bmatrix} -4.35 & -7.26 \\ -9.65 & -6.18 \end{bmatrix}$, $Y_{1,6} = \begin{bmatrix} -9.15 & -10.54 \\ -4.36 & -8.65 \end{bmatrix}$,
and $X_1 = X_2 = \text{diag}[1,\ 1]$, $X_3 = 1$, $\delta = 0.18$.

The following parameters are obtained by solving the LMIs in (7.5):

$$P_{1,1} = \begin{bmatrix} 40.2199 & 30.0436 \\ 30.0436 & 34.3470 \end{bmatrix}, \quad P_{1,2} = \begin{bmatrix} 31.8167 & 27.3989 \\ 27.3989 & 36.1007 \end{bmatrix},$$

$$T_1 = \begin{bmatrix} 52.0190 & 42.4215 \\ 42.4215 & 51.1644 \end{bmatrix}, \quad Y_{1,1} = \begin{bmatrix} -46.3069 & -19.0438 \\ -28.1585 & -43.0121 \end{bmatrix},$$

$$Y_{1,2} = \begin{bmatrix} 15.1552 & -5.7793 \\ -8.7843 & 21.6638 \end{bmatrix}, \quad Y_{1,3} = \begin{bmatrix} -24.7519 & -25.2701 \\ -24.1274 & -19.8668 \end{bmatrix},$$

$$K_1 = \begin{bmatrix} -2.5485 & -1.4118 \end{bmatrix}, \quad K_2 = \begin{bmatrix} -2.2405 & -2.4131 \end{bmatrix}.$$

The initial value of the system is $\varphi(t) = (1.8\ \ 0.5)^T$ for $t \in [-0.5, 0]$.

Fig. 7.1 presents the simulation results of the present GCNC method and illustrates the validity of Theorem 7.1. The trajectories of $x(t)$ are shown in Fig. 7.1. From the present method, we know that the state feedback controller is effective when the sum of sampling time and network-induced delay are less than δ. In this example, the sampling time is $t_s = 0.05$ and the network-induced delay is $\tau_D = 0.03 \times \text{rand}$, where rand is a random number between 0 and 1. The condition (7.3) is satisfied. When $t \to \infty$, the closed-loop system is stabilized with guaranteed cost performance.

If the above control system contains multiple time delays, we consider the following fuzzy model consisting of two rules:

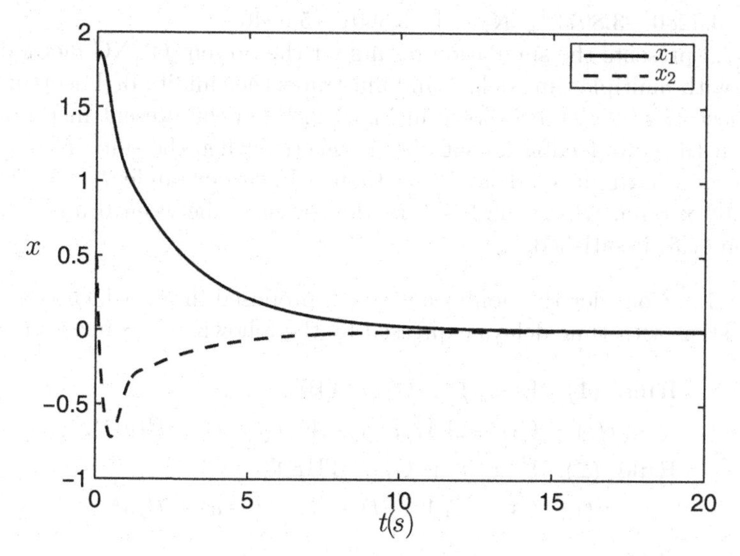

Fig. 7.1. Response of state x in the case ($d = 0.5,\ t_s = 0.05,\ \tau_D \leq 0.03$)

$$\textbf{Rule (1)} \text{ IF } (x_2(t)/0.5) \text{ is about } 0, \text{ THEN}$$

$$\dot{x}(t) = A_1 x(t) + \sum_{\tilde{m}=1}^{q} A_{d_{1\tilde{m}}} x(t - d_{\tilde{m}}) + B_1 u(t);$$

$$\textbf{Rule (2)} \text{ IF } (x_2(t)/0.5) \text{ is about } \pi \text{ or } -\pi, \text{ THEN}$$

$$\dot{x}(t) = A_2 x(t) + \sum_{\tilde{m}=1}^{q} A_{d_{2\tilde{m}}} x(t - d_{\tilde{m}}) + B_2 u(t);$$

where we assume that $q = 2$, $A_{d_{11}} = A_{d_{12}} = A_{d_{21}} = A_{d_{22}} = A_d$, $d_1 = 0.5$ and $d_2 = 0.2$, and other parameters in this system are the same as in the above model. Furthermore, we also assume that the membership functions and the initial value of the system will be the same as in the above model.

Next, we select

$$Y_{2,4} = \begin{bmatrix} -19.56 & -5.36 \\ -4.51 & -18.26 \end{bmatrix}, \; Y_{2,5} = \begin{bmatrix} -10.23 & -5.85 \\ -4.48 & -4.65 \end{bmatrix}, \; Y_{2,6} = \begin{bmatrix} -5.19 & -4.37 \\ -4.11 & -4.89 \end{bmatrix},$$

and $X'_1 = X'_2 = \text{diag}[1,\ 1]$, $X'_3 = 1$, $\delta = 0.130$.

The following parameters are obtained by solving the LMIs in (7.17):

$$P_{2,1} = \begin{bmatrix} 39.3544 & 21.4661 \\ 21.4661 & 37.0110 \end{bmatrix}, \; P_{2,2} = \begin{bmatrix} 16.2859 & 23.5709 \\ 23.5709 & 61.0953 \end{bmatrix},$$

$$T_2 = \begin{bmatrix} 44.1279 & 35.2036 \\ 35.2036 & 42.3502 \end{bmatrix}, \; Y_{2,1} = \begin{bmatrix} -193.0917 & -139.0116 \\ -158.4406 & -211.5975 \end{bmatrix},$$

$$Y_{2,2} = \begin{bmatrix} 177.2587 & 124.6491 \\ 127.3184 & 182.6014 \end{bmatrix}, \; Y_{2,3} = \begin{bmatrix} -11.8147 & -11.1031 \\ -10.9848 & -10.8312 \end{bmatrix},$$

$$K_1 = \begin{bmatrix} -3.4280 & -3.8977 \end{bmatrix}, \; K_2 = \begin{bmatrix} -3.3901 & -5.0946 \end{bmatrix}.$$

Fig. 7.2 presents the simulation results for the present GCNC method for systems with multiple time delays and illustrates the validity of Theorem 7.2. The trajectories of $x(t)$ are shown in Fig. 7.2. From the present method, we know that the state feedback controller is effective when the sum of sampling time and network-induced delay is less than δ. In this example, $t_s = 0.05$ and $\tau_D = 0.03 \times \text{rand}$, where rand is a random number between 0 and 1. The condition (7.3) is satisfied. □

Example 7.2. Consider the nonlinear system proposed in [6], which is a nonlinear system with time delays expressed by the following T–S fuzzy model:

$$\textbf{Rule (1)} \text{ IF } x_2(t) \text{ is } G_{11}, \text{ THEN}$$

$$\dot{x}(t) = (A_1 + \Delta A_1)x(t) + A_{d_1} x(t - d) + B_1 u(t);$$

$$\textbf{Rule (2)} \text{ IF } x_2(t) \text{ is } G_{12}, \text{ THEN}$$

$$\dot{x}(t) = (A_2 + \Delta A_2)x(t) + A_{d_2} x(t - d) + B_2 u(t);$$

where $x(t) = \begin{bmatrix} x_1(t) & x_2(t) \end{bmatrix}^T$, the grade of membership function $G_{11}(x_2(t)) = 1 - x_2^2(t)/2.25$, $G_{12}(x_2(t)) = 1 - G_{11}(x_2(t)) = x_2^2(t)/2.25$, and $d = 0.5$.

We have

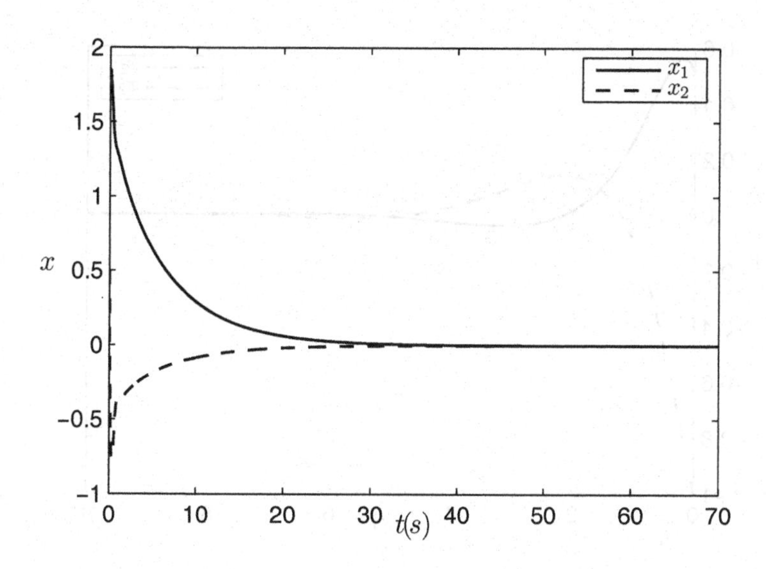

Fig. 7.2. Response of state x in the case ($d_1 = 0.5,\ d_2 = 0.2,\ t_s = 0.05,\ \tau_D \leq 0.03$)

$$A_1 = \begin{bmatrix} -0.1125 & -0.02 \\ 1 & 0 \end{bmatrix},\ A_{d_1} = \begin{bmatrix} -0.0125 & -0.005 \\ 0 & 0 \end{bmatrix},$$
$$A_2 = \begin{bmatrix} -0.1125 & -1.527 \\ 1 & 0 \end{bmatrix},\ A_{d_2} = \begin{bmatrix} -0.0125 & -0.23 \\ 0 & 0 \end{bmatrix},$$
$$\Delta A_1 = \Delta A_2 = \begin{bmatrix} -0.1125 \\ 0 \end{bmatrix} F(t) \begin{bmatrix} 1 & 0 \end{bmatrix},$$
$$B_1 = B_2 = \begin{bmatrix} 1 & 0 \end{bmatrix}^T,\ \Delta A_{d_i} = 0,\ \Delta B_i = 0,\ F(t) = \sin(t).$$

Next, we select

$$Y_{3,4} = \begin{bmatrix} -5.65 & 1.23 \\ -5.19 & -6.57 \end{bmatrix},\ Y_{3,5} = \begin{bmatrix} -2.46 & -1.37 \\ -6.05 & -6.23 \end{bmatrix},\ Y_{3,6} = \begin{bmatrix} -10.26 & 8.46 \\ -9.16 & -7.89 \end{bmatrix},$$

and $\varepsilon = 1$, $X_1 = X_2 = \mathrm{diag}[1,\ 1]$, $X_3 = 1$, $\delta = 0.34$.

Applying Theorem 7.3, the feasible solutions of (7.21) are given as follows:

$$P_{3,1} = \begin{bmatrix} 21.1872 & 9.4223 \\ 9.4223 & 17.5293 \end{bmatrix},\ P_{3,2} = \begin{bmatrix} 2.7158 & 0.1562 \\ 0.1562 & 3.8771 \end{bmatrix},$$
$$T_3 = \begin{bmatrix} 26.6336 & -0.5859 \\ -0.5859 & 15.1782 \end{bmatrix},\ Y_{3,1} = \begin{bmatrix} -78.6574 & 1.7293 \\ 1.7639 & -44.8555 \end{bmatrix},$$
$$Y_{3,2} = \begin{bmatrix} 78.6616 & -1.5901 \\ -1.7608 & 44.8578 \end{bmatrix},\ Y_{3,3} = \begin{bmatrix} 0.0222 & 0.0074 \\ 0.0298 & 0.0712 \end{bmatrix},$$
$$K_1 = \begin{bmatrix} -1.2811 & -0.5676 \end{bmatrix},\ K_2 = \begin{bmatrix} -1.3391 & 0.7855 \end{bmatrix}.$$

The initial value of the system is $\varphi(t) = (0.5, -1)^T$ for $t \in [-0.5, 0]$.

Fig. 7.3 presents the simulation results for the present GCNC method under uncertain conditions and illustrates the validity of Theorem 7.3. The trajectories of $x(t)$ are shown in Fig. 7.3. From the present method, we know

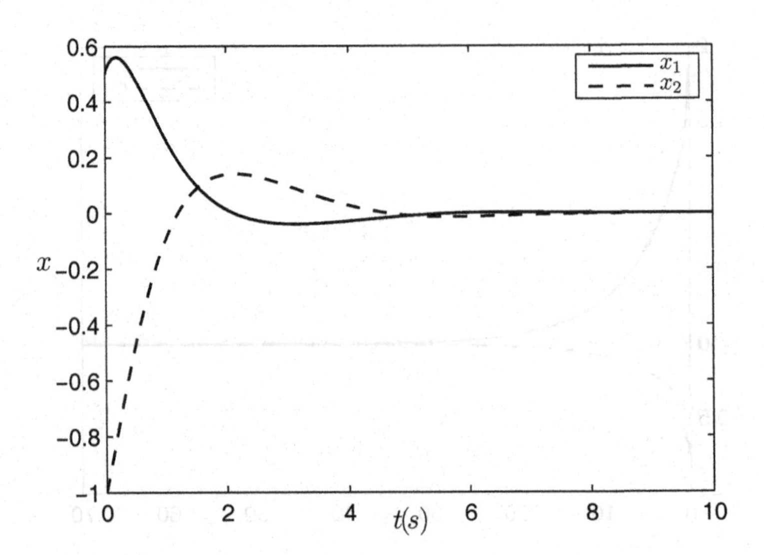

Fig. 7.3. Response of state x in the case $(d = 0.5,\ t_s = 0.05,\ \tau_D \leq 0.03)$

that the state feedback controller is effective when the sum of sampling time and network-induced delay is less than δ. In this example, the sampling time is $t_s = 0.05$ and the network-induced delay is $\tau_D = 0.03 \times \text{rand}$, where rand is a random number between 0 and 1. The condition (7.3) is satisfied. When $t \to \infty$, the closed-loop system is stabilized with guaranteed cost performance.

If in the above control system there exist multiple time delays, we consider the following fuzzy model consisting of two rules:

Rule (1) IF $x_2(t)$ is G_{11}, THEN

$$\dot{x}(t) = (A_1 + \Delta A_1)x(t) + \sum_{\tilde{m}=1}^{q} A_{d_{1\tilde{m}}} x(t - d_{\tilde{m}}) + B_1 u(t);$$

Rule (2) IF $x_2(t)$ is G_{12}, THEN

$$\dot{x}(t) = (A_2 + \Delta A_2)x(t) + \sum_{\tilde{m}=1}^{q} A_{d_{2\tilde{m}}} x(t - d_{\tilde{m}}) + B_2 u(t);$$

where we assume that $q = 2$, $A_{d_{11}} = A_{d_{21}} = A_{d_1}$, $A_{d_{12}} = A_{d_{22}} = A_{d_2}$, $\Delta A_{d_{i\tilde{m}}} = 0$, $d_1 = 0.5$ and $d_2 = 0.3$, and other parameters in this system are the same as in the above model. Furthermore, we also assume that the membership functions and the initial value of the system are the same as in the above model.

Next, we select

$$Y_{4,4} = \begin{bmatrix} -5.21 & 1.12 \\ -4.26 & -6.27 \end{bmatrix},\ Y_{4,5} = \begin{bmatrix} -3.46 & -1.09 \\ -5.38 & -4.89 \end{bmatrix},\ Y_{4,6} = \begin{bmatrix} -9.16 & 8.64 \\ -9.31 & -6.85 \end{bmatrix},$$

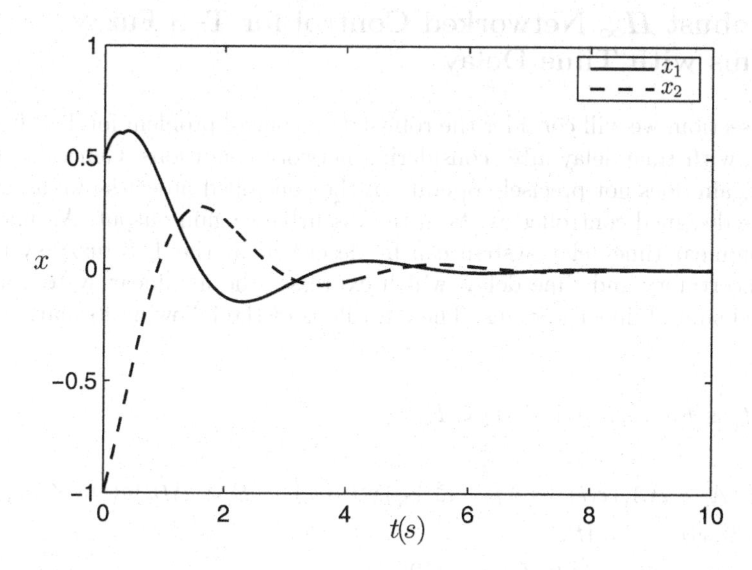

Fig. 7.4. Response of state x in the case ($d_1 = 0.5,\ d_2 = 0.3,\ t_s = 0.05,\ \tau_D \leq 0.03$)

$\varepsilon = 1$, $X_1' = X_2' = \text{diag}[1,\ 1]$, $X_3' = 1$, and $\delta = 0.21$.

Applying Theorem 7.4, the feasible solutions to (7.23) are given as follows:

$P_{4,1} = \begin{bmatrix} 20.3294 & 9.8239 \\ 9.8239 & 20.9635 \end{bmatrix}$, $P_{4,2} = \begin{bmatrix} 2.9968 & 0.8524 \\ 0.8524 & 4.3160 \end{bmatrix}$,

$T_4 = \begin{bmatrix} 29.3726 & 3.6414 \\ 3.6414 & 20.0449 \end{bmatrix}$, $Y_{4,1} = \begin{bmatrix} -139.9368 & -17.7953 \\ -17.9992 & -95.9921 \end{bmatrix}$,

$Y_{4,2} = \begin{bmatrix} 140.0012 & 17.9368 \\ 18.0651 & 96.1143 \end{bmatrix}$, $Y_{4,3} = \begin{bmatrix} 0.0351 & 0.0536 \\ -0.0449 & -0.0497 \end{bmatrix}$,

$K_1 = \begin{bmatrix} -1.3971 & -1.0254 \end{bmatrix}$, and $K_2 = \begin{bmatrix} -1.4113 & 0.4490 \end{bmatrix}$.

Fig. 7.4 presents the simulation results for the present GCNC method under uncertain and multiple time delay conditions and illustrates the validity of Theorem 7.4. The trajectories of $x(t)$ are shown in Fig. 7.4. From the present method, we know that the state feedback controller is effective when the sum of sampling time and network-induced delay is less than δ. In this example, $t_s = 0.05$ and $\tau_D = 0.03 \times \text{rand}$, rand is a random number between 0 and 1. The condition (7.3) is satisfied.

It is clear that the fuzzy system is asymptotically stable, which implies that the feature of guaranteed cost control in the fuzzy system can be guaranteed. Moreover, utilizing our scheme, under various network conditions, satisfactory results can be obtained. □

7.4 Robust H_∞ Networked Control for T–S Fuzzy Systems with Time Delay

In this section, we will consider the robust H_∞ control problem for T–S fuzzy systems with time delay after considering network conditions. Generally, the control gain does not precisely operate in the controlled object so we assume that the designed controller exists in the disturbance gain output. An uncertain nonlinear time-delay system can be described by the T–S fuzzy system with uncertainty and time delay, which expresses the nonlinear system as a weighted sum of linear systems. The ith rule is of the following format:

Rule i:
IF $\theta_1(t)$ is $F_{i1},\ldots,$ and $\theta_{\bar{n}}(t)$ is $F_{i\bar{n}}$
THEN

$$\begin{aligned}
&\dot{x}(t) = (A_i + \Delta A_i)x(t) + (A_{d_i} + \Delta A_{d_i})x(t-\tau) + (B_i + \Delta B_i)u(t) + C_i w(t),\\
&z(t) = D_i x(t) + E_i u(t),\\
&x(t) = \varphi(t),\ \ -\bar{\tau} \le t < 0,\ \text{ for } i = 1,2,\ldots,r,
\end{aligned}$$

where $i = 1,2,\ldots,r$ is the number of fuzzy rules, $x \in \mathbb{R}^n$ and $z \in \mathbb{R}^q$ denote the state vector and measurement output vector, $w(t) \in \mathbb{R}^p$ is disturbance input vector, $u(t) \in \mathbb{R}^m$ is the control input, A_i and $A_{d_i} \in \mathbb{R}^{n\times n}$ are the known system matrix, $B_i \in \mathbb{R}^{n\times m}$ is the input matrix, $C_i \in \mathbb{R}^{n\times p}$ is the disturbance input matrix, $D_i \in \mathbb{R}^{q\times n}$, $E_i \in \mathbb{R}^{q\times m}$, respectively, of the ith subsystem, We assume that the admissible uncertainties satisfy $\Delta A_i = M_{1i}F(t)N_{1i}$, $\Delta A_{d_i} = M_{2i}F(t)N_{2i}$, $\Delta B_i = M_{3i}F(t)N_{3i}$, where $M_{k_1 i}$ $(k_1 = 1,2,3)$, $N_{k_2 i}$ $(k_2 = 1,2,3)$ and $F^T(t)$ are real matrices with appropriate dimensions, and satisfy $F^T(t)F(t) \le I$. τ is the constant bounded time delay in the state and it is assumed to be $0 < \tau \le \bar{\tau}$, $\theta_1(t)$, $\theta_2(t)$, ..., $\theta_{\bar{n}}(t)$ are premise variables, and F_{ig} is a fuzzy set $(g = 1,2,\ldots,\bar{n})$. The inferred system is described by

$$\begin{aligned}
\dot{x}(t) = \sum_{i=1}^{r} h_i(\theta(t))[&(A_i + \Delta A_i)x(t) + (A_{d_i} + \Delta A_{d_i})x(t-\tau)\\
&+ (B_i + \Delta B_i)u(t) + C_i w(t)],
\end{aligned}$$

where

$$\mu_i(\theta(t)) = \prod_{g=1}^{\bar{n}} F_{ig}(\theta_g(t)),\quad h_i(\theta(t)) = \frac{\mu_i(\theta(t))}{\sum_{i=1}^{r}\mu_i(\theta(t))},$$

and $F_{ig}(\theta_g(t))$ is the grade of membership of $\theta_g(t)$ in the fuzzy set F_{ig}. Notice the following facts:

$$\mu_i(\theta(t)) \geq 0 \ \text{ and } \ \sum_{i=1}^{r} \mu_i(\theta(t)) > 0,$$

for all t. Then, we can see that

$$h_i(\theta(t)) \geq 0 \text{ for } i = 1, 2, \ldots, r, \ \text{ and } \ \sum_{i=1}^{r} h_i(\theta(t)) = 1.$$

While considering network action, the state feedback controller is

$$u(t) = \sum_{i=1}^{r} h_i(\theta(t))(K_i + \Delta K_i)x(t_k),$$

where ΔK_i is disturbance control gain and $\Delta K_i = \bar{D}_i F(t) \bar{E}_i$. We assume that $\bar{D}_i$, $\bar{E}_i$ and $F^T(t)$ are real matrices with appropriate dimensions, and satisfy $F^T(t)F(t) \leq I$. The inferred fuzzy system is reconstructed in the following form:

$$\begin{aligned}\dot{x}(t) = &\sum_{i=1}^{r}\sum_{j=1}^{r} h_i(\theta(t))h_j(\theta(t))[(A_i + \Delta A_i)x(t) + (A_{d_i} + \Delta A_{d_i})x(t-\tau)+ \\ &(B_i + \Delta B_i)(K_j + \Delta K_j)x(t_k) + C_i w(t)], \ \text{for } t \in [t_k + \tau_k, \ t_{k+1} + \tau_{k+1}),\end{aligned} \tag{7.24}$$

where t_k is the sampling instant, and $x(t_k)$ is the state vector of plant at the instant t_k, which is a piecewise constant function, by using the ZOH, τ_k denotes the network-induced delay $k = 0, 1, 2, \ldots$ $(\tau_0 = 0)$. $t_s = t_{k+1} - t_k$ is the sampling period.

The robust H_∞ networked control via state feedback will be designed according to (7.24). It is necessary that Assumptions 7.1–7.4 besides the following assumption will still hold in this section and the condition $p = 1$ is assumed.

Assumption 7.6. The overall closed-loop system is under zero initial condition. □

Lemma 7.3 ([5]). For any matrices $\mathbb{D} \in \mathbb{R}^{n \times n_f}$, $\mathbb{E} \in \mathbb{R}^{n_f \times n}$ and $\mathbb{F} \in \mathbb{R}^{n_f \times n_f}$, with $\|\mathbb{F}\| \leq 1$, and scalar $\varepsilon > 0$, the following inequality holds:

$$\mathbb{D}\mathbb{F}\mathbb{E} + \mathbb{E}^T\mathbb{F}^T\mathbb{D}^T \leq \varepsilon^{-1}\mathbb{D}\mathbb{D}^T + \varepsilon\mathbb{E}^T\mathbb{E}. \tag{7.25}$$

□

Theorem 7.5. If there exist matrices $P_1 = P_1^T > 0$, $P_2 = P_2^T > 0$, $T = T^T > 0$, matrices K_j $(j = 1, 2, \cdots, r)$, matrices Y_l $(l = 1, 2, 3)$ of appropriate dimensions and constant matrices Y_l $(l = 4, 5, 6)$ of appropriate dimensions such that the following LMIs hold for given scalars $\delta > 0$ and $\varepsilon > 0$, where δ

satisfies the condition (7.3), then the closed-loop system (7.24) with $w(t) \equiv 0$ is asymptotically stable,

$$\begin{aligned} &\Omega_{ii} < 0 \text{ for any } 1 \le i \le r, \\ &\Omega_{ij} + \Omega_{ji} < 0 \text{ for any } 1 \le i < j \le r, \end{aligned} \tag{7.26}$$

where

$$\Omega_{ij} = \begin{bmatrix} \Pi_{1,1} & \Pi_{1,2} & \Pi_{1,3} & \Pi_{1,4} & \Pi_{1,5} & 0 \\ * & \Pi_{2,2} & \Pi_{2,3} & \Pi_{2,4} & \Pi_{2,5} & 0 \\ * & * & \Pi_{3,3} & \Pi_{3,4} & 0 & 0 \\ * & * & * & \Pi_{4,4} & \Pi_{4,5} & \Pi_{4,6} \\ * & * & * & * & \Pi_{5,5} & 0 \\ * & * & * & * & * & \Pi_{6,6} \end{bmatrix},$$

$$\begin{aligned}
\Pi_{1,1} &= P_2 + Y_1 + Y_1^T - Y_4 A_i - A_i^T Y_4^T + \varepsilon^{-1} Y_4 (M_{1i} M_{1i}^T + M_{2i} M_{2i}^T \\
&\quad + M_{3i} M_{3i}^T) Y_4^T + (\varepsilon + 2\varepsilon^{-1}) N_{1i}^T N_{1i} + \varepsilon Y_4 B_i \bar{D}_j \bar{D}_j^T B_i^T Y_4^T, \\
\Pi_{1,2} &= P_1 + Y_3^T + Y_4 - A_i^T Y_6^T, \\
\Pi_{1,3} &= -Y_4 A_{d_i}, \\
\Pi_{1,4} &= -Y_1 + Y_2^T - Y_4 B_i K_j - A_i^T Y_5^T, \\
\Pi_{1,5} &= \delta Y_1, \\
\Pi_{2,2} &= \delta T + Y_6 + Y_6^T + \varepsilon Y_6 M_{1i} M_{1i}^T Y_6^T + \varepsilon^{-1} Y_6 (M_{2i} M_{2i}^T + M_{3i} M_{3i}^T) Y_6^T \\
&\quad + \varepsilon Y_6 B_i \bar{D}_j \bar{D}_j^T B_i^T Y_6^T, \\
\Pi_{2,3} &= -Y_6 A_{d_i}, \\
\Pi_{2,4} &= -Y_3 + Y_5^T - Y_6 B_i K_j, \\
\Pi_{2,5} &= \delta Y_3, \\
\Pi_{3,3} &= -P_2 + (2\varepsilon + \varepsilon^{-1}) N_{2i}^T N_{2i}, \\
\Pi_{3,4} &= -A_{d_i}^T Y_5^T, \\
\Pi_{4,4} &= -Y_2 - Y_2^T - Y_5 B_i K_j - K_j^T B_i^T Y_5^T + \varepsilon Y_5 (M_{1i} M_{1i}^T + M_{2i} M_{2i}^T) Y_5^T \\
&\quad + \varepsilon^{-1} Y_5 M_{3i} M_{3i}^T Y_5^T + (3\varepsilon^{-1} + 3) \bar{E}_j^T \bar{E}_j + \varepsilon Y_5 B_i \bar{D}_j \bar{D}_j^T B_i^T Y_5^T, \\
\Pi_{4,5} &= \delta Y_2, \\
\Pi_{4,6} &= \sqrt{3\varepsilon} K_j^T N_{3i}^T, \\
\Pi_{5,5} &= -\delta T, \\
\Pi_{6,6} &= -I + \varepsilon N_{3i} \bar{D}_j \bar{D}_j^T N_{3i}^T.
\end{aligned}$$

Proof. Consider a Lyapunov functional

$$V(t) = x^T(t) P_1 x(t) + \int_{t-\tau}^{t} x^T(s) P_2 x(s) ds + \int_{t-\delta}^{t} \int_{s}^{t} \dot{x}^T(v) T \dot{x}(v) dv ds,$$

where $P_1 = P_1^T > 0$, $P_2 = P_2^T > 0$ and $T = T^T > 0$.

We can see that the following equations hold for matrices Y_l $(l = 1, 2, 3, 4, 5, 6)$ of appropriate dimensions where $w(t) = 0$.

$$(x^T(t)Y_1 + x^T(t_k)Y_2 + \dot{x}^T(t)Y_3) \times \left[x(t) - x(t_k) - \int_{t_k}^{t} \dot{x}(s)ds\right] = 0, \quad (7.27)$$

$$(x^T(t)Y_4 + x^T(t_k)Y_5 + \dot{x}^T(t)Y_6) \times \Bigg[- \sum_{i=1}^{r}\sum_{j=1}^{r} h_i(\theta(t))h_j(\theta(t))[\bar{A}_i x(t)$$

$$+\bar{A}_{d_i} x(t-\tau) + \bar{B}_i(K_j + \Delta K_j)x(t_k)] + \dot{x}(t)\Bigg] = 0, \quad (7.28)$$

where $\bar{A}_i = A_i + \Delta A_i$, $\bar{A}_{d_i} = A_{d_i} + \Delta A_{d_i}$ and $\bar{B}_i = B_i + \Delta B_i$.

Combining (7.24), (7.27) and (7.28), the corresponding time derivative of $V(t)$, for $t \in [t_k + \tau_k,\ t_{k+1} + \tau_{k+1})$, is given by

$$\dot{V}(t) = 2\dot{x}^T(t)P_1 x(t) + x^T(t)P_2 x(t) - x^T(t-\tau)P_2 x(t-\tau) + 2(x^T(t)Y_1$$
$$+ x^T(t_k)Y_2 + \dot{x}^T(t)Y_3)(x(t) - x(t_k) - \int_{t_k}^{t} x(s)ds) + 2(x^T(t)Y_4$$
$$+ x^T(t_k)Y_5 + \dot{x}^T(t)Y_6)\Bigg(- \sum_{i=1}^{r}\sum_{j=1}^{r} h_i(\theta(t))h_j(\theta(t))[\bar{A}_i x(t)$$
$$+ \bar{A}_{d_i} x(t-\tau) + \bar{B}_i(K_j + \Delta K_j)x(t_k)] + \dot{x}(t)\Bigg) + \delta\dot{x}^T(t)T\dot{x}(t)$$
$$- \int_{t-\delta}^{t} \dot{x}^T(s)T\dot{x}(s)ds, \quad (7.29)$$

where Y_l $(l = 1, 2, 3, 4, 5, 6)$ are matrices of appropriate dimensions.

From (7.3), (7.4) and (7.9), we obtain, for $t \in [t_k + \tau_k,\ t_{k+1} + \tau_{k+1})$,

$$\int_{t_k}^{t} \dot{x}^T(s)T\dot{x}(s)ds \le \int_{t-\delta}^{t} \dot{x}^T(s)T\dot{x}(s)ds, \quad (7.30)$$

and

$$-2(x^T(t)Y_1 + x^T(t_k)Y_2 + \dot{x}^T(t)Y_3)\int_{t_k}^{t} x(s)ds$$
$$\le \delta\Lambda^T(t)\bar{Y}T^{-1}\bar{Y}^T\Lambda(t) + \frac{1}{\delta}\left[\int_{t_k}^{t} x(s)ds\right]^T T\left[\int_{t_k}^{t} x(s)ds\right]$$
$$\le \delta\Lambda^T(t)\bar{Y}T^{-1}\bar{Y}^T\Lambda(t) + \frac{t - t_k}{\delta}\int_{t_k}^{t} \dot{x}^T(s)T\dot{x}(s)ds$$
$$\le \delta\Lambda^T(t)\bar{Y}T^{-1}\bar{Y}^T\Lambda(t) + \int_{t-\delta}^{t} \dot{x}^T(s)T\dot{x}(s)ds, \quad (7.31)$$

where $\bar{Y}^T = [Y_1^T\ \ Y_3^T\ \ 0\ \ Y_2^T]^T$ and $\Lambda^T = [x^T(t)\ \ \dot{x}^T(t)\ \ x^T(t-\tau)\ \ x^T(t_k)]^T$.

Combining (7.29)–(7.31), we obtain

$$\dot{V}(t) \leq \sum_{i=1}^{r}\sum_{j=1}^{r} h_i h_j \left[x^T(t)\ \dot{x}^T(t)\ x^T(t-\tau)\ x^T(t_k)\right] \Xi_{ij}$$
$$\times \left[x^T(t)\ \dot{x}^T(t)\ x^T(t-\tau)\ x^T(t_k)\right]^T, \quad t \in [t_k + \tau_k,\ t_{k+1} + \tau_{k+1}) \tag{7.32}$$

where

$$\Xi_{ij} = \begin{bmatrix} \Phi_{1,1} & \Phi_{1,2} & \Phi_{1,3} & \Phi_{1,4} \\ * & \Phi_{2,2} & \Phi_{2,3} & \Phi_{2,4} \\ * & * & \Phi_{3,3} & \Phi_{3,4} \\ * & * & * & \Phi_{4,4} \end{bmatrix} + \delta \bar{Y} T^{-1} \bar{Y}^T,$$

$\Phi_{1,1} = P_2 + Y_1 + Y_1^T - Y_4 \bar{A}_i - \bar{A}_i^T Y_4^T,$
$\Phi_{1,2} = P_1 + Y_3^T + Y_4 - \bar{A}_i^T Y_6^T,$
$\Phi_{1,3} = -Y_4 \bar{A}_{d_i},$
$\Phi_{1,4} = -Y_1 + Y_2^T - Y_4 \bar{B}_i (K_j + \Delta K_j) - \bar{A}_i^T Y_5^T,$
$\Phi_{2,2} = \delta T + Y_6 + Y_6^T,$
$\Phi_{2,3} = -Y_6 \bar{A}_{d_i},$
$\Phi_{2,4} = -Y_3 + Y_5^T - Y_6 \bar{B}_i (K_j + \Delta K_j),$
$\Phi_{3,3} = -P_2,$
$\Phi_{3,4} = -\bar{A}_{d_i}^T Y_5^T,$
$\Phi_{4,4} = -Y_2 - Y_2^T - Y_5 \bar{B}_i (K_j + \Delta K_j) - (K_j + \Delta K_j)^T \bar{B}_i^T Y_5^T.$

Utilizing (7.25), the following result will hold,

$$\Xi_{ij} \leq \begin{bmatrix} \Phi'_{1,1} & \Phi'_{1,2} & \Phi'_{1,3} & \Phi'_{1,4} \\ * & \Phi'_{2,2} & \Phi'_{2,3} & \Phi'_{2,4} \\ * & * & \Phi'_{3,3} & \Phi'_{3,4} \\ * & * & * & \Phi'_{4,4} \end{bmatrix} + \delta \bar{Y} T^{-1} \bar{Y}^T \tag{7.33}$$

where

$\Phi'_{1,1} = P_2 + Y_1 + Y_1^T - Y_4 A_i - A_i^T Y_4^T + \varepsilon^{-1} Y_4 (M_{1i} M_{1i}^T + M_{2i} M_{2i}^T$
$\quad + M_{3i} M_{3i}^T) Y_4^T + (\varepsilon + 2\varepsilon^{-1}) N_{1i}^T N_{1i},$
$\Phi'_{1,2} = P_1 + Y_3^T + Y_4 - A_i^T Y_6^T,$
$\Phi'_{1,3} = -Y_4 A_{d_i},$
$\Phi'_{1,4} = -Y_1 + Y_2^T - Y_4 B_i (K_j + \Delta K_j) - A_i^T Y_5^T,$
$\Phi'_{2,2} = \delta T + Y_6 + Y_6^T + \varepsilon Y_6 M_{1i} M_{1i}^T Y_6^T + \varepsilon^{-1} Y_6 (M_{2i} M_{2i}^T + M_{3i} M_{3i}^T) Y_6^T,$
$\Phi'_{2,3} = -Y_6 A_{d_i},$
$\Phi'_{2,4} = -Y_3 + Y_5^T - Y_6 B_i (K_j + \Delta K_j),$
$\Phi'_{3,3} = -P_2 + (2\varepsilon + \varepsilon^{-1}) N_{2i}^T N_{2i},$
$\Phi'_{3,4} = -A_{d_i}^T Y_5^T,$
$\Phi'_{4,4} = -Y_2 - Y_2^T - Y_5 B_i (K_j + \Delta K_j) - (K_j + \Delta K_j)^T B_i^T Y_5^T$
$\quad + \varepsilon Y_5 (M_{1i} M_{1i}^T + M_{2i} M_{2i}^T) Y_5^T + \varepsilon^{-1} Y_5 M_{3i} M_{3i}^T Y_5^T$
$\quad + 3\varepsilon (K_j + \Delta K_j)^T N_{3i}^T N_{3i} (K_j + \Delta K_j).$

From (7.32), we can see $\dot{V} < 0$ for any nonzero $\Lambda(t)$ if the right side in (7.33) is negative. According to the Schur complement [2], the conditions in Theorem 7.5 can be obtained and the proof is completed. □

In the following derivation process, we will consider the robust stability of (7.24) with H_∞ performance index.

In order to attenuate the external disturbance of the fuzzy system (7.24), we introduce the H_∞ performance index

$$\int_{t_0}^{\infty} z^T(t)z(t)dt \le \gamma^2 \int_{t_0}^{\infty} w^T(t)w(t)dt, \tag{7.34}$$

where $\gamma > 0$ denotes the prescribed attenuation level.

Theorem 7.6. If there exist matrices $\bar{P}_1 = \bar{P}_1^T > 0$, $\bar{P}_2 = \bar{P}_2^T > 0$, $\bar{T} = \bar{T}^T > 0$, matrices K_j $(j = 1, 2, \cdots, r)$, matrices $\bar{Y}_l$ $(l = 1, 2, 3)$ of appropriate dimensions and constant matrices $\bar{Y}_l$ $(l = 4, 5, 6)$ of appropriate dimensions such that the following LMIs hold, for given scalars $\delta > 0$, $\varepsilon > 0$ and $\gamma > 0$, where δ satisfies the condition (7.3), then the closed-loop system (7.24) is robustly stable with H_∞ performance index (7.34),

$$\begin{aligned} &\Omega'_{ii} < 0 \quad \text{for any } 1 \le i \le r, \\ &\Omega'_{ij} + \Omega'_{ji} < 0 \quad \text{for any } 1 \le i < j \le r, \end{aligned} \tag{7.35}$$

where

$$\Omega'_{ij} = \begin{bmatrix} \Pi'_{1,1} & \Pi'_{1,2} & \Pi'_{1,3} & \Pi'_{1,4} & \Pi'_{1,5} & \Pi'_{1,6} & 0 & 0 \\ * & \Pi'_{2,2} & \Pi'_{2,3} & \Pi'_{2,4} & \Pi'_{2,5} & \Pi'_{2,6} & 0 & 0 \\ * & * & \Pi'_{3,3} & \Pi'_{3,4} & 0 & 0 & 0 & 0 \\ * & * & * & \Pi'_{4,4} & \Pi'_{4,5} & \Pi'_{4,6} & \Pi'_{4,7} & \Pi'_{4,8} \\ * & * & * & * & \Pi'_{5,5} & 0 & 0 & 0 \\ * & * & * & * & * & \Pi'_{6,6} & 0 & 0 \\ * & * & * & * & * & * & \Pi'_{7,7} & 0 \\ * & * & * & * & * & * & * & \Pi'_{8,8} \end{bmatrix},$$

$$\begin{aligned}
\Pi'_{1,1} = {}& \bar{P}_2 + \bar{Y}_1 + \bar{Y}_1^T - \bar{Y}_4 A_i - A_i^T \bar{Y}_4^T + \varepsilon^{-1}\bar{Y}_4(M_{1i}M_{1i}^T + M_{2i}M_{2i}^T \\
&+ M_{3i}M_{3i}^T)\bar{Y}_4^T + (\varepsilon + 2\varepsilon^{-1})N_{1i}^T N_{1i} + D_i^T D_i + \varepsilon \bar{Y}_4 B_i \bar{D}_j \bar{D}_j^T B_i^T \bar{Y}_4^T \\
&+ \varepsilon D_i^T E_i \bar{D}_j \bar{D}_j^T E_i^T D_i, \\
\Pi'_{1,2} = {}& \bar{P}_1 + \bar{Y}_3^T + \bar{Y}_4 - A_i^T \bar{Y}_6^T, \\
\Pi'_{1,3} = {}& -\bar{Y}_4 A_{d_i}, \\
\Pi'_{1,4} = {}& -\bar{Y}_1 + \bar{Y}_2^T - \bar{Y}_4 B_i K_j - A_i^T \bar{Y}_5^T + D_i^T E_i K_j, \\
\Pi'_{1,5} = {}& \delta \bar{Y}_1, \\
\Pi'_{1,6} = {}& -\bar{Y}_4 C_i, \\
\Pi'_{2,2} = {}& \delta \bar{T} + \bar{Y}_6 + \bar{Y}_6^T + \varepsilon \bar{Y}_6 M_{1i} M_{1i}^T \bar{Y}_6^T + \varepsilon^{-1}\bar{Y}_6(M_{2i}M_{2i}^T + M_{3i}M_{3i}^T)\bar{Y}_6^T
\end{aligned}$$

$$
\begin{aligned}
&+\varepsilon\bar{Y}_6 B_i \bar{D}_j \bar{D}_j^T B_i^T \bar{Y}_6^T,\\
&\Pi'_{2,3} = -\bar{Y}_6 A_{d_i},\\
&\Pi'_{2,4} = -\bar{Y}_3 + \bar{Y}_5^T - \bar{Y}_6 B_i K_j,\\
&\Pi'_{2,5} = \delta\bar{Y}_3,\\
&\Pi'_{2,6} = -\bar{Y}_6 C_i,\\
&\Pi'_{3,3} = -\bar{P}_2 + (2\varepsilon + \varepsilon^{-1}) N_{2i}^T N_{2i},\\
&\Pi'_{3,4} = -A_{d_i}^T \bar{Y}_5^T,\\
&\Pi'_{4,4} = -\bar{Y}_2 - \bar{Y}_2^T - \bar{Y}_5 B_i \bar{K}_j - \bar{K}_j^T B_i^T \bar{Y}_5^T + \varepsilon\bar{Y}_5 (M_{1i}M_{1i}^T + M_{2i}M_{2i}^T)\bar{Y}_5^T\\
&\qquad +\varepsilon^{-1}\bar{Y}_5 M_{3i} M_{3i}^T \bar{Y}_5^T + (5\varepsilon^{-1} + 3)\bar{E}_j^T \bar{E}_j + \varepsilon\bar{Y}_5 B_i \bar{D}_j \bar{D}_j^T B_i^T \bar{Y}_5^T,\\
&\Pi'_{4,5} = \delta\bar{Y}_2,\\
&\Pi'_{4,6} = -\bar{Y}_5 C_i,\\
&\Pi'_{4,7} = K_j^T E_i^T,\\
&\Pi'_{4,8} = \sqrt{3\varepsilon} K_j^T N_{3i}^T,\\
&\Pi'_{5,5} = -\delta\bar{T},\\
&\Pi'_{6,6} = -\gamma^2 I,\\
&\Pi'_{7,7} = -I + \varepsilon E_i \bar{D}_j \bar{D}_j^T E_i^T,\\
&\Pi'_{8,8} = -I + \varepsilon N_{3i} \bar{D}_j \bar{D}_j^T N_{3i}^T.
\end{aligned}
$$

Proof. Consider a Lyapunov functional given by

$$V(t) = x^T(t)\bar{P}_1 x(t) + \int_{t-\tau}^{t} x^T(s)\bar{P}_2 x(s)ds + \int_{t-\delta}^{t}\int_{s}^{t} \dot{x}^T(v)\bar{T}\dot{x}(v)dvds,$$

where $\bar{P}_1 = \bar{P}_1^T > 0$, $\bar{P}_2 = \bar{P}_2^T > 0$ and $\bar{T} = \bar{T}^T > 0$. The corresponding time derivative of $V(t)$ after considering the disturbance term in the system, for $t \in [t_k + \tau_k,\ t_{k+1} + \tau_{k+1})$ is given by

$$
\begin{aligned}
\dot{V}(t) &= 2\dot{x}^T(t)\bar{P}_1 x(t) + x^T(t)\bar{P}_2 x(t) - x^T(t-\tau)\bar{P}_2 x(t-\tau) + 2(x^T(t)\bar{Y}_1\\
&\quad + x^T(t_k)\bar{Y}_2 + \dot{x}^T(t)\bar{Y}_3)(x(t) - x(t_k) - \int_{t_k}^{t} \dot{x}(s)ds) + 2(x^T(t)\bar{Y}_4\\
&\quad + x^T(t_k)\bar{Y}_5 + \dot{x}^T(t)\bar{Y}_6)\Big(- \sum_{i=1}^{r}\sum_{j=1}^{r} h_i(\theta(t))h_j(\theta(t))[\bar{A}_i x(t)\\
&\quad + \bar{A}_{d_i} x(t-\tau) + \bar{B}_i(K_j + \Delta K_j)x(t_k) + C_i w(t)] + \dot{x}(t)\Big)\\
&\quad + \delta\dot{x}^T(t)\bar{T}\dot{x}(t) - \int_{t-\delta}^{t} \dot{x}^T(s)\bar{T}\dot{x}(s)ds,
\end{aligned}
$$

where $\bar{Y}_l$ $(l = 1, 2, 3, 4, 5, 6)$ are matrices of appropriate dimensions.

We let

$$J_k = \int_{t_k+\tau_k}^{t_{k+1}+\tau_{k+1}} \big(z^T(t)z(t) - \gamma^2 w^T(t)w(t)\big)dt$$
$$= \int_{t_k+\tau_k}^{t_{k+1}+\tau_{k+1}} \big(z^T(t)z(t) - \gamma^2 w^T(t)w(t) + \dot{V}(t)\big)dt - V(t)|_{t_k+\tau_k}^{t_{k+1}+\tau_{k+1}}, \quad (7.36)$$

and

$$J = \lim_{N\to\infty} \sum_{k=0}^{N} J_k$$
$$= \lim_{N\to\infty} \sum_{k=0}^{N} \int_{t_k+\tau_k}^{t_{k+1}+\tau_{k+1}} \big(z^T(t)z(t) - \gamma^2 w^T(t)w(t)\big)dt$$
$$= \int_{t_0}^{\infty} \big(z^T(t)z(t) - \gamma^2 w^T(t)w(t) + \dot{V}(t)\big)dt - V(t)|_{t_0}^{\infty}. \quad (7.37)$$

Similar to the proof of Theorem 7.5, (7.36) becomes

$$J_k \le \int_{t_k+\tau_k}^{t_{k+1}+\tau_{k+1}} \Bigg[\sum_{i=1}^{r}\sum_{j=1}^{r} h_i(\theta(t))h_j(\theta(t))\big(\Gamma^T(t)\Xi_{ij}^{'}\Gamma(t)$$
$$+ \delta\Gamma^T(t)\tilde{Y}\bar{T}^{-1}\tilde{Y}^T\Gamma(t)\big)\Bigg]dt - V(t)\Bigg|_{t_k+\tau_k}^{t_{k+1}+\tau_{k+1}}, \quad (7.38)$$

where

$$\Xi_{ij}^{'} = \begin{bmatrix} \bar{\Phi}_{1,1} & \bar{\Phi}_{1,2} & \bar{\Phi}_{1,3} & \bar{\Phi}_{1,4} & \bar{\Phi}_{1,5} \\ * & \bar{\Phi}_{2,2} & \bar{\Phi}_{2,3} & \bar{\Phi}_{2,4} & \bar{\Phi}_{2,5} \\ * & * & \bar{\Phi}_{3,3} & \bar{\Phi}_{3,4} & 0 \\ * & * & * & \bar{\Phi}_{4,4} & \bar{\Phi}_{4,5} \\ * & * & * & * & \bar{\Phi}_{5,5} \end{bmatrix},$$

$$\begin{aligned}
\bar{\Phi}_{1,1} &= P_2 + Y_1 + Y_1^T - Y_4\bar{A}_i - \bar{A}_i^T Y_4^T + D_i^T D_i, \\
\bar{\Phi}_{1,2} &= P_1 + Y_3^T + Y_4 - \bar{A}_i^T Y_6^T, \\
\bar{\Phi}_{1,3} &= -Y_4\bar{A}_{d_i}, \\
\bar{\Phi}_{1,4} &= -Y_1 + Y_2^T - Y_4\bar{B}_i(K_j + \Delta K_j) - \bar{A}_i^T Y_5^T + D_i^T E_i(K_j + \Delta K_j), \\
\bar{\Phi}_{1,5} &= -\bar{Y}_4 C_i, \\
\bar{\Phi}_{2,2} &= \delta T + Y_6 + Y_6^T, \\
\bar{\Phi}_{2,3} &= -Y_6\bar{A}_{d_i}, \\
\bar{\Phi}_{2,4} &= -Y_3 + Y_5^T - Y_6\bar{B}_i(K_j + \Delta K_j), \\
\bar{\Phi}_{2,5} &= -\bar{Y}_6 C_i, \\
\bar{\Phi}_{3,3} &= -P_2, \\
\bar{\Phi}_{3,4} &= -\bar{A}_{d_i}^T Y_5^T, \\
\bar{\Phi}_{4,4} &= -Y_2 - Y_2^T - Y_5\bar{B}_i(K_j + \Delta K_j) - (K_j + \Delta K_j)^T \bar{B}_i^T Y_5^T \\
&\quad +(K_j + \Delta K_j)^T E_i^T E_i(K_j + \Delta K_j), \\
\bar{\Phi}_{4,5} &= -\bar{Y}_5 C_i,
\end{aligned}$$

$\bar{\Phi}_{5,5} = -\gamma^2 I,$
$\tilde{Y}^T = [\bar{Y}_1^T \ \bar{Y}_3^T \ 0 \ \bar{Y}_2^T \ 0],$
$\Gamma^T = [x^T(t) \ \dot{x}^T(t) \ x^T(t-\tau) \ x^T(t_k) \ w^T(t)].$
According to (7.38), the following inequality can be shown,

$$\begin{aligned} J &= \lim_{N\to\infty} \sum_{k=0}^{N} J_k \\ &\le \lim_{N\to\infty} \sum_{k=0}^{N} \Bigg[\int_{t_k+\tau_k}^{t_{k+1}+\tau_{k+1}} \Bigg[\sum_{i=1}^{r}\sum_{j=1}^{r} h_i(\theta(t))h_j(\theta(t)) \\ &\quad \times \left(\Gamma^T(t)\Xi_{ij}^{'}\Gamma(t) + \delta\Gamma^T(t)\tilde{Y}\bar{T}^{-1}\tilde{Y}^T\Gamma(t)\right)\Bigg] dt - V(t)\big|_{t_k+\tau_k}^{t_{k+1}+\tau_{k+1}} \Bigg]. \end{aligned} \tag{7.39}$$

Combining (7.37) and (7.39), the following result is obtained,

$$\begin{aligned} &\int_{t_0}^{\infty} \left(z^T(t)z(t) - \gamma^2 w^T(t)w(t) + \dot{V}(t)\right) dt \le \\ &\lim_{N\to\infty} \sum_{k=0}^{N} \int_{t_k+\tau_k}^{t_{k+1}+\tau_{k+1}} \Bigg[\sum_{i=1}^{r}\sum_{j=1}^{r} h_i(\theta(t))h_j(\theta(t))\Gamma^T(t)(\Xi_{ij}^{'} + \delta\tilde{Y}\bar{T}^{-1}\tilde{Y}^T)\Gamma(t)\Bigg] dt. \end{aligned} \tag{7.40}$$

It is explicit that $z^T(t)z(t) - \gamma^2 w^T(t)w(t) + V(\infty) - V(t_0) \le 0$ if $\Xi_{ij}^{'} + \delta\tilde{Y}\bar{T}^{-1}\tilde{Y}^T < 0$ for any nonzero $\Gamma(t)$. According to the zero initial condition, we know that the H_∞ performance index is satisfied. According to the Schur complement [2], the conditions in Theorem 7.6 can be obtained and the proof is completed. □

7.5 Simulation Results

In this section, an example is presented to show the validity of our control scheme. We apply the above method to design a robust H_∞ networked controller for the following nonlinear systems.

Example 7.3. Consider the following nonlinear system proposed in [6].

The nonlinear system with time delay can be expressed by the following T–S fuzzy model,

$$\begin{aligned} &\textbf{Rule (1)} \text{ IF } x_2(t) \text{ is } \mathrm{N}_{11}, \text{ THEN } \dot{x}(t) = (A_1 + \Delta A_1)x(t) \\ &\qquad\qquad + A_{d_1}x(t-\tau) + B_1 u(t) + C_1 w(t); \\ &\qquad z(t) = D_1 x(t) + E_1 u(t); \\ &\textbf{Rule (2)} \text{ IF } x_2(t) \text{ is } \mathrm{N}_{12}, \text{ THEN } \dot{x}(t) = (A_2 + \Delta A_2)x(t) \\ &\qquad\qquad + A_{d_2}x(t-\tau) + B_2 u(t) + C_1 w(t); \\ &\qquad z(t) = D_2 x(t) + E_2 u(t); \end{aligned}$$

where $x(t) = \begin{bmatrix} x_1(t) \ x_2(t) \end{bmatrix}^T$, $N_{11}(x_2(t)) = 1 - x_2^2(t)/2.25$ and $N_{12}(x_2(t)) = 1 - N_{11}(x_2(t)) = x_2^2(t)/2.25$.

Similar to [6], we have

$$A_1 = \begin{bmatrix} -0.1125 & -0.02 \\ 1 & 0 \end{bmatrix}, \quad A_{d_1} = \begin{bmatrix} -0.0125 & -0.005 \\ 0 & 0 \end{bmatrix},$$
$$A_2 = \begin{bmatrix} -0.1125 & -1.527 \\ 1 & 0 \end{bmatrix}, \quad A_{d_2} = \begin{bmatrix} -0.0125 & -0.23 \\ 0 & 0 \end{bmatrix},$$
$$\Delta A_1 = \Delta A_2 = \begin{bmatrix} -0.1125 \\ 0 \end{bmatrix} \mathbb{F}(t) \begin{bmatrix} 1 \ 0 \end{bmatrix}, \quad \Delta K_1 = \Delta K_2 = \begin{bmatrix} 0.1 \ 0.1 \end{bmatrix} \mathbb{F}(t) \times 0.1,$$
$$B_1 = B_2 = \begin{bmatrix} 1 & 0 \end{bmatrix}^T, \quad C_1 = C_2 = \begin{bmatrix} 0.01 & 0 \end{bmatrix}^T, \quad D_1 = D_2 = \begin{bmatrix} 0 & 1 \end{bmatrix},$$
$$E_1 = E_2 = 0, \ \Delta A_{d_i} = 0, \ \ \Delta B_i = 0, \ \ \mathbb{F}(t) = \sin(t), \ \ w(t) = 0.1 \sin(t) e^{-0.1t}.$$

Next, we select

$$\bar{Y}_4 = \begin{bmatrix} -12.3 & -4.3 \\ -3.8 & -6.3 \end{bmatrix}, \quad \bar{Y}_5 = \begin{bmatrix} -3.4 & -4.3 \\ -3.0 & -3.6 \end{bmatrix}, \quad \bar{Y}_6 = \begin{bmatrix} -4.7 & -3.7 \\ -3.3 & -3.7 \end{bmatrix},$$
$\varepsilon = 1$, $\gamma = 0.65$, $\delta = 0.15$.

Applying Theorem 7.6, the feasible solutions to (7.35) are given as follows:

$$\bar{P}_1 = \begin{bmatrix} 13.2898 & 9.6762 \\ 9.6762 & 13.4112 \end{bmatrix}, \quad \bar{P}_2 = \begin{bmatrix} 1.2242 & 0.5533 \\ 0.5533 & 1.0490 \end{bmatrix},$$
$$\bar{T} = \begin{bmatrix} 11.7451 & 4.9364 \\ 4.9364 & 6.4083 \end{bmatrix}, \quad \bar{Y}_1 = \begin{bmatrix} -78.3072 & -32.9036 \\ -32.9034 & -42.7124 \end{bmatrix},$$
$$\bar{Y}_2 = \begin{bmatrix} 78.3042 & 32.9059 \\ 32.9039 & 42.7126 \end{bmatrix}, \quad \bar{Y}_3 = \begin{bmatrix} 0.0018 & -0.0004 \\ 0.0037 & -0.0012 \end{bmatrix},$$
$$K_1 = \begin{bmatrix} -1.7797 & -0.5432 \end{bmatrix}, \quad K_2 = \begin{bmatrix} -3.1463 & 0.0921 \end{bmatrix}.$$

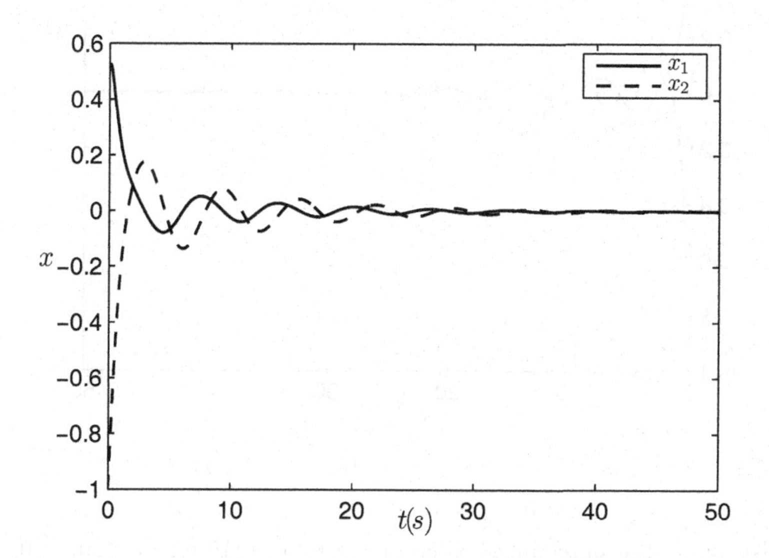

Fig. 7.5. Response of state x in Case I ($\tau = 0.5,\ t_s = 0.05,\ \tau_D \leq 0.03$)

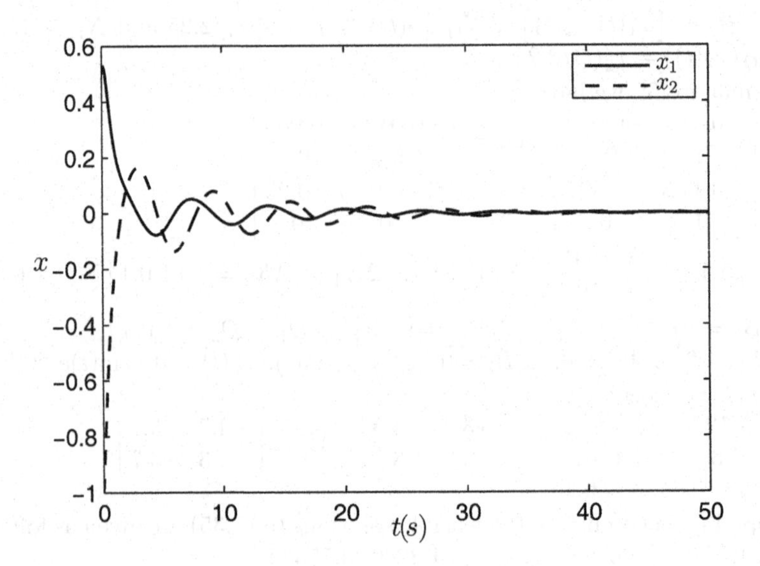

Fig. 7.6. Response of state x in Case II ($\tau = 1,\ t_s = 0.1,\ \tau_D \leq 0.01$)

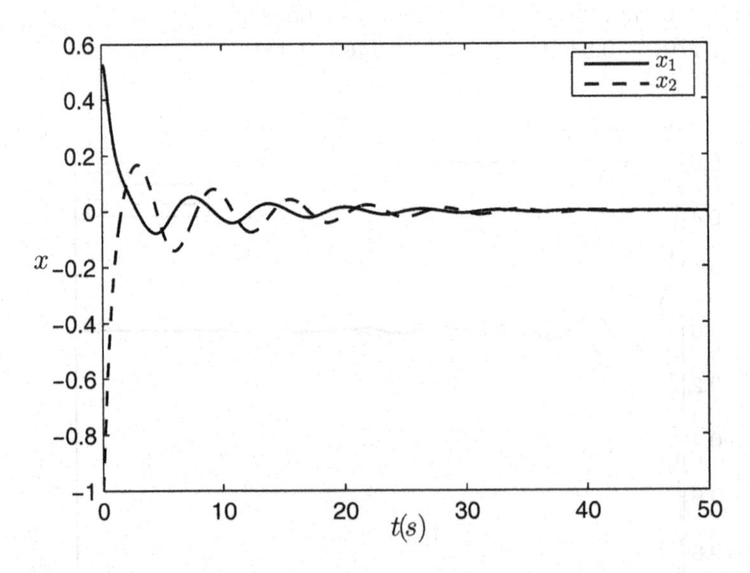

Fig. 7.7. Response of state x in Case III ($\tau = 3,\ t_s = 0.1,\ \tau_D \leq 0.01$)

Next, under the same initial value $\varphi(t) = (0.5\ \ -1)^T$ for $t \in [-0.5,\ \ 0]$, we show the results with different network conditions.

Case I: Sampling period $t_s = 0.05$, network-induced delay $\tau_D \leq 0.03$ and system state delay time $\tau = 0.5$ are given according to system demand.

Fig. 7.5 presents the simulation result for the proposed robust H_∞ networked control method with uncertain condition. The trajectory of x is shown in Fig. 7.5. From the method presented, we know that the state feedback controller is effective when the sum of the sampling period and network-induced delay is less than δ. In this example, $t_s = 0.05$ and $\tau_D \leq 0.03$, the above condition is satisfied. When $t \to \infty$, the state of this system can be robustly stabilized.

Case II: Sampling period $t_s = 0.1$, network-induced delay $\tau_D \leq 0.01$ and system state delay time $\tau = 1$ are given according to system demand.

The trajectory of x is shown in Fig. 7.6. In this example, $t_s = 0.1$ and $\tau_D \leq 0.01$, the condition is satisfied. When $t \to \infty$, the state of this system can be robustly stabilized.

Case III: Sampling period $t_s = 0.1$, network-induced delay $\tau_D \leq 0.01$ and system state delay time $\tau = 3$ are given according to system demand.

The trajectory of x is shown in Fig. 7.7. In this example, $t_s = 0.1$ and $\tau_D \leq 0.01$, the condition is satisfied. When $t \to \infty$, the state of this system can be robustly stabilized.

It is very explicit that the uncertain fuzzy system is robustly stable with H_∞ performance after considering network action. A satisfactory result can be obtained. □

7.6 Conclusions

In this chapter, we propose the GCNC method and robust H_∞ networked control method for T–S fuzzy systems with time delays, respectively. NCS theory is used to design system controller. Both network-induced delay and packet dropout are addressed in a uniform framework. Simulation results show the validity of the control schemes presented.

References

1. Almutairi NB, Chow MY (2003) Stabilization of networked PI control system using fuzzy logic modulation. In: Proc American Control Conference, Denver, CO, 975–980
2. Boyd S, Ghauoi LE, Feron E, Balakrishan V (1994) Linear matrix inequalities in system and control theory. SIAM, Philadelphia
3. Cao YY, Frank PM (2000) Analysis and synthesis of nonlinear time-delay systems via fuzzy control approach. IEEE Trans Fuzzy Systems 8:200–211
4. Cao YY, Frank PM (2001) Stability analysis and synthesis of nonlinear time-delay systems via linear Takagi–Sugeno fuzzy models. Fuzzy Sets and Systems 124:213–219

5. Chen WH, Guan ZH, Lu XM (2004) Delay-dependent output feedback guaranteed cost control for uncertain time-delay systems. Automatica 40:1263–1268
6. Chen B, Liu XP (2005) Fuzzy guaranteed cost control for nonlinear systems with time-varying delay. IEEE Trans Fuzzy Systems 13:238–249
7. Chen B, Liu XP (2005) Delay-dependent robust H_∞ control for T–S fuzzy systems with time delay. IEEE Trans Fuzzy Systems 13:544–556
8. Gu K (2000) An integral inequality in the stability problem of time-delay systems. In: Proc IEEE Conference on Decision and Control, Sydney, Australia, 2805–2810
9. Guan XP, Chen CL (2004) Delay-dependent guaranteed cost control for T–S fuzzy systems with time delays. IEEE Trans Fuzzy Systems 12:236–249
10. Lee KR, Kim JH, Jeun ET, Park HB (2000) Output feedback robust H_∞ control of uncertain fuzzy dynamic systems with time-varying delay. IEEE Trans Fuzzy Systems 8:657–664
11. Lee KC, Lee S, Lee MH (2003) Remote fuzzy logic control of networked control system via Profibus-DP. IEEE Trans Industrial Electronics 50:784–792
12. Liang Q, Karnik NN, Mendal JM (2000) Connection admission control in ATM networks using survey-based type-2 fuzzy logic systems. IEEE Trans Systems Man Cybernetics–Part C: Applications and Reviews 30:329–339
13. Nešić D, Teel AR (2004) Input-to-state stability of networked control systems. Automatica 40:2121–2128
14. Nešić D, Teel AR (2004) Input-output stability properties of networked control systems. IEEE Trans Automatic Control 49:1650–1667
15. Takagi T, Sugeno M (1985) Fuzzy identification of systems and its applications to modeling and control. IEEE Trans Systems Man Cybernetics 15:116–132
16. Wals GC, Beldiman O, Bushnell LG (2001) Asymptotic behavior of nonlinear networked control systems. IEEE Trans Automatic Control 46:1093–1097
17. Wang RJ, Lin WW, Wang WJ (2004) Stabilizability of linear quadratic state feedback for uncertain fuzzy time-delay systems. IEEE Trans Systems Man Cybernetics–Part B: Cybernetics 34:1288–1292
18. Wang HO, Tanaka K, Griffin MF (1996) An approach to fuzzy control of nonlinear systems: stability and design issues. IEEE Trans Fuzzy Systems 4:14–23
19. Xu SY, Lam J (2005) Robust H_∞ control for uncertain discrete-time-delay fuzzy systems via output feedback controllers. IEEE Trans Fuzzy Systems 13:82–93
20. Yi Z, Heng PA (2002) Stability of fuzzy control systems with bounded uncertain delays. IEEE Trans Fuzzy Systems 10:92–97
21. Yue D, Han QL, Chen P (2004) State feedback controller design of networked control systems. IEEE Trans Circuits and Systems II: Express Briefs 51:640–644
22. Yue D, Han QL, Lam J (2005) Network-based robust control of systems with uncertainty. Automatica 41:999–1007
23. Zhang W, Branicky M, Phillips S (2001) Stability of networked control systems. IEEE Control Systems Magazine 21:84–99
24. Zhang HG, Cai LL (2002) Decentralized nonlinear control of a HVAC system IEEE Trans Systems Man Cybernetics–Part C: Applications and Reviews 32:493–498
25. Zhang HG, Quan YB (2001) Modeling, identification and control of a class of nonlinear system. IEEE Trans Fuzzy Systems 9:349–354

8

A Discrete-time Jump Fuzzy System Approach to NCS Design

Fuchun Sun and Fengge Wu

Tsinghua University, Beijing 100084, P. R. China
fcsun@mail.tsinghua.edu.cn, wfg02@mails.tsinghua.edu.cn

Abstract. A discrete-time jump fuzzy system is proposed in this chapter for the modeling and control of a class of nonlinear networked control systems (NCS) with random but bounded communication delays and packets dropout. Above all, a guaranteed cost control with state feedback is developed by constructing a sub-optimal performance controller for the discrete-time jump fuzzy systems in such a way that a piecewise quadratic Lyapunov function (PQLF) can be used to establish the global stability of the resulting closed-loop fuzzy control system. A homotopy-based iterative algorithm solving for linear matrix inequality (LMI) is developed to get the feedback gains. When not all states are available, an output feedback controller is designed. For the NCS based on the mixed networks, a neuro-fuzzy controller is develped, which is composed of three parts: a guaranteed cost state-feedback controller, an adaptive neuro-fuzzy inference system (ANFIS) predictor and a fuzzy controller. The ANFIS predictor is used to improve the performance of the NCS when network delay is longer. Simulation examples are carried out to show the effectiveness of the proposed approaches.

Keywords. Discrete-time jump fuzzy systems, guaranteed cost control, LMI, Markovian jumping parameters, networked control systems.

8.1 Introduction

Over the past five years, networked control systems (NCSs) with feedback loops closed through networks, have received considerable attention in the literature, as illustrated by recent articles [1, 7, 8, 12, 14, 15, 17, 21, 23, 24, 27, 29], due to the enormous advantages, such as low cost, reduced power, simple maintenance and wide applications to novel teleoperating areas.

8.1.1 Fundamental Issues in NCS

An NCS exhibits issues which traditionally have not been taken into account in control system design because control loops are closed through a real-time network. Regardless of the type of network used, these special issues degrade the system dynamic performance and are a source of potential instability. So NCS issues should be investigated.

(i) A network-induced delay occurs while exchanging data among devices connected to the shared medium. The sensor data or control signal arrive at the controller or actuator of the NCS randomly due to network-induced delays.
(ii) The node of the network may discard some of the received packets if it is overloaded. Packets dropout renders the NCS data incomplete.

Compared with traditional control systems, an NCS does not possess data with two different characteristics, namely fixity and integrality. As a result, network delay and packets dropout should be considered simultaneously rather than separately when an NCS is modeled. Most researchers regarded an NCS as a time-delay control system or control system with packet dropout [1, 7, 12, 15, 29]. In addition, most existing literature reports consider only stabilization of linear NCSs whereas nonlinear NCSs have received little attention [1, 8]. Therefore, advanced approaches for nonlinear NCSs are required.

8.1.2 Previous Work

Usually, distributed linear feedback control systems with random network induced delay are modeled as Markovian jump linear control systems [8, 17, 21, 23], in which random variation of system delays corresponds to randomly varying structure of the state-representation. When the Markovian jump system changes abruptly from one mode to another [6, 16, 19, 22, 23, 28], the switching between modes is governed by a Markov process with discrete and finite state space. Markovian jump systems have been studied extensively because jumping systems have been a subject of great practical importance.

Fuzzy systems have been used in recent years for the control of nonlinear processes [5, 10, 11, 18, 20]. Fuzzy system theory enables us to utilize qualitative, linguistic information about a highly complex nonlinear system to construct a mathematical model. And a fuzzy linear model can be used to approximate global behaviors of a highly complex nonlinear system. Local dynamics in different state space regions are represented by local linear systems in this fuzzy linear model. The overall model of the system is obtained by “blending” these linear models through nonlinear fuzzy membership functions. Unlike conventional modeling, which uses a single model to describe the global behavior of a system, fuzzy modeling is essentially a multi-model approach in which simple submodels (a set of linear models) are combined to describe the global behavior of the system. From the middle of the 1980s, there have appeared a number of analysis/synthesis problems for Takagi–Sugeno (T–S) fuzzy systems [18]. Based on the T–S fuzzy systems, Palm and Driankov [16], Choi and Park [6], and Tanaka [19] introduced new switching fuzzy systems for more complicated nonlinear systems.

Motivated by these approaches, a discrete-time jump fuzzy system is proposed to model NCS with random but bounded delay and packet dropout in this chapter. Then new stability theorems and new controller design methods are developed for discrete-time jump fuzzy systems. The chapter is organized as follows. The discrete-time jump fuzzy system and the modeling of NCS are proposed in Section 8.2. In Section 8.3, the LMI-based design of a guaranteed cost state feedback fuzzy controller is presented. The fuzzy output feedback controller is developed in Section 8.4. The neuro-fuzzy controller is provided in Section 8.5. Finally, Section 8.6 summarizes some conclusions.

In this chapter, $\mathcal{Z}$, $\mathbb{R}^n$ and $\mathbb{R}^{m\times n}$ denote, respectively, the set of integer numbers, the n-dimensional Euclidean space and the set of all $m \times n$ real matrices. As usual, $P > 0$ ($\geqslant$, $<$, $\leqslant$, respectively) will denote that the matrix P is symmetric and positive definite (positive semi-definite, negative definite, negative semi-definite). I_n represents $n \times n$ identity matrix and diag$\{\cdots\}$ represents block diagonal matrix. The symmetric items in symmetric matrices are represented by "$*$". $E[\cdot]$ stands for the mathematical expectation.

8.2 Modeling NCS

The general NCS configuration is illustrated in Fig. 8.1, which is composed of a controller and a remote system containing a physical plant, sensors and actuators. The controller and the plant are physically located at different locations and are directly linked by a data network in order to perform remote closed-loop control. Most networked control methodologies use the discrete-time formulation [22].

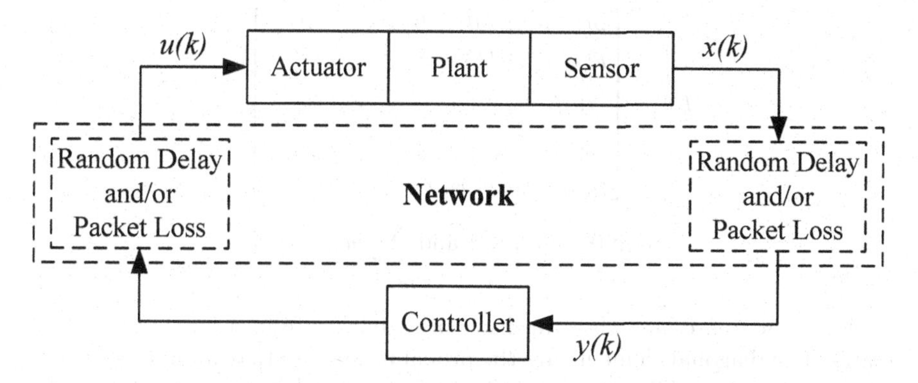

Fig. 8.1. The general NCS configuration

8.2.1 Markov Characteristics of NCS

Suppose that $r(k)$ is the network induced delay at time k with $0 \leqslant r(k) \leqslant d < \infty$, and d is the finite delay bound. When the data is transmitted in turn from sensor to controller or from controller to actuator through the network, the transition probability of $r(k+1)$ is determined only by $r(k)$ and not by $r(0), r(1), \ldots, r(k-1)$ or the time at which it reached the present state. Hence $\{r(k), k \in \mathcal{Z}\}$ is a homogeneous Markov chain. The transition probability is defined as follows:

$$
\begin{aligned}
pr_{ij} &= \mathrm{Prob}\{r(k+1) = j | r(k) = i\}, \\
pr_i &= \mathrm{Prob}(r(k) = i), \\
i, j &\in \mathcal{S} = \{0, 1, \ldots, d\}.
\end{aligned} \tag{8.1}
$$

Here $pr_{ij} \geqslant 0$ for $i, j \in \mathcal{S}$, and

$$\sum_{j=0}^{d} pr_{ij} = 1.$$

In real-time control systems, the newest data is the best data [27]. The assumption here means that the controller will always use the most recent data. That is, the data at step k is available for feedback when there is no new information coming in at step $k+1$ (data could be lost or there is a longer delay). So in the model of the NCS, the delay $r(k)$ can increase at most by 1 each step [17]. We develop a new controller for the set $\mathcal{S}$ denoting the possible jump state. In this case, we have

$$\mathrm{Prob}\{r(k+1) > r(k) + 1\} = 0. \tag{8.2}$$

Hence the structured transition probability matrix Pr is

$$
Pr = \begin{bmatrix}
pr_{00} & pr_{01} & 0 & 0 & \cdots & 0 \\
pr_{10} & pr_{11} & pr_{12} & 0 & \cdots & 0 \\
\vdots & \vdots & \vdots & \vdots & \ddots & \vdots \\
\vdots & \vdots & \vdots & \vdots & \vdots & pr_{d-1,d} \\
pr_{d0} & pr_{d1} & pr_{d2} & pr_{d3} & \cdots & pr_{d,d}
\end{bmatrix},
\quad \text{with } 0 \leqslant pr_{ij} \leqslant 1 \text{ and } \sum_{j=0}^{d} pr_{ij} = 1. \tag{8.3}
$$

Each row represents the transition probabilities from a fixed state to all states. The diagonal elements are the probabilities of data coming in sequence with equal delays. The elements above the diagonal indicate that data encounter longer delays, and the elements below the diagonal describe packet dropout.

8.2.2 Discrete-time Jump Fuzzy System

Many nonlinear dynamic systems can be represented by T–S fuzzy models. In fact, it is proved that T–S fuzzy models are universal approximators. So we shall introduce a discrete-time jump fuzzy system to model a class of nonlinear NCSs such as:

$$x_{k+1} = f_{r(k)}(x_k, u_k), \tag{8.4}$$

where $x_k \in \mathbb{R}^n$ is the state vector, $u_k \in \mathbb{R}^m$ is the input vector. Here $f_{r(k)}$ is a local fuzzy function. The models in two-level forms are inferred as follows:

$$\begin{aligned}
&\text{IF } r(k) = i\\
&\text{THEN local plant rule } l:\\
&\qquad \text{IF } z_{k,1} \text{ is } M_{il,1} \text{ and } \cdots \text{ and } z_{k,p} \text{ is } M_{il,p},\\
&\qquad \text{THEN } x_{k+1} = A_{il}x_k + B_{il}u_k,\\
&\qquad\qquad x_0 = x(0),\ l = 1, \ldots, t(i).
\end{aligned} \tag{8.5}$$

Here, $z_{k,1}, \ldots, z_{k,p}$ are the local premise variables, $M_{il,1}, \ldots, M_{il,p}$ are the local fuzzy sets, $t(i)$ is the number of IF-THEN rules when $r(k) = i$, $\{r(k), k \in \mathcal{Z}\}$ is a discrete-time homogeneous Markov chain taking values in a finite set $\mathcal{S} = 0, 1, \ldots, d$, with the transition probability from mode i at time k to mode j at time $k+1, i, j \in \mathcal{S}$, $k \in \mathcal{Z}$.

By the following local fuzzy weighting functions $h_{il}(z_k)$, which are determined by a local premise variable vector $z_k = [z_{k,1}\ z_{k,2}\ \cdots\ z_{k,p}]^T$, the final representation of the discrete-time jump fuzzy system is as follows:

$$x_{k+1} = \sum_{l=1}^{t(i)} h_{il}(z_k)\{A_{il}x_k + B_{il}u_k\}, \tag{8.6}$$

where

$$h_{il}(z_k) = \frac{\prod_{j=1}^{p} M_{il,j}(z_{k,j})}{\sum_{l=1}^{t(i)} \prod_{j=1}^{p} M_{il,j}(z_{k,j})}, \tag{8.7}$$

and $M_{il,j}(z_{k,j})$ is the grade of membership of $z_{k,j}$ in $M_{il,j}$.

To simplify the presentation, the discrete-time jump fuzzy system (8.5) can be represented as follows:

$$x_{k+1} = A_i(H_i(z_k))x_k + B_i(H_i(z_k))u_k, \tag{8.8}$$

where

$$[A_i(H_i(z_k)) \quad B_i(H_i(z_k))] \triangleq \sum_{l=1}^{t(i)} h_{il}(z_k)[A_{il} \quad B_{il}]. \tag{8.9}$$

8.3 State-feedback Controller Design

8.3.1 The Closed-loop Model of an NCS

According to the direction of data transfers, network delays and packets dropout in the NCS can be categorized as sensor-to-controller and controller-to-actuator. When the control or sensor data travel across one type of network, the data has the same transmission characteristic. So the simple NCS configuration in which the network exists only between the sensors and controller is illustrated in Fig. 8.2.

When $r(k) = i$, the mode-dependent jump state feedback control law is:

$$u_k = K_i(H_i(z_k))x_{k-i}, \tag{8.10}$$

where

$$K_i(H_i(z_k)) = \sum_{l=1}^{t(i)} h_{il}(z_k)K_{il}.$$

If we augment the state variable

$$X_k = [x_k^T \; x_{k-1}^T \; \cdots \; x_{k-d}^T]^T, \tag{8.11}$$

where $X_k \in \mathbb{R}^{(d+1)n}$, then the closed-loop system is:

$$\begin{aligned} X_{k+1} &= \left(\tilde{A}_i\left(H_i\left(z_k\right)\right) + \tilde{B}_i\left(H_i\left(z_k\right)\right) K_i\left(H_i\left(z_k\right)\right) \tilde{G}_{r(k)}\right) X_k, \\ X_0 &= [x_0^T \; x_{-1}^T \; \cdots \; x_{-d}^T]^T, \end{aligned} \tag{8.12}$$

where

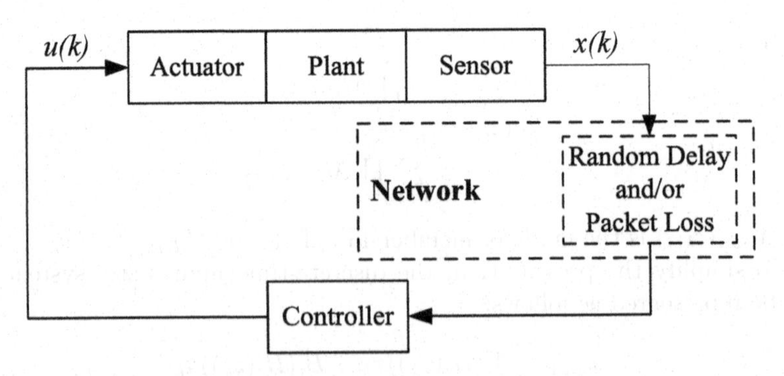

Fig. 8.2. The simple NCS configuration

$$\tilde{A}_i(H_i(z_k)) = \begin{bmatrix} A_i(H_i(z_k)) & 0 & \cdots & 0 & 0 \\ I & 0 & \cdots & 0 & 0 \\ 0 & I & \cdots & 0 & 0 \\ \vdots & \vdots & \ddots & \vdots & \vdots \\ 0 & 0 & \cdots & I & 0 \end{bmatrix},$$

$$\tilde{B}_i(H_i(z_k)) = \begin{bmatrix} B_i(H_i(z_k)) \\ 0 \\ 0 \\ \vdots \\ 0 \end{bmatrix},$$

$$\tilde{G}_{r(k)} = \begin{bmatrix} 0 \cdots 0 \; I \; 0 \cdots 0 \end{bmatrix},$$

and $\tilde{G}_r(k)$ has all elements zero except for the $r(k)$th block, which is an identity matrix. Equation (8.12) corresponds to a discrete-time jump fuzzy system.

8.3.2 Guaranteed Cost Controller Design

Now we will consider a guaranteed cost controller. For the performance criterion, an upper bound of LQ cost associated with states and inputs in the global systems called guaranteed cost is described as follows:

$$\min \max_{h_i(z_k)\in\mathcal{H}} \mathcal{E}\left\{\sum_{k=0}^{\infty}(X_k^T Q_{r(k)} X_k + u_k^T R_{r(k)} u_k)\right\}, \tag{8.13}$$

where $Q_{r(k)} > 0$, $R_{r(k)} > 0$, and $\mathcal{H}$ is defined as a set of all possible fuzzy weighting functions. In this chapter, the LQ cost is a function of the grades $h_i(z_k)$.

Definition 1.1 of [3] is extended, and we have the following definitions.

Definition 8.1. For System (8.6) with $u_k \equiv 0$ and $r(k) = i \in \mathcal{S}$, the equilibrium point 0 is said to be stochastically stable, if for every initial state $(X_0, r(0))$, there exists a finite $W > 0$ such that the following holds:

$$\mathcal{E}\left\{\sum_{k=0}^{\infty}\|X_k(X_0, r(0))\|^2 \,\middle|\, X_0, r(0)\right\} < X_0^T W X_0. \tag{8.14}$$

□

Lemma 8.1. The closed-loop system in (8.12) is stochastically stable if and only if there exists a set of symmetric matrices $P_i > 0$, $i \in \mathcal{S}$ satisfying the following coupled matrix inequalities:

$$L_i = \sum_{j=0}^{d} pr_{ij} [\tilde{A}_i(H_i(z_k)) + \tilde{B}_i(H_i(z_k)) K_i(H_i(z_k)) \tilde{G}_{r(k)}]^T P_j \times [\tilde{A}_i(H_i(z_k)) + \tilde{B}_i(H_i(z_k)) K_i(H_i(z_k)) \tilde{G}_{r(k)}] - P_i < 0. \tag{8.15}$$

Proof. Sufficiency. For the closed-loop system in (8.6), consider piecewise quadratic Lyapunov stability with the following PQLF candidate $V(X_k)$ mapping from $\mathbb{R}^n$ to $\mathbb{R}$:

$$V(X_k, r(k) = i) = V(X_k, i) = X_k^T P_i X_k > 0. \tag{8.16}$$

The weak infinitesimal operator $\tilde{\mathcal{A}}V(X, i)$ [2, 4] of the stochastic process (X, i) is defined by:

$$\begin{aligned} \tilde{\mathcal{A}}V(X,i) &= \mathcal{E}\{V(X_{k+1}, r(k+1)) \,|X_k, r(k) = i\} - V(X_k, i) \\ &= X_k^T \Bigg\{ [\tilde{A}_i(H_i(z_k)) + \tilde{B}_i(H_i(z_k)) K_i(H_i(z_k)) \tilde{G}_{r(k)}]^T \\ &\quad \times \left[\sum_{j=0}^{d} pr_{ij} P_j\right] \Big[\tilde{A}_i(H_i(z_k)) + \tilde{B}_i(H_i(z_k)) \\ &\quad \times K_i(H_i(z_k)) \tilde{G}_{r(k)}\Big] - P_i \Bigg\} X_k. \end{aligned} \tag{8.17}$$

Thus, if $L_i < 0$, then:

$$\begin{aligned} \tilde{\mathcal{A}}V(X,i) &= \mathcal{E}\{V(X_{k+1}, r(k+1)) \,|X_k, r(k) = i\} \\ &\quad - V(X_k, i) \leqslant -\lambda_{\min}(L_i) X_k^T X_k \leqslant -\beta X_k^T X_k \\ &= -\beta \|X_k\|^2, \end{aligned} \tag{8.18}$$

where

$$\beta = \inf\{\lambda_{\min}(-L_i), i \in \mathcal{S}\} > 0.$$

From (8.18), we can see that for any $T \geqslant 1$

$$\mathcal{E}\{V(X_{T+1}, r(T+1))\} - \mathcal{E}\{V(X_0, r(0))\} \leqslant -\beta \mathcal{E}\left\{\sum_{t=0}^{T} \|X_t\|^2\right\}.$$

Then,

$$\begin{aligned} \mathcal{E}\left\{\sum_{t=0}^{T} \|X_t\|^2\right\} &\leqslant \frac{1}{\beta} (\mathcal{E}\{V(X_0, r(0))\} - \mathcal{E}\{V(X_{T+1}, r(T+1))\}) \\ &\leqslant \frac{1}{\beta} \mathcal{E}\{V(X_0, r(0))\}. \end{aligned}$$

From Definition 8.1, the stochastic stability is obtained.

Necessity. Let us assume that the closed-loop system in (8.18) is stochastically stable. That is, we have

$$\mathcal{E}\left\{\sum_{k=0}^{\infty}\|X_k(X_0,r(0))\|^2\,|X_0,r(0)\right\} < X_0^T W X_0. \tag{8.19}$$

Consider the following function:

$$X_t^T\tilde{P}_{T-t,r(t)}X_t \triangleq \mathcal{E}\left\{\sum_{k=t}^{T}X_k^T O_{r(k)}X_k\,|X_t,r(t)\right\}, \tag{8.20}$$

with $O_{r(k)}>0$. Assume that $X_k\neq 0$. Since $O_{r(k)}>0$, as T increases, either $X_t^T\tilde{P}_{T-t,r(t)}X_t$ is monotonically increasing or it increases monotonically until

$$\mathcal{E}\left\{X_k^T O_{r(k)}X_k\,|X_k,r(k)\right\}=0$$

for all $k\geqslant k_1\geqslant t$. It is shown in (8.19) that $X_t^T\tilde{P}_{T-t,r(t)}X_t$ is bounded above, and thus, its limit is given by

$$\begin{aligned} X_t^T P_i X_t &\triangleq \lim_{T\to\infty} X_t^T\tilde{P}_{T-t,r(t)}X_t \\ &\triangleq \lim_{T\to\infty}\mathcal{E}\left\{\sum_{k=t}^{T}X_k^T O_{r(k)}X_k\,|X_t,r(t)=i\right\}. \end{aligned} \tag{8.21}$$

Since this is valid for any X_t, we have

$$P_i=\lim_{T\to\infty}\tilde{P}_{T-t,r(t)}. \tag{8.22}$$

According to (8.21), $P_i>0$ since $O_{r(k)}>0$. We get

$$\begin{aligned} &\mathcal{E}\left\{X_t^T\tilde{P}_{T-t,r(t)}X_t - X_{t+1}^T\tilde{P}_{T-t-1,r(t+1)}X_{t+1}\right|\left. X_t,r(t)=i\right\} \\ &= X_t^T O_i X_t. \end{aligned} \tag{8.23}$$

Note that:

$$\begin{aligned} &\mathcal{E}\left\{X_{t+1}^T\tilde{P}_{T-t-1,r(t+1)}X_{t+1}\right|\left. X_t,r(t)=i\right\} \\ &= X_t^T\sum_{j=0}^{d}pr_{ij}\left(\tilde{A}_i(H_i(z_k))+\tilde{B}_i(H_i(z_k))K_i(H_i(z_k))\tilde{G}_{r(k)}\right)^T \\ &\times \tilde{P}_{T-t-1,j}\left(\tilde{A}_i(H_i(z_k))+\tilde{B}_i(H_i(z_k))K_i(H_i(z_k))\tilde{G}_{r(k)}\right)X_t. \end{aligned} \tag{8.24}$$

This, together with (8.23), implies that for any X_t,

$$X_t^T \left[\tilde{P}_{T-t,r(t)} - \sum_{j=0}^{d} pr_{ij} \left(\tilde{A}_i(H_i(z_k)) + \tilde{B}_i(H_i(z_k))K_i(H_i(z_k))\tilde{G}_{r(k)} \right)^T \right.$$
$$\left. \times \tilde{P}_{T-t-1,j} \left(\tilde{A}_i(H_i(z_k)) + \tilde{B}_i(H_i(z_k))K_i(H_i(z_k))\tilde{G}_{r(k)} \right) \right] X_t$$
$$= X_t^T O_i X_t.$$

Letting $T \to \infty$ and noticing that (8.22) and $O_i > 0$, we obtain:

$$P_i - \sum_{j=0}^{d} pr_{ij} \left[\tilde{A}_i(H_i(z_k)) + \tilde{B}_i(H_i(z_k))K_i(H_i(z_k))\tilde{G}_{r(k)} \right]^T$$
$$\times P_j \left[\tilde{A}_i(H_i(z_k)) + \tilde{B}_i(H_i(z_k))K_i(H_i(z_k))\tilde{G}_{r(k)} \right] > 0.$$

□

Lemma 8.2. The closed-loop jump fuzzy system(8.12) is stochastically stable in the large and the cost (8.13) will be bounded by $x_0^T P_i x_0$ for any nonzero initial state $x_0 \in \mathbb{R}_i$, if there exist $P_i > 0$, $i \in \mathcal{S}$, and $K_i(H_i(z_k))$ satisfying the following conditions:

$$\begin{bmatrix} -\bar{P}_i & * & \cdots & * & * & * \\ \begin{pmatrix} \tilde{A}_i(H_i(z_k))\bar{P}_i \\ +\tilde{B}_i(H_i(z_k)) \\ \times K_i(H_i(z_k)) \\ \times \tilde{G}_{r(k)}\bar{P}_i \end{pmatrix} & \begin{pmatrix} -(pr_{i0})^{-1} \\ \times \bar{P}_0 \end{pmatrix} & \cdots & 0 & 0 & 0 \\ \vdots & \vdots & \ddots & \vdots & \vdots & \vdots \\ \begin{pmatrix} \tilde{A}_i(H_i(z_k))\bar{P}_i \\ +\tilde{B}_i(H_i(z_k)) \\ \times K_i(H_i(z_k)) \\ \times \tilde{G}_{r(k)}\bar{P}_i \end{pmatrix} & 0 & \cdots & \begin{pmatrix} -(pr_{id})^{-1} \\ \times \bar{P}_d \end{pmatrix} & 0 & 0 \\ \bar{P}_i & 0 & \cdots & 0 & -Q_i^{-1} & 0 \\ \begin{pmatrix} K_i(H_i(z_k)) \\ \times \tilde{G}_{r(k)}\bar{P}_i \end{pmatrix} & 0 & \cdots & 0 & 0 & -R_i^{-1} \end{bmatrix} < 0. \quad (8.25)$$

Furthermore, a sub-optimal guaranteed cost controller can be obtained via the following semi-definite programming:

$$\text{Minimize } \gamma \text{ subject to (8.25) and } \begin{bmatrix} \gamma & x_0^T \\ x_0 & \bar{P}_i \end{bmatrix} \leqslant 0. \quad (8.26)$$

Proof. Consider the cost (8.13) associated with states as follows:

$$\min_{h_i(z_k)\in\mathcal{H}} \max \sum_{k=0}^{\infty} \Big\{ X_k^T Q_i X_k + X_k^T (K_i(H_i(z_k))\tilde{G}_{r(k)})^T \times R_i K_i\left(H_i\left(z_k\right)\right)\tilde{G}_{r(k)} X_k \Big\}. \tag{8.27}$$

Then, the closed-loop system is stable via the guaranteed cost controller, if there exists positive-definite symmetric P_i and P_j such that for all X_k and $i, j \in \mathcal{S}$, the following condition holds:

$$\tilde{\mathcal{A}}V(X,i) X_k^T Q_i X_k + X_k^T \left(K_i(H_i(z_k))\tilde{G}_{r(k)}\right)^T \times R_i K_i(H_i(z_k))\tilde{G}_{r(k)} X_k < 0. \tag{8.28}$$

We obtain:

$$[ABK]^T \sum_{j=0}^{d} pr_{ij} P_j \times [ABK] - P_i + Q_i + (K_i(H_i(z_k))\tilde{G}_{r(k)})^T R_i K_i(H_i(z_k))\tilde{G}_{r(k)} X_k < 0,$$

where

$$ABK = \tilde{A}_i(H_i(z_k)) + \tilde{B}_i(H_i(z_k))K_i(H_i(z_k))\tilde{G}_{r(k)}.$$

Using Schur complements, we have the following matrix inequality:

$$\begin{bmatrix} -P_i & * & \cdots & * & * & * \\ ABK & -(pr_{i0}P_0)^{-1} & & 0 & 0 & 0 \\ \vdots & \vdots & \ddots & \vdots & \vdots & \vdots \\ ABK & 0 & \cdots & -(pr_{id}P_d)^{-1} & 0 & 0 \\ I_n & 0 & \cdots & 0 & -Q_i^{-1} & 0 \\ \begin{pmatrix} K_i(H_i(z_k)) \\ \times \tilde{G}_{r(k)} \end{pmatrix} & 0 & \cdots & 0 & 0 & -R_i^{-1} \end{bmatrix} < 0. \tag{8.29}$$

The left-hand side of the inequality (8.29) can be pre- and post-multiplied by J^T and J, respectively, where

$$J = \text{blockdiag}\left\{P_i^{-1}, I_n, I_n, I_m\right\},$$

which yields the following:

$$\begin{bmatrix} -P_i^{-1} & * & \cdots & * & * & * \\ ABKP & -(pr_{i0}P_0)^{-1} & \cdots & 0 & 0 & 0 \\ \vdots & \vdots & \ddots & \vdots & \vdots & \vdots \\ ABKP & 0 & \cdots & -(pr_{id}P_d)^{-1} & 0 & 0 \\ P_i^{-1} & 0 & \cdots & 0 & -Q_i^{-1} & 0 \\ \begin{pmatrix} K_i\left(H_i\left(z_k\right)\right) \\ \times \tilde{G}_{r(k)} P_i^{-1} \end{pmatrix} & 0 & \cdots & 0 & 0 & -R_i^{-1} \end{bmatrix} < 0, \tag{8.30}$$

with

$$ABKP = \tilde{A}_i(H_i(z_k))P_i^{-1} + \tilde{B}_i(H_i(z_k))K_i(H_i(z_k))\tilde{G}_{r(k)}P_i^{-1}.$$

Let $\bar{P}_i \triangleq P_i^{-1}$. We obtain:

$$\begin{bmatrix} -\bar{P}_i & * & \cdots & * & * & * \\ \begin{pmatrix} \tilde{A}_i(H_i(z_k))\bar{P}_i \\ +\tilde{B}_i(H_i(z_k)) \\ \times K_i(H_i(z_k)) \\ \times \tilde{G}_{r(k)}\bar{P}_i \end{pmatrix} & -(pr_{i0})^{-1}\bar{P}_0 & \cdots & 0 & 0 & 0 \\ \vdots & \vdots & \ddots & \vdots & \vdots & \vdots \\ \begin{pmatrix} \tilde{A}_i(H_i(z_k))\bar{P}_i \\ +\tilde{B}_i(H_i(z_k)) \\ \times K_i(H_i(z_k)) \\ \times \tilde{G}_{r(k)}\bar{P}_i \end{pmatrix} & 0 & \cdots & -(pr_{id})^{-1}\bar{P}_d & 0 & 0 \\ \bar{P}_i & 0 & \cdots & 0 & -Q_i^{-1} & 0 \\ \begin{pmatrix} K_i(H_i(z_k)) \\ \times \tilde{G}_{r(k)}\bar{P}_i \end{pmatrix} & 0 & \cdots & 0 & 0 & -R_i^{-1} \end{bmatrix} < 0. \tag{8.31}$$

So, if the condition (8.25), $\bar{P}_i > 0$, $P_j > 0$ hold for all $i, j \in \mathcal{S}$, then

$$\tilde{A}V(X, i) < 0 \text{ at } x_k \neq 0.$$

Owing to continuity, there exists $M_i > 0$ such that

$$\tilde{A}V(X, i) - M_i < 0.$$

Based on Lemma 8.1, the closed-loop jump fuzzy system is stochastically stable.

When the condition (8.25) holds, the cost (8.13) will be bounded for any nonzero initial state $X_0 \in \mathbb{R}_i$:

$$\max_{h_l(z_k)\in\mathcal{H}} \sum_{k=0}^{\infty} \{X_k^T Q_i X_k + u_k^T R_i u_k\} < X_0^T P_i X_0.$$

Since any feasible solutions γ, $\bar{P}_i$, P_j, and $K_i(H_i(z_k))$ yielding (8.25) will also satisfy

$$\max_{h_l(z_k)\in\mathcal{H}} \sum_{k=0}^{\infty} \{X_k^T Q_i X_k + u_k^T R_i u_k\} < X_0^T P_i X_0 \leq \gamma,$$

for any h_l and nonzero $X_0 \in \mathbb{R}_i^{d+1}$, we can use (8.26) to minimize $X_0^T P_i X_0$ for known nonzero initial states. The proof is completed. □

8.3.3 Homotopy Algorithm

The design problem to determine the state feedback gains K_i for (8.26) can be defined as follows: Find P with the constraints (8.25) and K_i such that (8.26) are satisfied.

However, in general, the inequalities (8.25) cannot be transformed equivalently to LMIs and we will utilize the homotopy method [13] to solve it in an iterative manner.

The homotopy algorithm uses a continuous deformation to embed difficult problems into a family of related problems. As a result, once the solution to an "easy to solve" problem in this family is obtained, a continuous path may be followed in solution space to obtain the desired solution to the original problem. To construct a homotopy path, we introduce a real number λ varying from 0 to 1, and define:

$$\begin{bmatrix} -\bar{P}_i & * & \cdots & * & * & * \\ \left(\tilde{A}_i(H_i(z_k))\bar{P}_i + \tilde{B}_i(H_i(z_k))\left((1-\lambda)K_0 + \lambda K_i(H_i(z_k))\right)\tilde{G}_{r(k)}\bar{P}_i\right) & -(pr_{i0})^{-1}\bar{P}_0 & \cdots & 0 & 0 & 0 \\ \vdots & \vdots & \ddots & \vdots & \vdots & \vdots \\ \left(\tilde{A}_i(H_i(z_k))\bar{P}_i + \tilde{B}_i(H_i(z_k))\left((1-\lambda)K_0 + \lambda K_i(H_i(z_k))\right)\tilde{G}_{r(k)}\bar{P}_i\right) & 0 & \cdots & -(pr_{id})^{-1} & 0 & 0 \\ \bar{P}_i & 0 & \cdots & 0 & -Q_i^{-1} & 0 \\ \left((1-\lambda)K_0 + \lambda K_i(H_i(z_k))\right)\tilde{G}_{r(k)}\bar{P}_i & 0 & \cdots & 0 & 0 & -R_i^{-1} \end{bmatrix} < 0, \quad i,j \in \mathcal{S}. \tag{8.32}$$

Then the homotopy algorithm can be summarized as follows:

Step 1: Initialization: set $k = 0$, select N and $N_{\max}$. Compute the initial values K_0 and P_0.

Step 2: Set $k = k + 1$ and $k = k/N$, set P to P_{k-1}.

If the LMIs (8.32) are feasible,
Then denote the feasible solution as ${K_i}^k$, set $P_k = P_{k-1}$, and go to Step 4,
Else go to Step 3.

Step 3: Set K_i to ${K_i}^{k-1}$,

If the LMIs (8.32) are feasible,
Then solve the minimization problem:
min trace(P) subject to (8.32),
denote the feasible solutions as P_k, and set $K_i = {K_i}^{k-1}$, then go to Step 4,

Else set $N = 2N$,
If $N > N_{\max}$, then the algorithm fails in giving feasible solution,
Else set $k = 0$, go to Step 2.

Step 4: If $k < N$, go to Step 2. If $k = N$, the obtained solutions ${K_i}^k$ and P_k are a set of feasible solutions of (8.25) and (8.26). □

8.4 Output Feedback Controller Synthesis of an NCS

8.4.1 Fuzzy Observer Design

Suppose not all state variables are available, the following fuzzy observer is considered:

$$\begin{aligned} \hat{x}_{k+1} &= A_i(H_i(z_k))\hat{x}_k + B_i(H_i(z_k))u_k \\ &\quad + \hat{L}_i(H_i(z_k))(\hat{y}_k - y_k), \\ \hat{y}_k &= C_i(H_i(z_k))\hat{x}_k + D_i(H_i(z_k))u_k. \end{aligned} \tag{8.33}$$

We wish to find observer gains $\hat{L}_i\left(H_i\left(z_k\right)\right)$ such that $e_k = \hat{x}_k - x_k \to 0$ asymptotically as $k \to \infty$. Define the fuzzy error system as:

$$e_{k+1} = \hat{x}_{k+1} - x_{k+1} = A_i^{cL}\left(H_i\left(z_k\right)\right) e_k, \tag{8.34}$$

where $A_i^{cL}\left(H_i(z_k)\right) = A_i\left(H_i(z_k)\right) + \hat{L}_i\left(H_i(z_k)\right) C_i\left(H_i(z_k)\right)$.

Then we come to the result for piecewise fuzzy jump observer synthesis.

Lemma 8.3. The closed-loop fuzzy error system (8.34) is stochastically stable, if for any given set of matrices $N_i > 0$, $i \in \mathcal{S}$, there exists a set of matrices E_i, F_i, and a set of symmetric matrices $X_i > 0$, $i \in \mathcal{S}$, satisfying the following matrix inequality:

$$\begin{bmatrix} -X_i + N_i & * & \cdots & * \\ O_i & pr_{i0}X_0 - E_i - {E_i}^T & \cdots & 0 \\ \vdots & \vdots & \ddots & \vdots \\ O_i & 0 & \cdots & pr_{id}X_d - E_i - {E_i}^T \end{bmatrix} < 0, \tag{8.35}$$

where

$$O_i = E_i A_i\left(H_i\left(z_k\right)\right) + F_i C_i\left(H_i\left(z_k\right)\right),\ i \in \mathcal{S}.$$

In addition, the observer gain for each subspace is given by:

$$\hat{L}_i = E_i^{-1} F_i,\ i \in \mathcal{S}. \tag{8.36}$$

Proof. Based on Definition 8.1 and Lemma 8.1, the closed-loop fuzzy error system (8.34) is stochastically stable if there exists a set of symmetric positive definite matrices $P_i > 0$, satisfying the following inequalities,

$$\sum_{j=0}^{d} pr_{ij} \left[A_i^{cL}\left(H_i\left(z_k\right)\right)\right]^T P_j \times \left[A_i^{cL}\left(H_i\left(z_k\right)\right)\right] - P_i + N_i < 0. \tag{8.37}$$

With $F_i = E_i \hat{L}_i$, the LMI (8.35) is equivalent to:

$$\begin{bmatrix} -X_i + N_i & * & \cdots & * \\ E_i A_i^{cL}\left(H_i\left(z_k\right)\right) & pr_{i0} X_0 - E_i - E_i^T & \cdots & 0 \\ \vdots & \vdots & \ddots & \vdots \\ E_i A_i^{cL}\left(H_i\left(z_k\right)\right) & 0 & \cdots & pr_{id} X_d - E_i - E_i^T \end{bmatrix} < 0. \tag{8.38}$$

The left-hand side of the inequality (8.38) can be pre-multiplied by T_1 and post-multiplied by $T_2 = T_1^T$ to yield the inequality (8.37), where

$$T_1 = \big[I \;\; \underbrace{\left(A_i^{cL}(H_i(z_k))\right)^T \cdots \left(A_i^{cL}\left(H_i\left(z_k\right)\right)\right)^T}_{d+1}\big].$$

Thus, LMI (8.35) implies the inequality (8.37). It can be concluded that the fuzzy error system (8.34) is stochastically stable. □

8.4.2 Output Feedback Controller Design

The output feedback fuzzy controller design presented above with sub-optimal guaranteed cost performance is based on the sub-optimal state feedback fuzzy controller and fuzzy observer in each subspace. When $r(k) = i$, the observer equation is defined in (8.33) and the output feedback jump fuzzy control law is:

$$u_k = \hat{K}_i\left(H_i\left(z_k\right)\right) \hat{x}_{k-i}, \tag{8.39}$$

where

$$\hat{K}_i\left(H_i\left(z_k\right)\right) = \sum_{l=1}^{t(i)} h_{il}\left(z_k\right) \hat{K}_{il}.$$

If we augment the variable as

$$\begin{aligned} \tilde{\hat{x}}_k &= \left[\hat{x}_k^T \;\; \hat{x}_{k-1}^T \cdots \hat{x}_{k-d}^T\right]^T, \;\; \tilde{\hat{x}}(k) \in^{(d+1)n}, \\ \tilde{e}_k &= \left[e_k^T \; e_{k-1}^T \cdots e_{k-d}^T\right]^T, \end{aligned} \tag{8.40}$$

then the closed-loop system becomes

$$\tilde{\hat{x}}_{k+1} = \left(\tilde{A}_i(H_i(z_k)) + \tilde{B}_i(H_i(z_k))\hat{K}_i(H_i(z_k))\tilde{G}_{r(k)}\right)\tilde{\hat{x}}_k + \hat{L}_i(H_i(z_k))\tilde{C}_i(H_i(z_k))\tilde{e}_k, \tag{8.41}$$
$$\tilde{e}_{k+1} = \left(\tilde{A}_i(H_i(z_k)) + \hat{L}_i(H_i(z_k))\tilde{C}_i(H_i(z_k))\right)\tilde{e}_k,$$

where

$$\tilde{C}_i(H_i(z_k)) = \begin{bmatrix} C_i(H_i(z_k)) & 0 \cdots 0 \end{bmatrix},$$
$$\tilde{L}_i(H_i(z_k)) = \begin{bmatrix} L_i(H_i(z_k)) \\ 0 \\ 0 \\ \vdots \\ 0 \end{bmatrix}.$$

Here for simplicity, the closed-loop output feedback jump fuzzy system dynamics can be described by

$$\bar{x}_{k+1} = \bar{A}_i(H_i(z_k))\bar{x}_k \tag{8.42}$$

where

$$\bar{x}_k = \begin{bmatrix} \tilde{\hat{x}}_k & \tilde{e}_k \end{bmatrix}^T,$$
$$\bar{A}_i(H_i(z_k)) = \begin{bmatrix} \tilde{A}_i^{CK} & \hat{L}_i(H_i(z_k))\tilde{C}_i(H_i(z_k)) \\ 0 & \tilde{A}_i(H_i(z_k)) + \hat{L}_i(H_i(z_k))\tilde{C}_i(H_i(z_k)) \end{bmatrix},$$
$$\tilde{A}_i^{CK} = \tilde{A}_i(H_i(z_k)) + \tilde{B}_i(H_i(z_k))\hat{K}_i(H_i(z_k))\tilde{G}_{r(k)}.$$

Then the output feedback fuzzy controller is obtained.

Lemma 8.4. The closed-loop output feedback jump fuzzy system (8.39) is stochastically stable if for any given set of symmetric matrices $W_i > 0$, $i \in \mathcal{S}$, there exists a set of symmetric matrices $\tilde{P}_i > 0$, $i \in \mathcal{S}$ satisfying the following matrix inequality:

$$\sum_{j=0}^{d} pr_{ij}\bar{A}_i^T(H_i(z_k))\tilde{P}_j\bar{A}_i(H_i(z_k)) - \tilde{P}_i + W_i < 0. \tag{8.43}$$

Proof. The result directly follows from Lemma 8.1. □

Lemma 8.4 is only useful for checking the closed-loop stability of the discrete-time jump fuzzy control system when the output feedback fuzzy controller is already available. Note that the matrix inequality (8.43) contains product terms involving $\hat{K}_i$, $\hat{L}_i$, $\tilde{P}_i$ and W_i. Nonlinear matrix inequality (NMI) technique is required to generate the output feedback jump fuzzy controller. Luckily, we have the following theorem by extending Theorem 5 of [25].

Lemma 8.5. The matrix inequality (8.43) has feasible solutions if LMIs (8.25)–(8.26) and (8.35)–(8.36) do.

Proof. Based on Theorem 5 of [25], we will show that if the feasible solutions to LMIs (8.25)–(8.26) and (8.35)–(8.36) can be found then there always exists a positive scalar α, such that:

$$\tilde{P}_i = \begin{bmatrix} \bar{P}_i & 0 \\ 0 & \alpha X_i \end{bmatrix}, \; i \in \mathcal{S}, \tag{8.44}$$

satisfies (8.43). And α can be obtained from the following inequality:

$$\begin{aligned} \alpha \left[\sum_{j=0}^{d} pr_{ij} \left(A_i^{cL} \right)^T X_i A_i^{cL} - X_i \right] &< \left(B_i \left(H_i \left(z_k \right) \right) \hat{K}_i \left(H_i \left(z_k \right) \right) \right)^T \\ &\times \left[\sum_{j=0}^{d} pr_{ij} \left(A_i^{CK} \right)^T \bar{P}_i^{-1} A_i^{CK} - \bar{P}_i^{-1} \right]^{-1} \\ &\times \left(B_i \left(H_i \left(z_k \right) \right) \hat{K}_i \left(H_i \left(z_k \right) \right) \right), \end{aligned} \tag{8.45}$$

where

$$A_i^{CK} = A_i \left(H_i \left(z_k \right) \right) + B_i \left(H_i \left(z_k \right) \right) \hat{K}_i \left(H_i \left(z_k \right) \right).$$

□

Based on Lemmas 8.4 and 8.5, the output feedback jump fuzzy controller can be designed.

8.4.3 Simulation Example

To illustrate the proposed theoretical results, a numerical example is considered.

Different networks vary in network-induced delay bounds and the rate of data loss. The induced delays and data dropout of typical networks are shown in Figs. 8.3 and 8.4 by simulations using OPNET software.

In our example, local area network (LAN) including Ethernet, token ring, etc., in which the induced delay is low and rate of data dropout is nearly zero, is used as the communication network in the NCS. With the purpose of defining $v(r(k))$, NCS experiments with fixed constant delays bounded by the LAN delay are presented. If the low delay is considered, a good output result is shown in Fig. 8.5 when delay is less than 0.001. On the contrary, if the delay is high, the system is out of control when delay is larger than 0.007.

Based on the experiment, the states $r(k) = 0,\ 1,\ 2$ denote that the network induced-delay is low, medium and high, respectively, and have the following transition probability matrix:

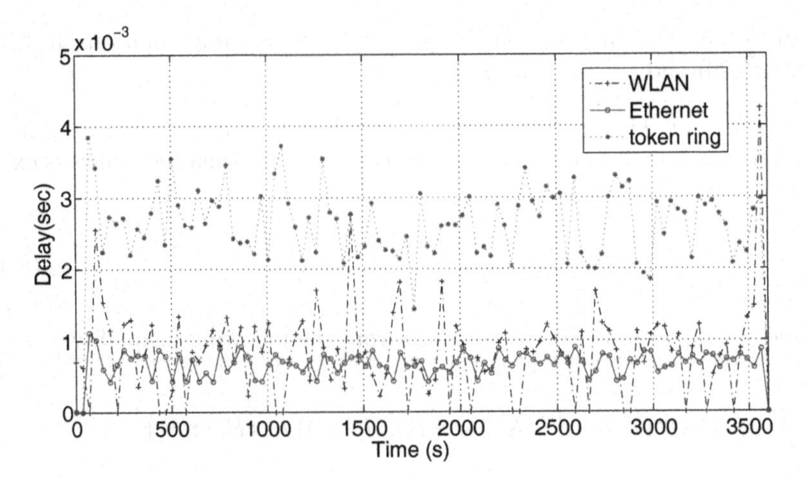

Fig. 8.3. Induced delay of typical networks

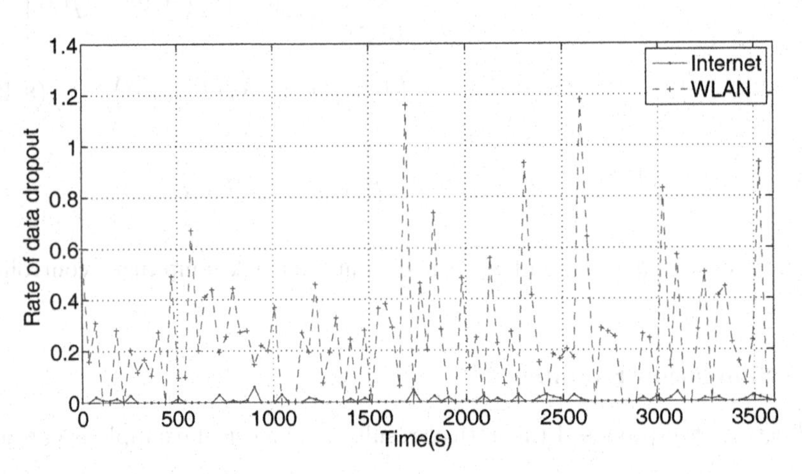

Fig. 8.4. Data dropout rate of typical networks

$$Pr = \begin{bmatrix} 0.5 & 0.5 & 0 \\ 0.3 & 0.6 & 0.1 \\ 0.3 & 0.6 & 0.1 \end{bmatrix}.$$

The delays $v(r(k))$ corresponding to the three states are:

$$\begin{cases} v(r(k)) \in [0, 0.001], & r(k) = 0, \\ v(r(k)) \in [0.001, 0.007], & r(k) = 1, \\ v(r(k)) \in [0.007, 0.01], & r(k) = 2. \end{cases}$$

Consider the discrete-time jump fuzzy system given by

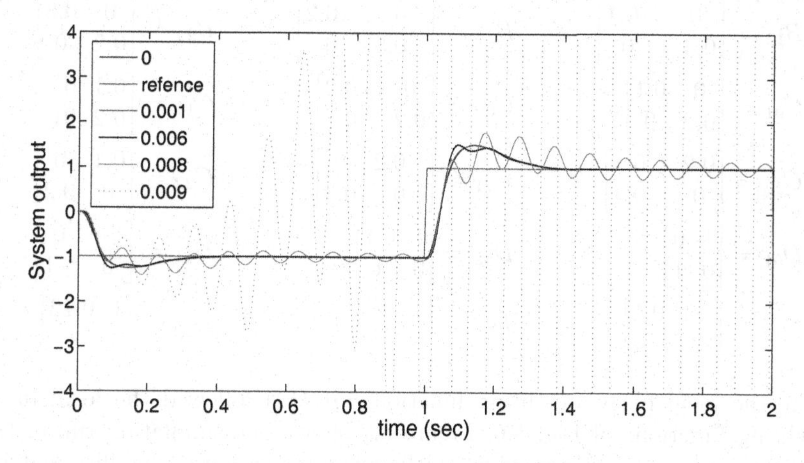

Fig. 8.5. System outputs under different fixed delays

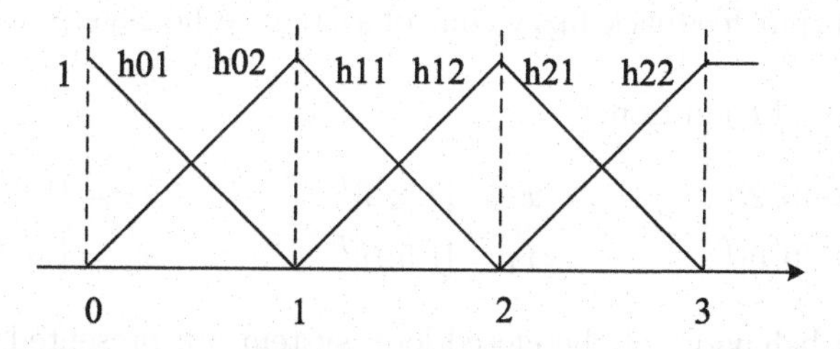

Fig. 8.6. The local fuzzy weighting functions of the example

$$x_{k+1} = \sum_{l=1}^{2} h_{il} \left(\tau_k \left(r\left(k\right)\right)\right) \{A_{il}x_k + B_{il}u_k\},$$

$$y_{k+1} = \sum_{l=1}^{2} h_{il} \left(i + pr_{(i-1)i}\right) \{C_{il}x_k + D_{il}u_k\},$$

with

$$A_{01} = \begin{bmatrix} 0.8 & 0.1 \\ -0.5 & 0.8 \end{bmatrix}, \quad A_{02} = \begin{bmatrix} 0.5 & 0.1 \\ -0.4 & 0.5 \end{bmatrix}, \quad A_{11} = \begin{bmatrix} 0.7 & 0.1 \\ -0.3 & 0.7 \end{bmatrix},$$

$$A_{12} = \begin{bmatrix} 0.5 & 0.1 \\ -0.1 & 0.5 \end{bmatrix}, \quad A_{21} = \begin{bmatrix} 1 & 0.1 \\ -0.4 & 1 \end{bmatrix}, \quad A_{22} = \begin{bmatrix} 0.5 & 0.1 \\ -0.3 & 0.5 \end{bmatrix},$$

$$B_{01} = \begin{bmatrix} 0 & 0.1 \\ 0.2 & 0 \end{bmatrix}, \quad B_{02} = \begin{bmatrix} 0 & -0.2 \\ 0.2 & 0 \end{bmatrix}, \quad B_{11} = \begin{bmatrix} 0 & 0.1 \\ 0.2 & 0 \end{bmatrix},$$

$$B_{12} = \begin{bmatrix} 0 & -0.1 \\ 0.4 & 0 \end{bmatrix}, \quad B_{21} = \begin{bmatrix} 0 & -0.2 \\ -0.5 & 0 \end{bmatrix}, \quad B_{22} = \begin{bmatrix} 0 & 0.3 \\ 0.2 & 0 \end{bmatrix},$$

$$C_{01} = \begin{bmatrix} 0 & 0.1 \\ 0.2 & 0 \end{bmatrix}, \quad C_{02} = \begin{bmatrix} 0 & -0.2 \\ 0.2 & 0 \end{bmatrix}, \quad C_{11} = \begin{bmatrix} 0.1 & 0 \\ 0.2 & 0 \end{bmatrix},$$

$$C_{12} = \begin{bmatrix} 0.2 & 0 \\ 0 & -0.4 \end{bmatrix}, \quad C_{21} = \begin{bmatrix} -0.2 & 0 \\ 0 & 0.3 \end{bmatrix}, \quad C_{22} = \begin{bmatrix} 0.1 & 0 \\ 0 & 0.2 \end{bmatrix},$$

$$D_{01} = \begin{bmatrix} 1 & 0 \\ 0 & 1 \end{bmatrix}, \quad D_{02} = \begin{bmatrix} 1 & 0.1 \\ 0 & 1 \end{bmatrix}, \quad D_{11} = \begin{bmatrix} -1 & 0 \\ 0 & -1 \end{bmatrix},$$

$$D_{12} = \begin{bmatrix} -1 & 0.1 \\ 0 & -1 \end{bmatrix}, \quad D_{21} = \begin{bmatrix} 1 & 0 \\ 0 & 1 \end{bmatrix}, \quad D_{22} = \begin{bmatrix} 1 & 0.12 \\ 0 & 1 \end{bmatrix},$$

where the local fuzzy weighting function $h_{il}(r(k))$ followed the local fuzzy weighting functions of Fig. 8.6. Figure 8.7 shows one simulation run of the Markovian jump delays according to the given transition probability matrix.

By using mincx() in MATLAB® LMI Toolbox, the minimal α for the closed-loop output feedback fuzzy control system to be asymptotically stable is 0.5438.

For the initial condition:

$$x_0 = [3,2]^T, \quad x_1 = [2.5,2]^T, \quad x_2 = [1,1]^T,$$
$$\hat{x}_0 = [0,0]^T, \quad \hat{x}_1 = [1.5,0]^T, \quad \hat{x}_2 = [1,1]^T.$$

The response behaviors of the closed-loop system are presented in Figs. 8.8 and 8.9 using the output feedback fuzzy controller. Fig. 8.8 shows the system state responses and their estimates, while Fig. 8.9 shows outputs and control variables curves and their corresponding estimates. It is shown from these two figures that the estimated variables can converge to the original ones asymptotically such that good system performance can be guaranteed.

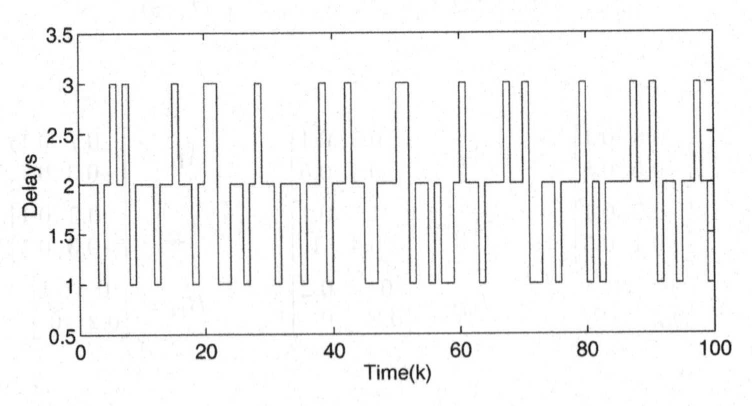

Fig. 8.7. Random delays

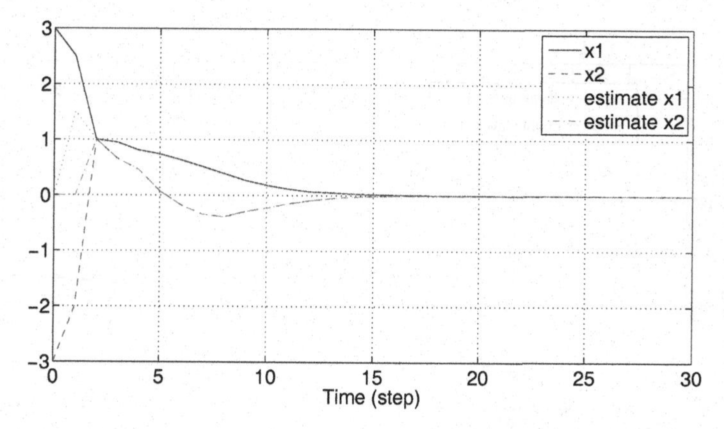

Fig. 8.8. The system state responses and their estimates

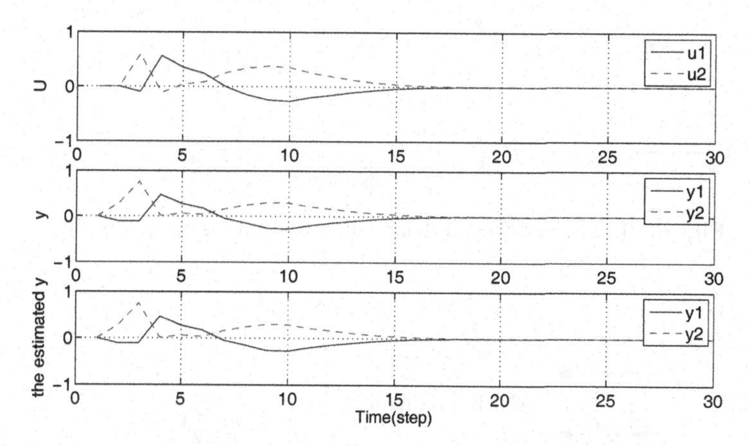

Fig. 8.9. The output and control variables curves and their corresponding estimates

8.5 Neuro-fuzzy Controller Design

Network type has a great effect on the characteristic of NCS. Fig. 8.5 shows the system is out of control when the delay is high. Satellite network [26] is a typical network with longer induced delay shown in Fig. 8.10 by simulation using OPNET software.

A new neuro-fuzzy controller is presented for the NCS based on the mixed network including terrestrial networks and satellite networks. It is constructed of three parts: the guaranteed cost controller, the ANFIS predictor and the fuzzy controller, presented in Fig. 8.11.

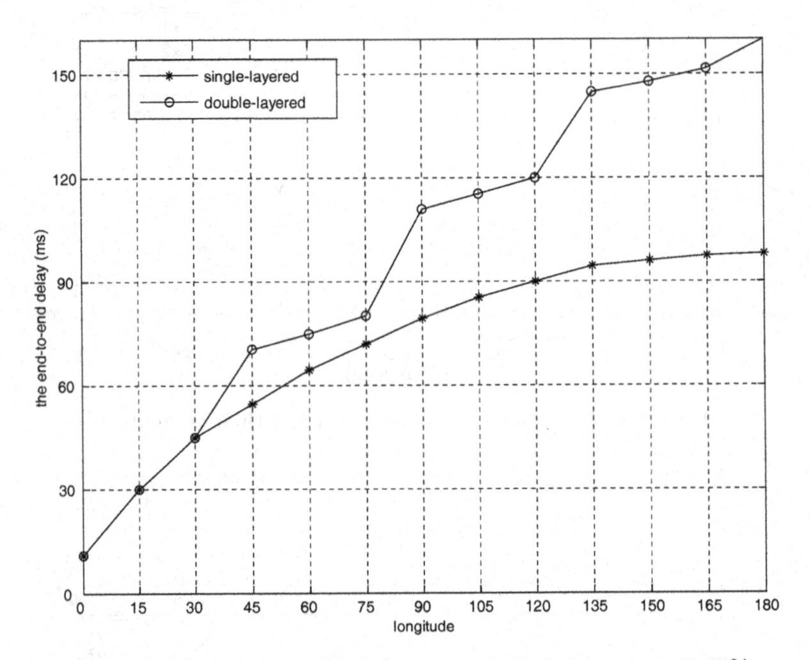

Fig. 8.10. The end-to-end delay when the link usage rate is 90%

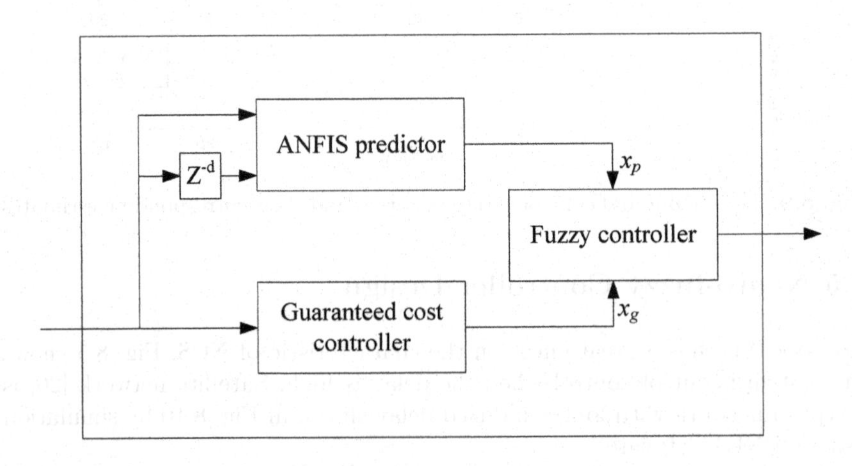

Fig. 8.11. The neuro-fuzzy controller of NCS

8.5.1 Neuro-fuzzy Predictor

In this subsection, a new method is proposed to improve the performance of the NCS by adding a predictor to estimate the plant state. The neuro-fuzzy predictor is computationally straightforward and has shown excellent prediction capabilities. So the decision is made to use the ANFIS [18]. The ANFIS predictor has two inputs: system states at time k and $k-d$, and produce the predicted state value at time $k+d$. The architecture of the ANFIS predictor is shown in Fig. 8.12, where d is the delay bound.

In the following description, ${u_i}^k$ denotes the ith input of a node in the kth layer, ${o_i}^k$ denotes the ith node output in layer k, and there are n input values.

The ANFIS predictor uses Gaussian functions for fuzzy sets. The reason is that a multidimensional Gaussian membership function can easily be decomposed into the product of one-dimensional membership functions. With this choice, the operation performed in this layer is

$$o_{ij}^2 = \exp\left\{-\frac{(u_i^2 - m_{ij}^2)}{\left(\delta_{ij}^2\right)^2}\right\},\ i = 1,\ 2,\ j = 1,\ 2,\ \ldots,\ m, \tag{8.46}$$

where u_i^2 and δ_{ij}^2 are, respectively, the center and the width of the Gaussian membership function. The ANFIS predictor uses Gaussian functions for fuzzy sets, linear functions for the outputs, and Sugeno's inference mechanism [16]. The parameters of the network are the mean and standard deviation of the membership functions and the coefficients of the output linear functions. The ANFIS predictor learning algorithm is then used to obtain these parameters.

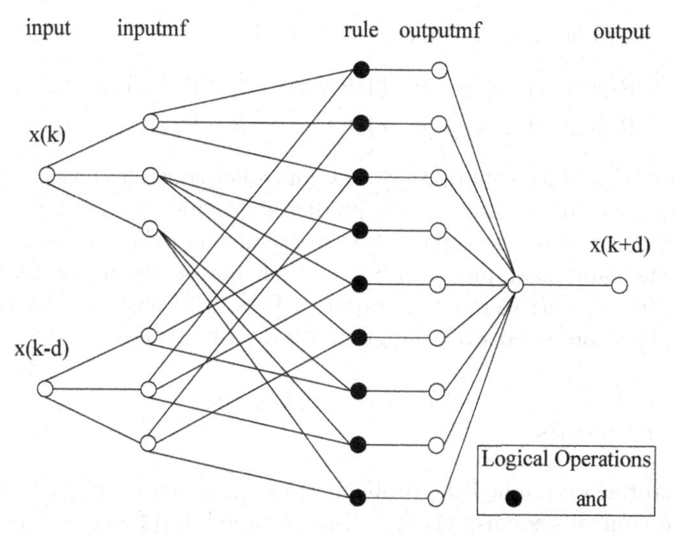

Fig. 8.12. The structure of the ANFIS predictor

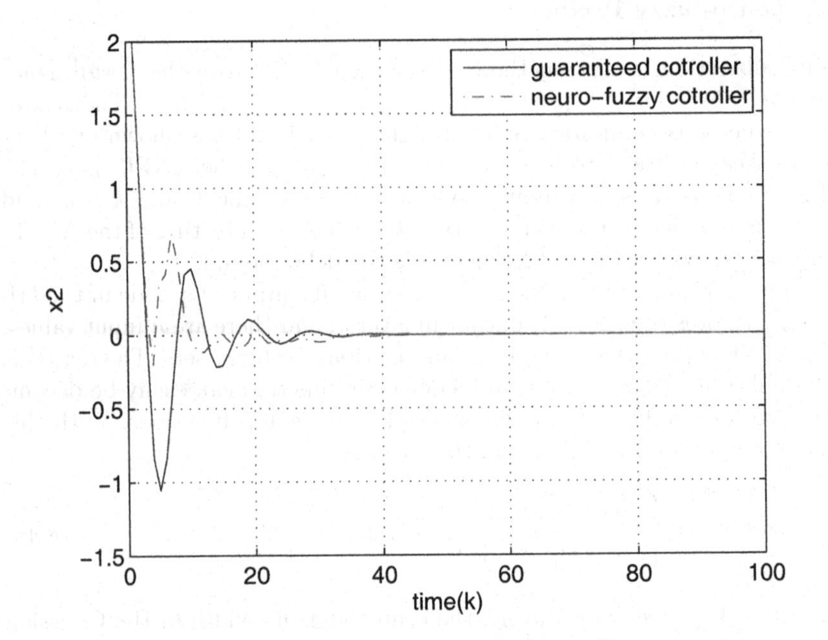

Fig. 8.13. State responses

The learning algorithm is a hybrid algorithm consisting of the gradient descent and the least squares estimate.

8.5.2 Fuzzy Controller

The fuzzy controller has two rules:

$$\text{Rule 1: IF } r(k) < d, \text{ THEN } u_k = K_i(H_i(z_k)) = x_g;$$
$$\text{Rule 2: IF } r(k) = d, \text{ THEN } u_k = K_i(H_i(z_k)) = x_p.$$

When $r(k) < d$, the guaranteed cost controller controls the system. When network delay is longer, the ANFIS predictor provides the predicted state at time $k+d$. In this way, the impact of network's longer delay can be moderated.

With the same example as in Section 8.4, better control performance of the neuro-fuzzy controller is illustrated in Figs. 8.13 and 8.14 for state and control inputs compared with the guaranteed cost controller.

8.6 Conclusions

In this chapter, we studied the problem of modeling and controller design for networked control systems, where a discrete-time jump fuzzy system is developed to model networked control systems with random but bounded delays

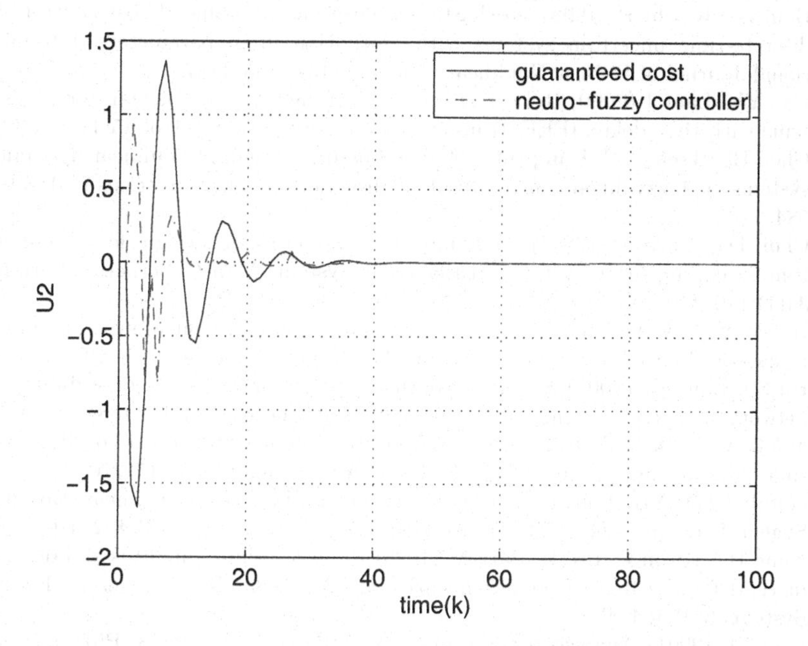

Fig. 8.14. Control inputs

and packets dropout. On the basis of the assumption that all state variables of an NCS are available, a state feedback controller is developed for the discrete-time jump system with sub-optimal guaranteed cost performance based on a piecewise quadratic Lyapunov function. It is shown that the state feedback sub-optimal fuzzy controller can be obtained by solving a set of LMIs using the homotopy approach. Because not all state variables are available in many practical cases, an output feedback fuzzy controller is proposed, which cannot only stabilize the system, but also meet certain desired sub-optimal system performance criteria. The LMI technique is used to effectively minimize the overall cost function and thus achieve the sub-optimal system. When the network is in a poor condition, a novel neuro-fuzzy controller including an ANFIS predictor is designed to deal with the problem. Finally, the effectiveness of the proposed approaches are verified by a numerical example.

References

1. Almutairi NB, Chow MY, Tipsuwan Y (2001) Network-based controlled DC motor with fuzzy compensation. In: Proceedings of the 27th Annual Conference of the IEEE Industrial Electronics Society, Denver, CO, 3:1844–1849
2. Benjelloun K, Boukas EK, Shi P (1997) Robust stochastic stability of discrete-time linear systems with markovian jump parameters. In: Proceedings of the 36th IEEE Conference on Decision and Control, San Diego, CA, 559–563

3. Boukas EK, Shi P (1998) Stochastic stability and guaranteed cost control of discrete-time uncertain systems with Markovian jump parameters. International Journal of Robust Nonlinear Control 8(13):1155–1167
4. Cao YY, Lam J (2000) Robust H^∞ control of uncertain markovian jump systems with time-delay. IEEE Transactions on Automatic Control 45(1):77–83
5. Chen BS, Tseng CS, Uang HJ (1999) Robustness design of nonlinear dynamic systems via fuzzy linear control. IEEE Transactions on Fuzzy Systems 7(5):571–584
6. Choi DJ, Park P (2004) Guaranteed cost controller design for discrete-time switching fuzzy systems. IEEE Trans Systems Man Cybernetics–Part B 34(1):110–119
7. Halevi Y, Ray A (1998) Integrated communication and control systems: Part I-analysis. Journal of Dynamic Systems, Measurement and Control 110:367–373
8. Hu SS, Zhu QX (2003) Stochastic optimal control and analysis of stability of networked control systems with long delay. Automatica 39:1877–1884
9. Ji YH, Chizeck J, Feng X, Loparo KA (1991) Stability and control of discrete-time jump linear systems. Control Theory Adv Technol 7(1):247–270
10. Joh J, Chen YH, Langari R (1998) On the stability issues of linear Takagi–Sugeno fuzzy models. IEEE Transactions on Fuzzy Systems 6(3):402–410
11. Kang HJ, Kwon C, Lee H, Park M (1998) Robust stability analysis and design method for the fuzzy feedback linearization regulator. IEEE Trans on Fuzzy Systems 6(4):464–472
12. Lian FL (2001) Analysis, design, modeling, and control of NCS. PhD Thesis, University of Michigan, USA
13. Liu HP, Sun FC, Hu YN (2005) H^∞ control for fuzzy singularly perturbed systems. Fuzzy Sets and Systems 155:272–291
14. Nesic D, Teel AR (2004) Input-to-state stability of networked control systems. Automatica 40:2121–2128
15. Nilson J (1998) Real-time control systems with delays. PhD Thesis, Lund Institute of Technology, Sweden
16. Palm R, Driankov D (1998) A fuzzy switched hybrid systems-modeling and identification. In: Proceedings of the IEEE ISIC/CIRA/ISAS Joint Conference, Gaithersburg, MD, 130–135
17. Peter VZ, Richard HM (2003) Networked control dedign for linear systems. Automatica 39:743–750
18. Takagi T, Sugeno M (1985) Fuzzy identification of systems and its applications to modeling and control. IEEE Trans Systems Man Cybernetics SMC-15:116–132
19. Tanaka K, Iwasaki M, Wang HO (2000) Stable switching fuzzy control and its application to a Hovercraft type vehicle. In: Proceedings of the Ninth IEEE International Conference on Fuzzy Systems, San Antonio, TX, 804–809
20. Tanaka K, Ikeda T, Wang HO (1996) Robust stabilization of a class of uncertain nonlinear systems via fuzzy control: Quadratic stabilizability, H^∞ control theory and linear matrix inequalities. IEEE Transactions on Fuzzy Systems 4(1):1–13
21. Tipsuwan Y, Chow MY (2001) Networked-based controller adaptation based on QoS negotiation and deterioration. In: Proceedings of the 27th Annual Conference of the IEEE Industrial Electronics Society, Denver, CO, 3:1794–1799
22. Tipsuwan Y, Chow MY (2003) Control methodologies in networked control systems. Control Engineering Practice 11:1099–1111

23. Walsh GC, Beldiman O, Bushnell LG (2001) Asymptotic behavior of non-linear networked control systems. IEEE Transactions on Automatic Control 46(7):1093–1097
24. Walsh GC, Ye H, Bushnell LG (2002) Stability analysis of networked control systems. IEEE Transactions on Control Systems Technology 10(3):438–446
25. Wang L, Feng G, Hesketh T (2003) Piecewise output feedback controller synthesis of discrete time fuzzy systems. In: Proceedings of the 42th Conference on Decision and Control, Las Vegas, NV, 4741–4746
26. Wang CJ (1995) Delivery time analysis of low earth orbit satellite work for seamless PCS. IEEE Journal on Selected Areas in Communications 13(2):389–396
27. Xiao L, Hassibi A, How JP (2000) Control with random communication delays via a discrete-time jump system approach. In: Proceedings of the American Control Conference, Chicago, IL, 3:2199–2204
28. Yu ZX, Chen HT, Wang YJ (2002) Research on Markov delay characteristic-based closed loop network control system (in Chinese). Control Theory and Applications 19(2):263–267
29. Zhang W, Branicky MS, Phillips SM (2001) Stability of networked control systems. IEEE Control Systems Magazine 21(1):84–99

9

Networked Boundary Control of Damped Wave Equations

YangQuan Chen

Utah State University, Logan, UT 84322, USA yqchen@ece.usu.edu

Abstract. This chapter considers the boundary control of damped wave equations using a boundary measurement in a networked control system (NCS) setting. In this networked boundary control system, the induced delays can be lumped as the boundary measurement delay. The Smith predictor is applied to the networked boundary control problem and the instability problem due to large delays is solved and the scheme is proved to be robust against a small difference between the assumed delay and the actual delay. In addition, we analyze the robustness of the time-fractional order wave equation with a fractional order boundary controller subject to delayed boundary measurement. Conditions are given to guarantee stability when the delay is small. For large delays, again the Smith predictor is applied to solve the instability problem and the scheme is proved to be robust against a small difference between the assumed delay and the actual delay. The analysis shows that fractional order controllers are better than integer order controllers in the robustness against delays in the boundary measurement.

Keywords. Boundary control, distributed parameter system, fractional order calculus, robustness, wave equation.

9.1 Introduction

In recent years, boundary control of flexible systems has become an active research area, due to the increasing demand on high precision control of many mechanical systems, such as spacecraft with flexible attachments or robots with flexible links, which are governed by PDEs (partial differential equations) rather than ODEs (ordinary differential equations) [2, 3, 4, 7, 8, 18, 19, 20, 21]. In this research area, the robustness of controllers against delays is an important topic and has been studied by many researchers [5, 6, 14, 15, 17], due to the fact that delays are unavoidable in practical engineering. All the available publications focus on the analysis of systems against a small delay, i.e., under what conditions a very small delay will not cause instability problems and can therefore be neglected. An equally important and very practical issue is, how to synthesize a boundary controller when the delay is large and makes the system unstable. To the best of our knowledge, publications studying this problem are very few. In this chapter, we solve the instability problem caused by large delays by applying the Smith predictor to the boundary control of the damped wave equation. The control scheme is shown to be stable and robust against a small difference between the actual delay and the assumed delay.

Fractional diffusion and wave equations are obtained from the classical diffusion and wave equations by replacing the first and second order time derivative term by a fractional derivative of an order satisfying $0 < \alpha \leq 1$ and $1 < \alpha \leq 2$, respectively. Since many of the universal phenomena can be modeled accurately using the fractional diffusion and wave equations (see [22]), there has been growing interest in investigating the solutions and properties of these evolution equations. Compared with the publications on control of integer order PDEs, results on control of fractional wave equations are relatively few [10, 11, 16].

In this chapter, we will also investigate two robust stabilization problems of the fractional wave equations subject to delayed boundary measurement. First, under what conditions a very small delay in boundary measurement will not cause instability problems. Second, how to stabilize the system when the delay is large and makes the system unstable.

9.2 A Brief Introduction to the Smith Predictor

The Smith predictor was proposed by Smith in [24] and is probably the most famous method for control of systems with time delays; see [9] and [25]. Consider a typical feedback control system with a time delay in Fig. 9.1, where $C(s)$ is the controller and $P(s)e^{-\theta s}$ is the plant with a time delay θ.

With the presence of the time delay, the transfer function of the closed-loop system relating the output $y(s)$ to the reference $r(s)$ becomes

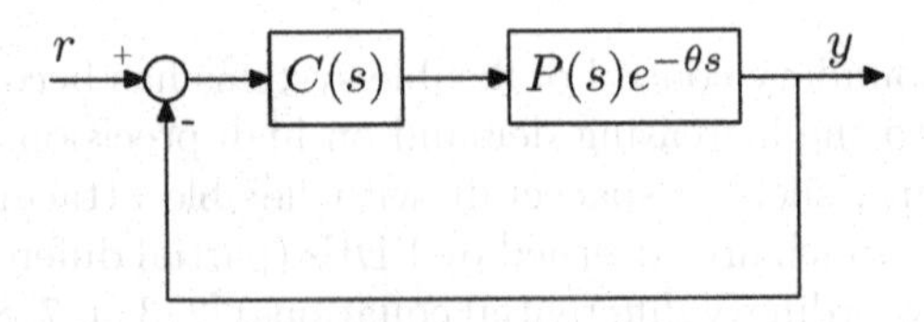

Fig. 9.1. A feedback control system with a time delay

$$\frac{y(s)}{r(s)} = \frac{C(s)P(s)e^{-\theta s}}{1 + C(s)P(s)e^{-\theta s}}. \tag{9.1}$$

Obviously, the time delay θ directly changes the closed-loop poles. Usually, the time delay reduces the stability margin of the control system, or more seriously, destabilizes the system.

The classical configuration of a system containing a Smith predictor is depicted in Fig. 9.2, where $\hat{P}(s)$ is the assumed model of $P(s)$ and $\hat{\theta}$ is the assumed delay. The block $C(s)$ combined with the block $\hat{P}(s) - \hat{P}(s)e^{-\hat{\theta}s}$ is called "the Smith predictor". If we assume perfect model matching, i.e., $\hat{P}(s) = P(s)$ and $\theta = \hat{\theta}$, the closed-loop transfer function becomes

$$\frac{y(s)}{r(s)} = \frac{C(s)P(s)e^{-\theta s}}{1 + C(s)P(s)}. \tag{9.2}$$

Now, it is clear what the underlying idea of the Smith predictor is. With perfect model matching, the time delay can be removed from the denominator of the transfer function, making the closed-loop stability independent of the time delay.

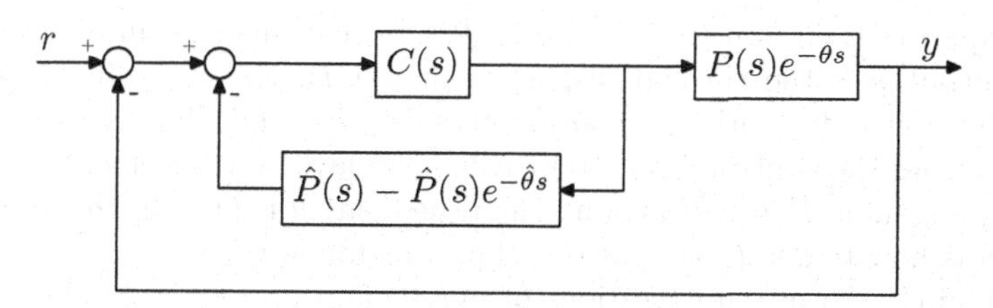

Fig. 9.2. The Smith predictor

9.3 Boundary Control of Damped Wave Equations with Large Delays

Consider a string clamped at one end and free at the other end. We denote the displacement of the string by $u(x,t)$, where $x \in [0,1]$ and $t \geq 0$. The string is controlled by a boundary control force at the free end. The governing equations are given as

$$u_{tt}(x,t) - u_{xx}(x,t) + 2au_t(x,t) + a^2u(x,t) = 0, \tag{9.3}$$

$$u(0,t) = 0, \tag{9.4}$$

$$u_x(1,t) = f(t), \tag{9.5}$$

where $a > 0$ is the damping constant and $f(t)$ is the boundary control force applied at the free end of the string.

It is known that the following boundary feedback controller stabilizes the system [2],

$$f(t) = -ku_t(1,t), \tag{9.6}$$

where $k > 0$ is the constant boundary control gain.

Now, we consider the presence of a time delay in the feedback loop, which is shown as follows:

$$f(t) = -ku_t(1,t-\theta), \tag{9.7}$$

where θ is the time delay.

In [6] and [15], it was shown that if k and a satisfy

$$k\frac{e^{2a}+1}{e^{2a}-1} < 1, \tag{9.8}$$

then the delayed feedback systems is stable for all sufficiently small delays.

In this chapter, we will solve the following problem: what if the time delay θ is large enough to make the system unstable? We will apply the Smith predictor to solve this problem.

Comparing Equation (9.7) with Fig. 9.2, we can see that in our case, the plant output y is the tip end displacement $u(1,t)$; the controller $C(s)$ is a derivative controller with the transfer function ks; and $P(s)$ is the transfer function from the control force $f(t)$ to the undelayed displacement of the tip end. If we assume $\hat{P}(s) = P(s)$ and the time delay θ is known, the remaining problem is how to get $P(s)$, which is shown as follows.

Assuming zero initial conditions of $u(x,0)$ and $u_t(x,0)$, take the Laplace transform of (9.3), (9.4), and (9.5) with respect to t, the original PDE of $u(x,t)$ with initial and boundary conditions can be transformed into the following ODE of $U(x,s)$ with boundary conditions:

$$\frac{d^2U(x,s)}{dx^2} - (s+a)^2U(x,s) = 0, \tag{9.9}$$

$$U(0,s) = 0, \tag{9.10}$$

$$U_x(1,s) = F(s), \tag{9.11}$$

where $U(x,s)$ is the Laplace transform of $u(x,t)$ and $F(s)$ is the Laplace transform of $f(t)$.

Solving the ODE (9.9), we have the following solution of $U(x,s)$ with two arbitrary constants C_1 and C_2 (s can be treated as a constant in this step),

$$U(x,s) = C_1e^{-(s+a)x} + C_2e^{(s+a)x}. \tag{9.12}$$

Substitute (9.12) into (9.10) and (9.11), we have the following two equations:

$$C_1 + C_2 = 0, \tag{9.13}$$

$$(-C_1 e^{-(s+a)} + C_2 e^{s+a})(s+a) = F(s). \tag{9.14}$$

Solving (9.13) and (9.14) simultaneously, we can obtain the exact values of C_1 and C_2

$$C_1 = \frac{-F(s)}{(s+a)(e^{-(s+a)} + e^{s+a})}, \tag{9.15}$$

$$C_2 = \frac{F(s)}{(s+a)(e^{-(s+a)} + e^{s+a})}. \tag{9.16}$$

Now we have obtained the solution of $U(x,s)$. Substituting $x = 1$ into $U(x,s)$, we obtain the following Laplace transform of the tip end displacement.

$$U(1,s) = \frac{F(s)(1 - e^{-2(s+a)})}{(s+a)\left(1 + e^{-2(s+a)}\right)}. \tag{9.17}$$

So the transfer function of the plant, which is $P(s)$ in Fig. 9.2, is obtained as

$$P(s) = \frac{U(1,s)}{F(s)} = \frac{1 - e^{-2(s+a)}}{(s+a)\left(1 + e^{-2(s+a)}\right)}. \tag{9.18}$$

Finally, we have the following expression for the boundary controller (the Smith predictor), denoted as $C_{sp}(s)$:

$$C_{sp}(s) = \frac{ks}{1 + ksP(s)(1 - e^{-\hat{\theta}s})}. \tag{9.19}$$

Notice that the controller (9.19) is physically implementable.

9.4 Stability and Robustness Analysis

In [2], the stability of the controller (9.6) was proved for the boundary control of the damped wave equation without delays. If the assumed delay is equal to the actual delay, the Smith predictor removes the delay term completely from the denominator of the closed-loop transfer function. This means the stability of the controller (9.19) is already proved.

Since the actual delay θ and the assumed delay $\hat{\theta}$ cannot be exactly the same, another important issue is the robustness of the controller (9.19), i.e., what if an unknown small difference ϵ between the assumed delay and the actual delay is introduced to the system, as shown in Fig. 9.3.

To study the robustness of the controller (9.19), we will first introduce a theorem presented in [14, 15].

Theorem 9.1. Let $H(s)$ be the open-loop transfer function as illustrated in Fig. 9.4, and $\mathfrak{D}_H$ the set of all its poles. Define two closed-loop transfer functions $G_0(s)$ and $G_\epsilon(s)$ as

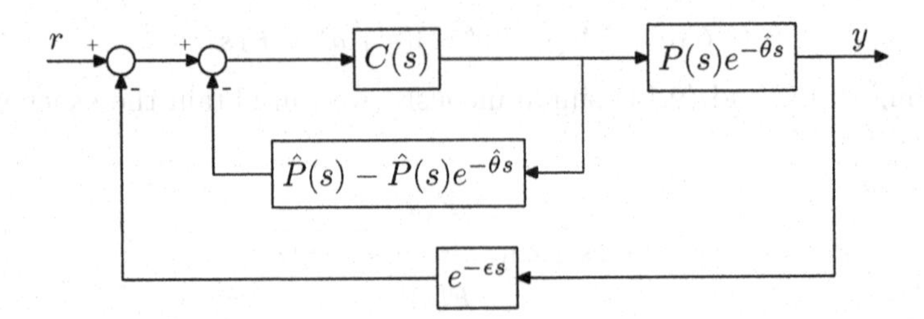

Fig. 9.3. System with mis-matched delays

$$G_0(s) = \frac{H(s)}{1 + H(s)},$$

and

$$G_\epsilon(s) = \frac{H(s)}{1 + e^{-\epsilon s} H(s)}.$$

Define again

$$\mathbb{C}_0 = \{s \in \mathbb{C} | \Re(s) > 0\},$$

and

$$\gamma(H(s)) = \limsup_{|s| \to \infty, s \in \mathbb{C}_0 \backslash \mathfrak{D}_H} |H(s)|.$$

Suppose G_0 is L^2-stable. If $\gamma(H) < 1$, then there exists ϵ^* such that G_ϵ is L^2-stable for all $\epsilon \in (0, \epsilon^*)$. □

The underlying idea of the above theorem is that the robustness of the closed-loop transfer function $G_0(s)$ against a small unknown delay can be determined by studying the open-loop transfer function $H(s)$. Now we can prove the robustness of the controller (9.19).

Claim. If $\hat{\theta}$ is chosen as the minimum value of the possible delay and k is chosen to satisfy

$$k \frac{e^{2a} + 1}{e^{2a} - 1} \leq \frac{1}{3}, \tag{9.20}$$

then the controller (9.19) is robust against a small difference ϵ between the assumed delay $\hat{\theta}$ and the actual delay $\theta = \hat{\theta} + \epsilon$.

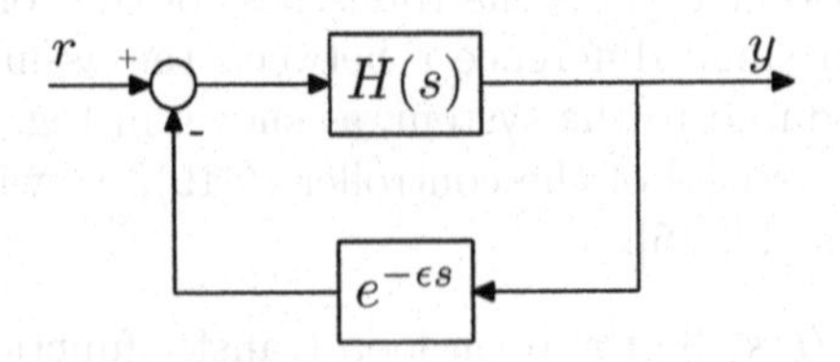

Fig. 9.4. Feedback system with delay

Proof. For

$$H(s) = C_{sp}(s)P(s)e^{-\hat{\theta}s} = \frac{ksP(s)e^{-\theta s}}{1 + ksP(s)(1 - e^{-\theta s})}.$$

Let $T(s) = ksP(s)$. Then

$$|H(s)| = \frac{1}{\left|\left(\frac{1}{T(s)} + 1\right) e^{\hat{\theta}s} - 1\right|}. \tag{9.21}$$

Let $Q(s) = \left(\frac{1}{T(s)} + 1\right) e^{\hat{\theta}s} - 1$. Then

$$\begin{aligned} |Q(s)| &= \left|\left(\frac{1}{T(s)} + 1\right) e^{\hat{\theta}s} - 1\right| \\ &\geq \left|\left|\left(\frac{1}{T(s)} + 1\right) e^{\hat{\theta}s}\right| - 1\right| \\ &\geq \left|\left|\frac{1}{T(s)} + 1\right| \left|e^{\hat{\theta}s}\right| - 1\right|. \end{aligned} \tag{9.22}$$

In [15], it was proved that

$$\limsup_{|s|\to\infty, s\in\mathbb{C}_0} |T(s)| = k\frac{e^{2a} + 1}{e^{2a} - 1}.$$

So, if

$$k\frac{e^{2a} + 1}{e^{2a} - 1} \leq \frac{1}{3},$$

for $|s|$ large enough,

$$\left|\frac{1}{T(s)} + 1\right| \geq \left|\left|\frac{1}{T(s)}\right| - 1\right| \geq 2. \tag{9.23}$$

Considering $|e^{\hat{\theta}s}| > 1$, we have

$$|Q(s)| > 1. \tag{9.24}$$

So

$$\limsup_{|s|\to\infty, s\in\mathbb{C}_0} |H(s)| < 1. \tag{9.25}$$

□

Remark 9.1. In Theorem 9.1, ϵ is positive. To satisfy this condition, $\hat{\theta}$ should be chosen as the minimal value of the possible delay.

The damping constant a plays a key role in making the controllers (both the original derivative controller ks and the Smith predictor) robust. if $a =$

0, the damped wave equation becomes the conservative wave equation, the transfer function of which is

$$P(s) = \frac{1 - e^{-2s}}{s(1 + e^{-2s})}. \tag{9.26}$$

We can see that $P(s)$ has an infinite number of poles on the imaginary axis. In order to make $\gamma(H(s)) < 1$, controllers must cancel these poles completely, which is impossible due to the uncertainty of the plant parameters. This means both the original derivative controller ks and the Smith predictor are not robust when applied to the boundary control of conservative wave equation. □

9.5 Fractional Order Case – Problem Formulation

We consider a cable made with special smart materials governed by the fractional wave equation, fixed at one end, and stabilized by a boundary controller at the other end. Omitting the mass of the cable, the system can be represented by

$$\frac{\partial^\alpha u}{\partial t^\alpha} = \frac{\partial^2 u}{\partial x^2}, \quad 1 < \alpha \leq 2, \quad x \in [0,1], \quad t \geq 0 \tag{9.27}$$

$$u(0,t) = 0, \tag{9.28}$$

$$u_x(1,t) = f(t), \tag{9.29}$$

$$u(x,0) = u_0(x), \tag{9.30}$$

$$u_t(x,0) = v_0(x), \tag{9.31}$$

where $u(x,t)$ is the displacement of the cable at $x \in [0,1]$ and $t \geq 0$, $f(t)$ is the boundary control force at the free end of the cable, $u_0(x)$ and $v_0(x)$ are the initial conditions of displacement and velocity, respectively.

The control objective is to stabilize $u(x,t)$, given the initial conditions (9.30) and (9.31).

We adopt the following Caputo definition for fractional derivative of order α of any function $f(t)$, because the Laplace transform of the Caputo derivative allows utilization of initial values of classical integer-order derivatives with known physical interpretations [1, 23]

$$\frac{d^\alpha f(t)}{dt^\alpha} = \frac{1}{\Gamma(\alpha - n)} \int_0^t \frac{f^{(n)}(\tau)d\tau}{(t-\tau)^{\alpha+1-n}}, \tag{9.32}$$

where n is an integer satisfying $n - 1 < \alpha \leq n$ and Γ is Euler's Gamma function.

In this chapter, we study the robustness of the controllers in the following format:

$$f(t) = -k\frac{\mathrm{d}^\mu u(1,t)}{\mathrm{d}t^\mu}, \quad 0 < \mu \leq 1 \tag{9.33}$$

where k is the controller gain, μ is the order of fractional derivative of the displacement at the free end of the cable.

Based on the definition in (9.32), the Laplace transform of the fractional derivative is [1, 23]:

$$\mathcal{L}\left\{\frac{\mathrm{d}^\alpha f}{\mathrm{d}t^\alpha}\right\} = s^\alpha F(s) - \sum_{k=0}^{n-1} f^k(0^+)s^{\alpha-1-k}. \tag{9.34}$$

In the following, the transfer function from the boundary controller $f(t)$ to the tip end displacement will be derived for later use.

Assuming zero initial conditions of $u(x,0)$ and $u_t(x,0)$, take the Laplace transform of (9.27), (9.28), and (9.29) with respect to t, making use of (9.34), the original PDE of $u(x,t)$ with initial and boundary conditions can be transformed into the following ODE of $U(x,s)$ with boundary conditions,

$$\frac{\mathrm{d}^2U(x,s)}{\mathrm{d}x^2} - s^\alpha U(x,s) = 0, \tag{9.35}$$

$$U(0,s) = 0, \tag{9.36}$$

$$U_x(1,s) = F(s), \tag{9.37}$$

where $U(x,s)$ is the Laplace transform of $u(x,t)$ and $F(s)$ is the Laplace transform of $f(t)$.

Solving the ODE (9.35), we have the following solution of $U(x,s)$ with two arbitrary constants C_1 and C_2 (s can be treated as a constant in this step),

$$U(x,s) = C_1 e^{xs^{\alpha/2}} + C_2 e^{-xs^{\alpha/2}}. \tag{9.38}$$

Substituting (9.38) into (9.36) and (9.37), we have the following two equations,

$$C_1 + C_2 = 0, \tag{9.39}$$

$$s^{\alpha/2}(C_1 e^{s^{\alpha/2}} - C_2 e^{-s^{\alpha/2}}) = F(s). \tag{9.40}$$

Solving (9.39) and (9.40) simultaneously, we can obtain the exact value of C_1 and C_2

$$C_1 = -C_2 = \frac{F(s)e^{s^{\alpha/2}}}{s^{\alpha/2}(e^{2s^{\alpha/2}} + 1)}. \tag{9.41}$$

Now we have obtained the solution of $U(x,s)$. Substituting $x = 1$ into $U(x,s)$ and dividing $U(x,s)$ by $F(s)$, we obtain the following transfer function of the fractional wave equation $P(s)$:

$$P(s) = \frac{U(1,s)}{F(s)} = \frac{1 - e^{-2s^{\alpha/2}}}{s^{\alpha/2}\left(1 + e^{-2s^{\alpha/2}}\right)}. \tag{9.42}$$

9.6 Fractional Order Case – Robustness of Boundary Stabilization

We consider the presence of a very small time delay θ in boundary measurement, shown as follows

$$f(t) = -ku_t^{(\mu)}(1, t - \theta), \tag{9.43}$$

where θ is the time delay.

The situation is also illustrated in Fig.9.1, where $P(s)$ is the transfer function of the plant and $C(s)$ is the Laplace transform of the controller. In our case, $P(s)$ is (9.42) and $C(s)$ is

$$C(s) = k\, s^{\mu}. \tag{9.44}$$

In [5, 6, 14, 15], it was shown that an arbitrarily small delay in boundary measurement causes the instability problem in boundary control of wave equations using integer order controllers $f(t) = -ku_t(1, t)$. Does this problem exist in boundary control of the fractional wave equation? Since fractional order controllers are chosen in this chapter, will this additional tuning knob bring us any benefits of robustness against the small delay? To answer these questions, we will use Theorem 9.1 in Section 9.4 [14, 15].

Again, the underlying idea of the above theorem is that the robustness of the closed-loop transfer function $G_0(s)$ against a small unknown delay can be determined by studying the open-loop transfer function $H(s)$. Notice that $H(s) = C(s)P(s)$ in our case.

Claim. If the derivative order μ of controller (9.33) and the fractional order α in the fractional wave equation (9.27) satisfy

$$\mu < \frac{\alpha}{2}, \tag{9.45}$$

then the system is stable for a delay θ small enough in boundary measurement.

Proof. For $s \in \mathbb{C}_0$,

$$\begin{aligned} |H(s)| &= |C(s)P(s)| \\ &= \left| \frac{ks^{\mu}(1 - e^{-2s^{\alpha/2}})}{s^{\alpha/2}\left(1 + e^{-2s^{\alpha/2}}\right)} \right| \\ &= \left| \frac{k(1 - e^{-2s^{\alpha/2}})}{s^{(\alpha/2-\mu)}\left(1 + e^{-2s^{\alpha/2}}\right)} \right| \\ &\le \frac{k|1 - e^{-2s^{\alpha/2}}|}{|s^{(\alpha/2-\mu)}||1 + e^{-2s^{\alpha/2}}|}. \end{aligned} \tag{9.46}$$

Since $\frac{\alpha}{2} > \mu$, $|s^{(\alpha/2-\mu)}| \to \infty$ for $|s| \to \infty$.

Since $\frac{1}{2} < \frac{\alpha}{2} < 1$, for $|s|$ large enough, $|1 - e^{-2s^{\alpha/2}}|$ is bounded and

$$|1 - e^{-2s^{\alpha/2}}| > \eta > 0,$$

where η is a positive number.

So

$$\limsup_{|s|\to\infty, s\in\mathbb{C}_0} |H(s)| = 0 < 1. \qquad \square$$

Following the above proof, it can easily be proved that an integer order controller $f(t) = -ku_t(1,t)$ is not robust against an arbitrarily small delay.

9.7 Fractional Order Case – Compensation of Large Delays in Boundary Measurement

In the last section, it is shown that a fractional order controller is robust against a small delay under the condition (9.45). In this section, we investigate the problem that the delay is large and makes the system unstable. We will apply the Smith predictor to solve this problem.

In Section 9.2, it is shown that if the assumed delay is equal to the actual delay, the Smith predictor removes the delay term completely from the denominator of the closed-loop. However, the actual delay is not exactly known. In this section, we will investigate what happens if an unknown small difference ϵ between the assumed delay and the actual delay is introduced to the system, as shown in Fig. 9.3.

Claim. If $\hat{\theta}$ is chosen as the minimum value of the possible delay and μ is chosen to satisfy (9.45), then the controller (9.19) is robust against a small difference ϵ between the assumed delay $\hat{\theta}$ and the actual delay $\theta = \hat{\theta} + \epsilon$.

Proof. For $s \in \mathbb{C}_0$,

$$\begin{aligned}
|H(s)| &= \left| \frac{ks^{\mu}P(s)e^{-\hat{\theta}s}}{1 + ks^{\mu}P(s)(1 - e^{-\hat{\theta}s})} \right| \\
&\le \frac{k|1 - e^{-2s^{\alpha/2}}||e^{-\theta s}|}{|s^{(\alpha/2-\mu)}(1 + e^{-2s^{\alpha/2}}) + k(1 - e^{-2s^{\alpha/2}})(1 - e^{-\theta s})|} \\
&< \frac{k|1 - e^{-2s^{\alpha/2}}|}{||s^{(\alpha/2-\mu)}(1 + e^{-2s^{\alpha/2}})| - k|(1 - e^{-2s^{\alpha/2}})(1 - e^{-\theta s})||}.
\end{aligned}$$

When $|s| \to \infty$,

$$|s^{(\alpha/2-\mu)}(1 + e^{-2s^{\alpha/2}})| \to \infty,$$

while both $|1 - e^{-2s^{\alpha/2}}|$ and $|(1 - e^{-2s^{\alpha/2}})(1 - e^{-\theta s})|$ are bounded.

So

$$\limsup_{|s|\to\infty, s\in\mathbb{C}_0} |H(s)| = 0 < 1. \qquad \square$$

Remark 9.2. In Theorem 9.1, ϵ is positive. To satisfy this condition, $\hat{\theta}$ should be chosen as the minimal value of the possible delay. □

9.8 Conclusions

For both integer order and fractional order cases, this chapter considers the boundary control of damped wave equations using a boundary measurement in a networked control system (NCS) setting. In this networked boundary control system, the induced delays can be lumped as the boundary measurement delay. The Smith predictor is applied to the networked boundary control problem and the instability problem due to large delays is solved and the scheme is proved to be robust against a small difference between the assumed delay and the actual delay. Our analysis shows that fractional order boundary controllers are better than integer order boundary controllers in terms of robustness against delays in the boundary measurement.

Future work includes studying the robustness of the controller against plant modeling errors and the controller performance of the Smith predictor.

Acknowledgement

I would like to thank Dr. Igor Podlubny, Dr. Jinsong Liang and Weiwei Zhang for their collaboration. This book chapter is partly based on [10, 12, 13].

References

1. Caputo M (1967) Linear models of dissipation whose q is almost frequency independent-II. Geophys J R Astronom Soc 13:529–539
2. Chen G (1979) Energy decay estimates and exact boundary value controllability for the wave equation in a bounded domain. J Math Pure Appl 58:249–273
3. Chen G, Delfour MC, Krall AM, Payre G (1987) Modelling, stabilization and control of serially connected beams. SIAM J Control Optimization 25:526–546
4. Conrad F, Morgül O (1998) On the stability of a flexible beam with a tip mass. SIAM J Control Optimization 36:1962–1986
5. Datko R (1993) Two examples of ill-posedness with respect to small time delays in stabilized elastic systems. IEEE Trans Automatic Control 38:163–166
6. Datko R, Lagnese J, Polis MP (1986) An example on the effect of time delays in boundary feedback stabilization of wave equations. SIAM J Control Optimization 24:152–156
7. Guo BZ (2001) Riesz basis approach to the stabilization of a flexible beam with a tip mass. SIAM J Control Optimization 39:1736–1747
8. Guo BZ (2002) Riesz basis property and exponential stability of controlled Euler-Bernoulli beam equations with variable coefficients. SIAM J Control Optimization 40:1905–1923

9. Levine W (1996) The control handbook. CRC Press, New York
10. Liang JS (2005) Control of linear time-invariant disturbed parameter systems - from integer order to fractional order. MS Thesis, Utah State University, Logan, Utah
11. Liang JS, Chen YQ, Fullmer R (2004) Simulation studies on the boundary stabilization and disturbance rejection for fractional diffusion-wave equation. In: Proceedings of the 2004 IEEE American Control Conference, Boston, MA, 5010–5015
12. Liang JS, Zhang WW, Chen YQ (2005) Robustness of boundary control of damped wave equations with large delays at boundary measurement. In: Proceedings of the IFAC World Congress, Prague, Czeck Republic
13. Liang JS, Zhang WW, Chen YQ, Podlubny I (2005) Robustness of boundary control of fractional wave equations with delayed boundary measurement using fractional order controller and the Smith predictor. In: Proceedings of the ASME DETC05/Symposium on Dynamics and Control of Time-Varying and Time-Delay, Long Beach, CA
14. Logemann H, Rebarber R (1998) PDEs with distributed control and delay in the loop: transfer function poles, exponential modes and robustness of stability. European J Control 4:333–344
15. Logemann H, Rebarber R, Weiss G (1996) Conditions for robustness and nonrobustness of the stability of feedback systems with respect to small delays in the feedback loop. SIAM J Contr Optimimiz 34:572–600
16. Matignon D, d'Andréa-Novel B (1995) Spectral and time-domain consequences of an integro-differential perturbation of the wave PDE. In: SIAM Proc of the Third International Conference on Mathematical and Numerical Aspects of Wave Propagation Phenomena, France, 769–771
17. Morgül O (1995) On the stabilization and stability robustness against small delays of some damped wave equations. IEEE Trans. Automatic Control 40:1626–1623
18. Morgül O (1998) Stabilization and disturbance rejection for the wave equation. IEEE Trans. Automatic Control 43:89–95
19. Morgül O (2001) Stabilization and disturbance rejection for the beam equation. IEEE Trans. Automatic Control 46:1913–1918
20. Morgül O (2002) An exponential stability result for the wave equation. Automatica 38:731–735
21. Morgül O (2002) On the boundary control of beam equation. In: Proc of the 15th IFAC World Congress on Automatic Control, Barcelona, Spain
22. Nigmatullin RR (1986) Realization of the generalized transfer equation in a medium with fractal geometry. Phys Stat Sol (b) 133:425–430
23. Podlubny I (1999) Fractional differential equations. Academic Press, New York
24. Smith OJM (1957) Closer control of loops with dead time. Chem Eng Progress 53:217–219
25. Wang QG, Lee TH, Tan KK (1999) Finite spectrum assignment for time-delay systems. Spring-Verlag, Berlin

10

Coordination of Multi-agent Systems Using Adaptive Velocity Strategy

Wei Li and Xiaofan Wang

Shanghai Jiao Tong University, Shanghai 200240, P. R. China
liwei1@sjtu.edu.cn, xfwang@sjtu.edu.cn

Abstract. Collective behaviors of biological swarms have attracted significant interest in recent years, but much attention and correlative effort has been focused on constant speed models in which all agents are assumed to move with the same constant speed. One limitation of the constant speed models without attraction functions is that it is quite difficult or even practically impossible for the swarm to form large biological cluster(s) if the speed is relatively fast or the sensory radius is small. In this chapter, we propose an adaptive velocity model with more reasonable assumptions in which every agent not only adjusts its moving direction but also adjusts its speed based on the degree of direction consensus among its local neighbors. It is also a nearest neighbor rule but much easier for swarm agents to form a giant cluster or only one cluster in the adaptive velocity model if each agent moves with a speed that is proportional to its local direction consensus, even though the steady-state speed is still fast. The adaptive velocity strategy also shows that attraction actions or dominant leaders of swarms are not necessities for swarm cohesion. Therefore, the adaptive velocity model provides a powerful mechanism for coordinated motion in biological and technological multi-agent systems.

Keywords. Graph, multi-agent, power-law, swarm.

10.1 Introduction

The emergence of biological swarms is a beauty and wonder of nature [3, 23, 24]. It is common to see huge herds of animals or flocks of birds or schools of fish moving as if they were a single living creature. These swarms often travel in the absence of any leader/leaders or external stimuli, and agents in these swarms usually do not share any global information. How do they form a congregation and move? What collective behaviors and properties do they have? In recent years efforts have been devoted to modeling and exploring the dynamic properties of such systems which can roughly be divided into three approaches: Lagrangian approach [4, 9, 10, 11, 12, 17, 22], Eulerian approach [18, 27, 28, 29], and discrete approach [2, 5, 6, 8, 13, 15, 16, 19, 25, 30].

In 1987, Reynolds introduced three heuristic rules - cohesion, separation and alignment - to create the first computer simulated model of flocking [25]. Later on, Vicsek *et al.* proposed a simplified minimal model, which focused mainly on the emergence of directional alignment in self-driven particle systems [30]. In recent years, the Vicsek model has been one of the most frequently investigated swarm models using nearest neighbor rules to imitate swarming behaviors. For example, effects of noise and scaling behavior of the model were considered in [8]. Intermittency and clustering in self-driven particles [15] and the onset of collective motion [13] were also studied. Stability analysis of swarms revealed the relationship between network connectivity and the stability property [16, 19]. There are some other models that capture the important rule of the directional alignment used in the Vicsek model. For example, Couzin *et al.* showed that the alignment actions together with attraction/repulsion functions between neighboring agents can lead to complex patterns of swarms and revealed the existence of major group-level behavioral transitions [5]. Effective leadership was investigated in [6], which indicated that information owned by a few agents can be transferred within the whole group. Self-driven many-particle systems with general network topologies such as the vectorial network model (VNM) were investigated in [2].

All these researchers assumed that all agents in a swarm move with the same constant speed (i.e., absolute value of the velocity). However, we believe that in natural swarms, it is a more reasonable assumption that agents may not only adjust their moving directions in the swarming evolution but also adjust their speed according to the behavior of their neighbors. Indeed, when an agent finds itself surrounded by scattered moving agents, it may naturally feel at a loss to follow any direction, and may hesitate to move; in this dilemma, it is safer for the agent to move with a slower speed. On the other hand, if a certain moving direction is dominant, the agent may take this direction without hesitation and thus moves relatively fast. Similar analogies are often found in human lives and politics: when several different proposals or choices have nearly the same support or weights, individuals (or organizations) may find themselves embarrassed to decide on and thus little progress will be made in this situation; but when consensus is reached by dominant or all individuals, rapid progress tends to be made immediately. Another human-scale example is the rhythmic clapping in a concert hall after a good performance, which is suggested to be formed by each individual who tend to adjust the natural clapping frequency lower or higher according to his/her hearing [20], just as biological swarms, humans sometimes tend to do what their neighbors do.

In this chapter, we propose an adaptive velocity model in which each agent adjusts its velocity (i.e., both direction and speed) simultaneously according to the behavior of its neighbors. The direction adjustment consideration follows the same rule that used in the Vicsek model. To design our speed adjustment rule, we introduce the concept of local order parameter to measure the local degree of direction consensus (or local polarity) among the neighbors of an agent. At each time step, each agent will move along the average direction of its neighbors with a speed which is taken as the maximum possible speed scaled by a power-law function of the magnitude of its local order parameter. The power-law exponent $\alpha \geq 0$ reflects the willingness of each agent to move faster or slower based on the local degree of direction consensus among its neighbors. If $\alpha = 0$, then the adaptive velocity model reduces to the constant speed Vicsek model and each agent always moves with the maximum constant speed. However, if $\alpha > 0$, then an agent will move with the maximum speed if and only if complete local direction consensus is achieved among its neighbors. A larger value of α implies that an agent will move with a slower speed in the face of a given level of non-complete local direction consensus, which results in higher convergence probability that a group of initially randomly distributed agents will finally move along a global consensus direction.

This chapter is organized as follows. In Section 10.2, we describe briefly the constant speed model proposed by Vicsek *et al.* and compare two order parameters to measure the phase transition phenomena of the swarm. In Section 10.3, we propose an adaptive velocity model with a tunable parameter α based on the concept of local order parameter. Simulation results and discussions are given in Section 10.4. Conclusions are given in Section 10.5.

10.2 The Constant Speed Vicsek Model

We first describe the original constant speed Vicsek model [30]. Consider N agents, labeled from 1 through N, all moving synchronously in a square shaped cell of linear size L with periodic boundary conditions. Each agent has the same absolute velocity v_0 but with different direction at different time steps. Originally, all agents' positions are randomly distributed in the cell with randomly distributed directions in $[0, 2\pi)$. At each time step, agent i adjusts its direction as the average moving direction $\langle\theta_i(k)\rangle_R$ of its neighbors with some random perturbation $\Delta\theta$ added:

$$\theta_i(k+1) = \langle\theta_i(k)\rangle_R + \Delta\theta. \tag{10.1}$$

Here, the neighbors of agent i are defined as those agents who fall in a circle of predefined sensory radius R centered at the current position of agent i. One characteristic of this homogeneous model is that only by local interactions it shows phase transition through spontaneous symmetry breaking of the rotational symmetry. The different pattern behaviors, such as large-scale emergence, convergence and disordered disperse motion, can be observed under different parameters using simulation [30]. This directional rule of local interactions together with constant speed motion of agents has considerable influences [2, 5, 6, 8, 13, 15, 16, 19, 25, 30]. The swarm model in [5] is another important constant speed model that consists of homogeneous agents with directional alignment, attraction and repulsion rules, the emergences are generated by spontaneous symmetry breaking. Certainly, attraction action between agents is another reasonable consideration to form gathering and to have considerable influence. We will show that attraction action is not a necessity for large swarm clusters.

The following order parameter has been widely adopted to measure the phase transition phenomena of the constant speed model from the initial zero net transport to emergence [2, 5, 7, 14, 15, 30]:

$$\Phi_v(k) = \frac{1}{Nv_0}\left|\sum_{i=1}^{N}\vec{v_i}(k)\right|, \ 0 \le \Phi_v(k) \le 1. \tag{10.2}$$

Here $\vec{v_i}(k)$ is the velocity of agent i with direction $\theta_i(k)$ and the constant speed $v_0 = |\vec{v_i}(k)|$ for all $i = 1, 2, \ldots, N$ at all steps k. $\Phi_v(k)$ is a univocal physical parameter by definition – a scaled average momentum of the whole system and emergent behavior can be observed if $\Phi_v(k) \gg 0$. $\Phi_v(k) = 0$ corresponds to the isotropy state of directional distribution and $\Phi_v(k) = 1$ implies convergent or linear coherent motion of all agents only on the prerequisite that all agents have the same fixed speed v_0.

Now suppose that different agents may have generally different speed at different time steps. Let v_0 be the average value of all agents' possible maximum speeds, that is, $v_0 = \frac{1}{N}\left|\sum_{i=1}^{N} v_{i0}\right|$, where v_{i0} is the maximum possible speed of agent i. In this general case, it is possible that $\Phi_v(k) > 1$ even if the moving directions of all agents are isotropic which corresponds to a non-emergence state. And $\Phi_v(k) = 1$ does not necessarily mean linear coherence, unless v_{i0} is the same value for all agents. Thus $\Phi_v(k)$ is not appropriate to measure the level of emergence.

Another order parameter that has been widely adopted, especially for synchronous characteristic in the networked phase oscillators, is defined as follows [13, 26]

$$\Phi_\theta(k) = \frac{1}{N}\left|\sum_{i=1}^{N} e^{i\theta_i(k)}\right|, \ \ 0 \le \Phi_\theta(k) \le 1. \tag{10.3}$$

This order parameter eliminates the influence of the agent's speed, but at the expense of having no physical meaning of scaled average momentum. For the constant speed Vicsek model, it is obvious that the two order parameters defined above are the same, i.e., $\Phi_v = \Phi_\theta$.

10.3 The Adaptive Velocity Model

In this section, we propose an adaptive velocity model in which each agent adjusts its direction and speed at different time steps simultaneously. To do so, we first define the complex-valued *local order parameter* of agent i at step $k+1$ as follows:

$$\phi_i(k+1)e^{i\theta_i(k+1)} = \frac{1}{n_i(k+1)} \sum_{j\in\Gamma_i(k+1)} e^{i\theta_j(k)}, \ i = 1,2,\dots,N; \ k = 0,1,\dots, \tag{10.4}$$

where $e^{i\theta_j(k)}$ is the unit directional vector and $\Gamma_i(k+1)$ is the set of $n_i(k+1)$ neighbors of agent i at step $(k+1)$. Magnitude (or local polarity) $\phi_i(k+1)$ of the local order parameter measures the local degree of direction consensus among the neighbors of agent i at step $(k+1)$. Obviously, $\phi_i(k+1)$ is a local form of the global order parameter (10.3) and $0 \leq \phi_i(k+1) \leq 1$. A larger value of $\phi_i(k+1)$ implies a higher degree of local direction consensus among neighbors of agent i (Fig. 10.1). Angle $\theta_i(k+1)$ is the corresponding moving direction of agent i at step $(k+1)$, which is the average directions of agents in set $\Gamma_i(k)$. Computations using this expression can also avoid some undesired directional problems mentioned in [16].

Denote $\mathbf{X}_i(k)$ as the position of agent i on the complex plane at step k. In our adaptive velocity model, each agent not only adjusts its moving direction, but also adjusts its speed according to the degree of local direction consensus among its neighbors, which is represented by its local polarity. Specifically, the speed of agent i at step k is scaled by a power-law function of its local polarity, i.e.,

$$s_i(\phi_i(k)) \triangleq \phi_i^{\alpha}(k) \tag{10.5}$$

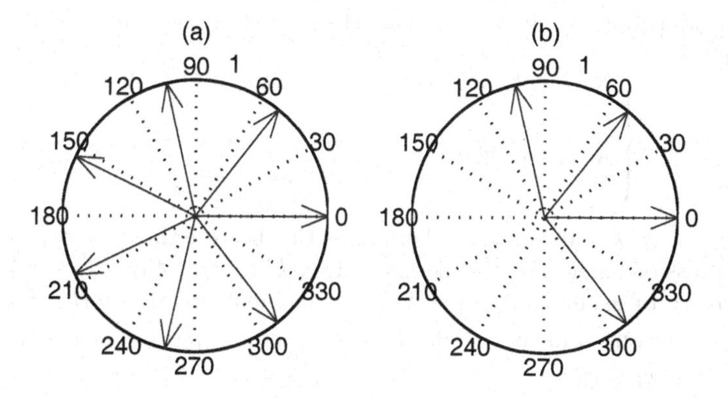

Fig. 10.1. Illustration of local polarity ϕ_i of agent i. The arrows show the moving directional vectors of neighboring agents of agent i. For simplicity, these modular vectors are plotted with the same starting points located in the center of a circle. (a) The collection of agents moving scattered in the plane with no dominant direction, the order parameter $\phi_i \approx 0$ for this situation. (b) The agents with a relatively strong dominant direction, $\phi_i \neq 0$ for this situation. The polarity $\phi_i = 0$ if and only if all the agents in set $\Gamma_i(k)$ move in the same direction.

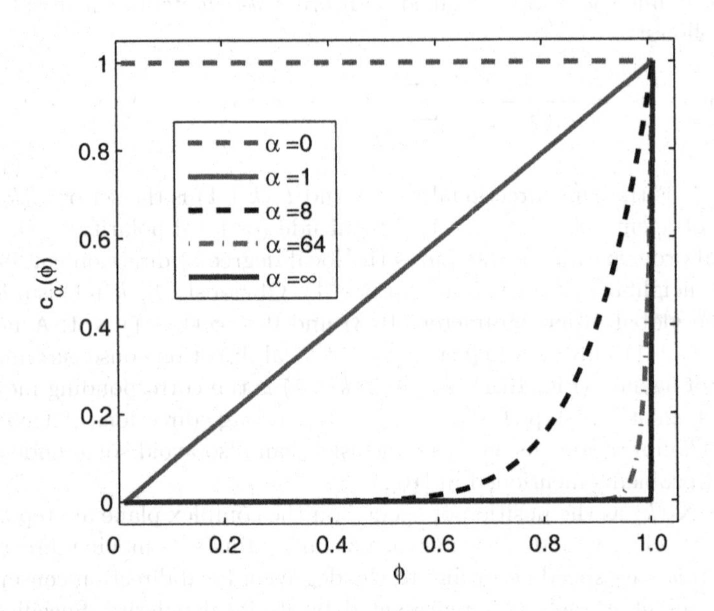

Fig. 10.2. Scaled speed coefficient $c_\alpha(\phi)$ as a power function of local polarity ϕ. For any value of α, $c_\alpha(\phi) = 1$ if $\phi = 1$. For $\alpha = 0$, $c_\alpha(\phi) \equiv 1$. For $0 < \alpha < \infty$, $0 < c_\alpha(\phi) < 1$ if $0 < \phi < 1$. For $\alpha = \infty$, $c_\alpha(\phi) = 0$ if $0 < \phi < 1$.

with an power-law exponent $\alpha \geq 0$ (Fig. 10.2).

The adaptive velocity model can then be described mathematically as follows:

$$\begin{cases} \mathbf{X}_i(k+1) = \mathbf{X}_i(k) + v_0 \times \phi_i^\alpha(k) e^{i\theta_i(k)} \times \Delta t \\ \\ \phi_i(k+1)e^{i\theta_i(k+1)} = \dfrac{1}{n_i(k+1)} \displaystyle\sum_{j \in \Gamma_i(k+1)} e^{i\theta_j(k)} \end{cases} \tag{10.6}$$

$i = 1, 2, \ldots, N$; $k = 0, 1, 2, \ldots$, where Δt is the discrete time interval, and here without loss of generality, we take $\Delta t = 1$. $\vec{v_i}(k) \equiv v_0 \times \phi_i^\alpha(k) e^{i\theta_i(k)}$ represents the velocity of agent i at step k with its moving direction $\theta_i(k)$. Since $0 \leq \phi_i^\alpha(k) \leq 1$ for any value of $\alpha \geq 0$, the corresponding speed $|\vec{v_i}(k)| = v_0 \times \phi_i^\alpha(k)$ satisfies $0 \leq \vec{v_i} \leq v_0$.

This adaptive speed is another important factor that contributes to emergence or swarming clusters that has been previously overlooked, especially for swarms in three or higher dimensions. This adaptive speed model also satisfies fundamental swarm's characteristics: no any leader/leaders, no external stimuli, only homogeneous agents, and only local interactions, but induces more intensified phase transition and symmetry-broken from disordered to ordered state than the constant speed model.

The power-law exponent $\alpha \geq 0$ reflects the willingness of each agent to move faster or lower along the average direction of its neighbors based on the local degree of direction consensus. If $\alpha = 0$, then $c_\alpha(\phi) \equiv 1$. The adaptive velocity model (10.6) reduces to the constant speed Vicsek model and each agent always moves with the maximum constant speed v_0 without any considerations about its local polarity. However, if $\alpha > 0$, then an agent will move with the maximum speed if and only if complete local direction consensus is achieved among its neighbors. In the case of $\alpha = 1$, the local order parameter of agent i is just the direct sum of directional vectors of agent i's neighbors. A larger value of α implies that an agent will move with a slower speed in the face of a given level of local direction consensus. In the limit case that $\alpha = \infty$, we have

$$c_\infty(\phi) = \phi^{+\infty} = \begin{cases} 0, & 0 \leq \phi \leq 1, \\ 1, & \phi = 1. \end{cases} \tag{10.7}$$

It means that each agent will not move unless complete local direction consensus is achieved among its neighbors.

The 2-dimensional adaptive velocity model (10.6) can easily be generalized to general M-dimensional Euclidean space case. Let $P_i = [p_{i1}, p_{i2}, \ldots, p_{iM}]^T$ represent position of agent i, $i = 1, 2, \ldots, N$. The motion direction of agent i is represented by a unitary vector $d_i = [d_{i1}, d_{i2}, \ldots, d_{iM}]^T$ which satisfies

$$\|d_i\| = 1, \; -1 \leq d_{ij} \leq 1, \; j = 1, 2, \ldots, M, \tag{10.8}$$

for all i. Agent i and agent j are neighbors if $\|p_i(k) - p_j(k)\| \leq R$.

Define the order parameter as

$$r_i(k+1) = \frac{1}{n_i(k+1)} \left\| \sum_{j \in \Gamma_i(k+1)} d_j(k) \right\|. \tag{10.9}$$

Of cause, $0 \leq r_i(k+1) \leq 1$. The M-dimensional adaptive velocity model can be described as:

$$P_i(k+1) = P_i(k) + v_0 \times r_i^\alpha(k) \times d_i(k) \times \Delta t, \; k = 0, 1, 2, \ldots, \tag{10.10}$$

$$d_i(k+1) = \Big(\sum_{j \in \Gamma_i(k+1)} d_j(k) \Big) \Big/ \left\| \sum_{j \in \Gamma_i(k+1)} d_j(k) \right\|, \; k = 0, 1, 2, \ldots. \tag{10.11}$$

10.4 Simulations and Discussions

It is a more natural assumption for swarms moving in the plane to ensure that they can evolve freely and sufficiently. To illustrate the effect of adaptive velocity strategy, we consider N agents moving in the complex plane for simulation instead of in a rectangle of open boundary or periodic boundary

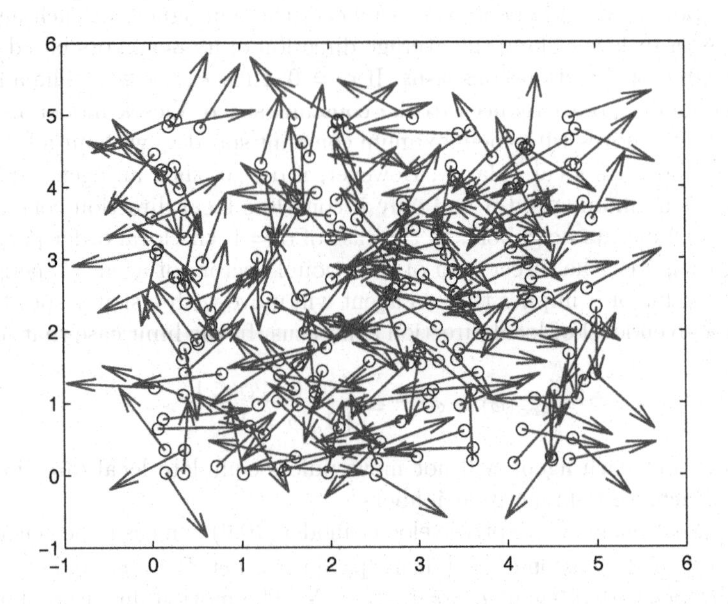

Fig. 10.3. Illustration of initially random distribution of positions and directions of agents, the arrows point to the initial directions of agents, the ends of arrows (denoted by blue circles) are positions of agents. Here the rectangle is 5×5 cell. The swarm evolves in the whole 2D plane.

conditions [30]. The N agents' positions and directions are initially randomly distributed on a rectangle of linear size L (Fig. 10.3). Denote the initially distributed directions and positions of agent i as θ_i and $P_i(0)$, respectively, $i = 1, 2, \ldots, N$. Note that the initial distribution of direction θ_i is not the initial moving direction $\theta_i(0)$. We compute the initial moving direction $\theta_i(0)$ and initial polarity $\phi_i(0)$ of agent i according to local order parameter formula

$$\phi_i(0)e^{i\theta_i(0)} = \frac{1}{n_i(0)} \sum\nolimits_{j \in \Gamma_i(0)} e^{i\theta_j}.$$

This means that each agent moves with adaptive velocity strategy in the very beginning of its evolution. This beginning step is denoted as step $k = 0$ with the corresponding initial speed $v_0 \times \phi_i(0)$.

In simulations, we take the parameters $N = 300$, $L = 5$ and $R = 2$. All estimates are the results of averaging over 400 realizations, if without special mention. We first investigate the influence of power-law exponent α in the adaptive velocity model on the convergence probability p, which is defined as the probability that a group of N initially randomly distributed agents will finally all move along a global consensus direction with the same maximum speed v_0.

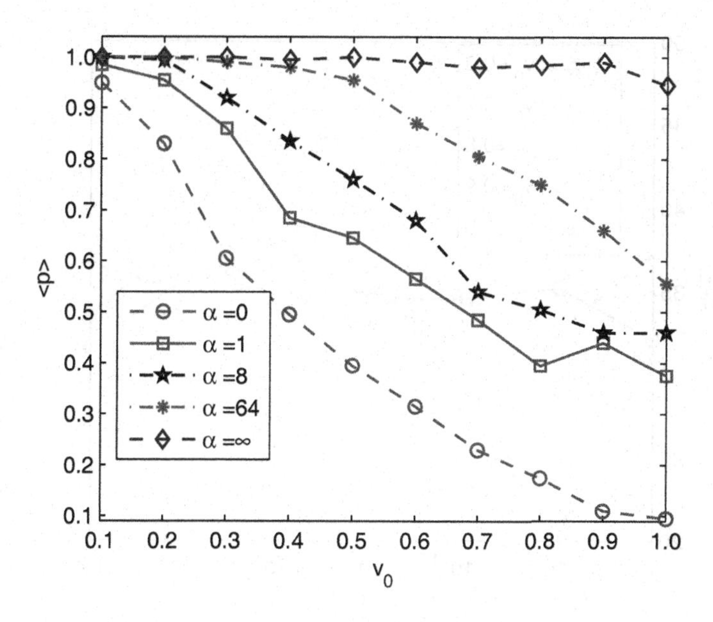

Fig. 10.4. Convergent probability p as a function of the maximum speed v_0 for five different values of α

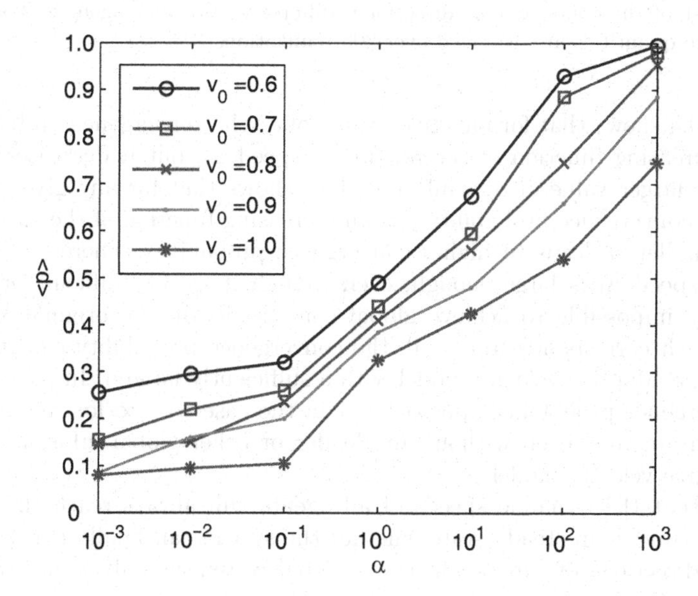

Fig. 10.5. Convergent probability p as a function of the exponent α for five different values of v_0

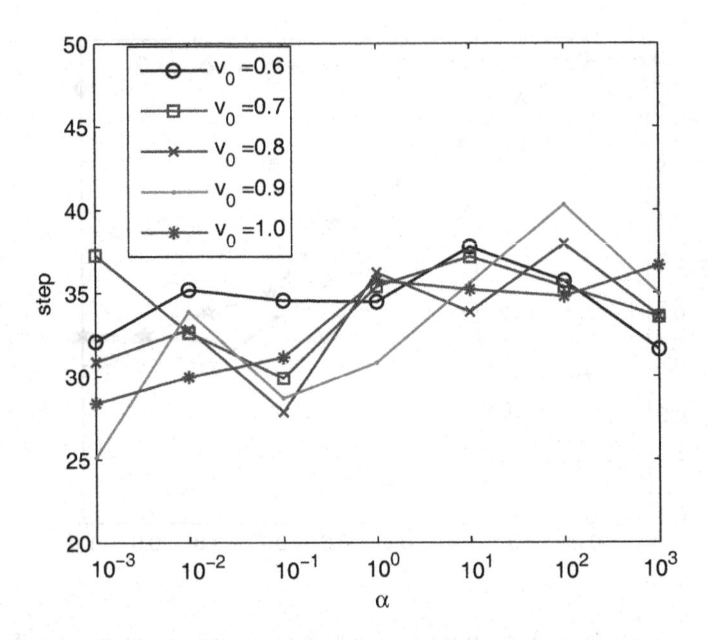

Fig. 10.6. Transient time step as a function of the exponent α. The termination condition for steady state of the swarms is that, the standard deviation of N vectors that consist of the conterminous directional differences of every agent is less than 0.0001. The quantities are averaged over 200 realizations.

Fig. 10.4 shows that for any given value of α, the convergence probability p is a decreasing function of the maximum speed v_0, but it decreases more slowly for larger value of α; while Fig. 10.5 shows that for any given value of v_0, the convergence probability p is an increasing function of the exponent α, and smaller v_0 leads to higher convergence probability. Therefore, if the constant speed v_0 is large enough, even though it is very difficult or even practically impossible to achieve global convergence in the original Vicsek model which corresponds to $\alpha = 0$, the convergence probability can still be high for the adaptive velocity model with a sufficiently large α. In particular, the convergence probability approaches 1 in the case $\alpha = \infty$ for the present system parameters, even without any leader or other global information in the adaptive velocity model.

Note that the dynamic speeds of all agents will always reach the same maximal value v_0 in steady state whether the swarm can finally converge or not; but directions of agents will reach global consensus only under certain conditions. Generally, speeds of agents in the adaptive velocity model are varied over transient time and the average speed $v_{\text{ave}}(k)$ of all agents in the swarm increases monotonically until steady state is achieved. Since a larger value of α implies that an agent will move with a slower speed in the face of a

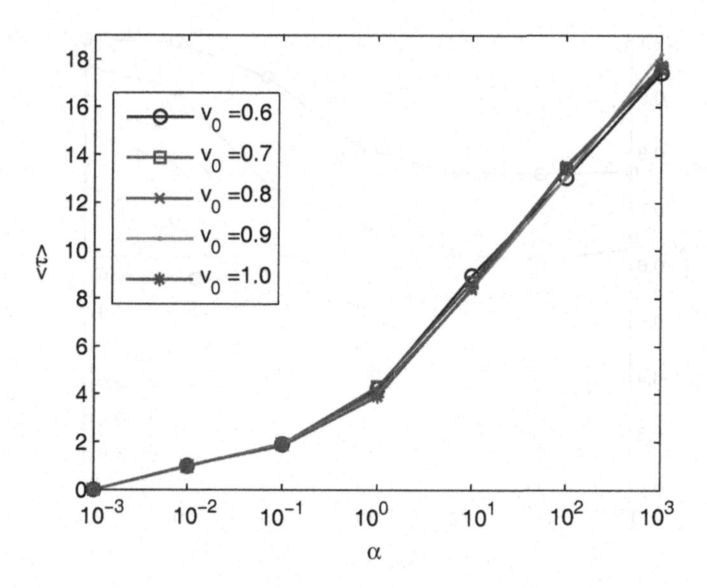

Fig. 10.7. The time step τ required for the average speed of all agents to reach 98% of maximum speed v_0 as a function of the exponent α. All estimates are the results of averaging over 200 realizations.

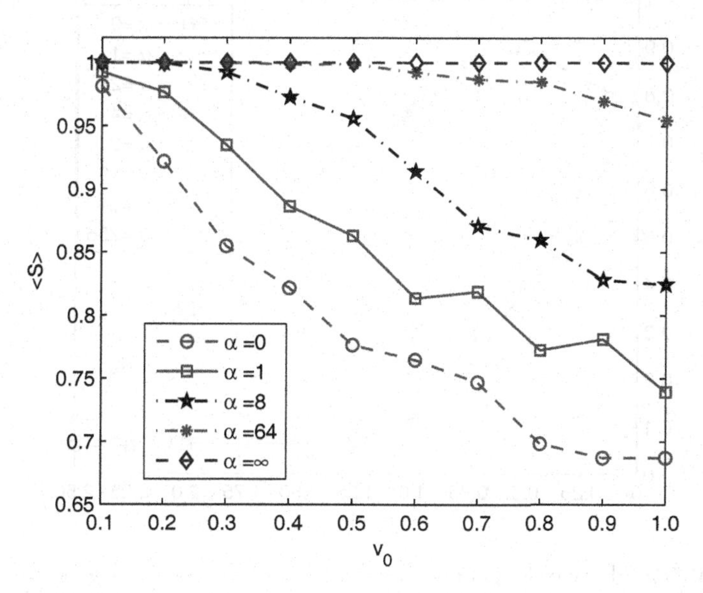

Fig. 10.8. MCSG in steady state as a function of the maximum speed v_0 for five different values of α

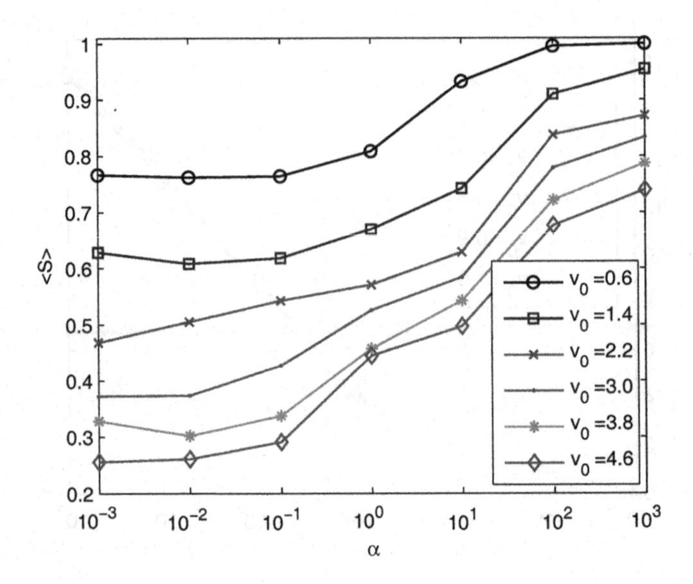

Fig. 10.9. MCSG in steady state as a function of the exponent α for six different values of v_0

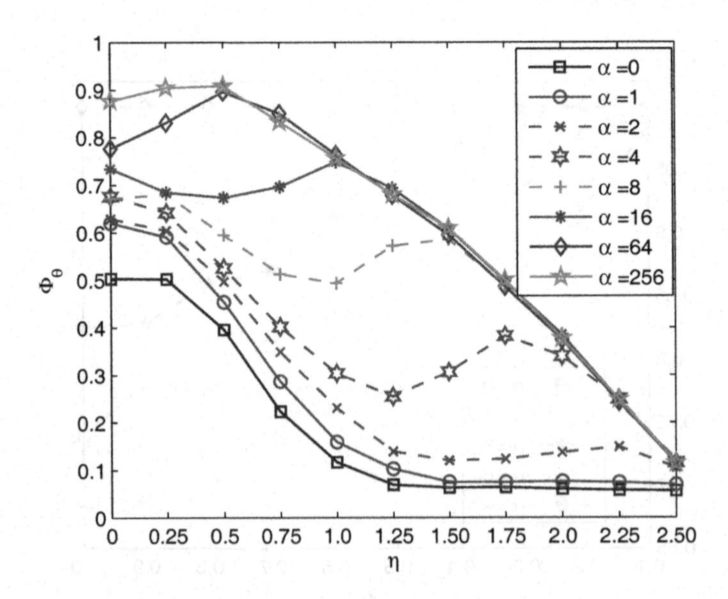

Fig. 10.10. The global order parameter Φ_θ of swarm as a function of noise amplitude η. For large η and α, Φ_θ decreases linearly. All estimates are the results of averaging over 200 realizations.

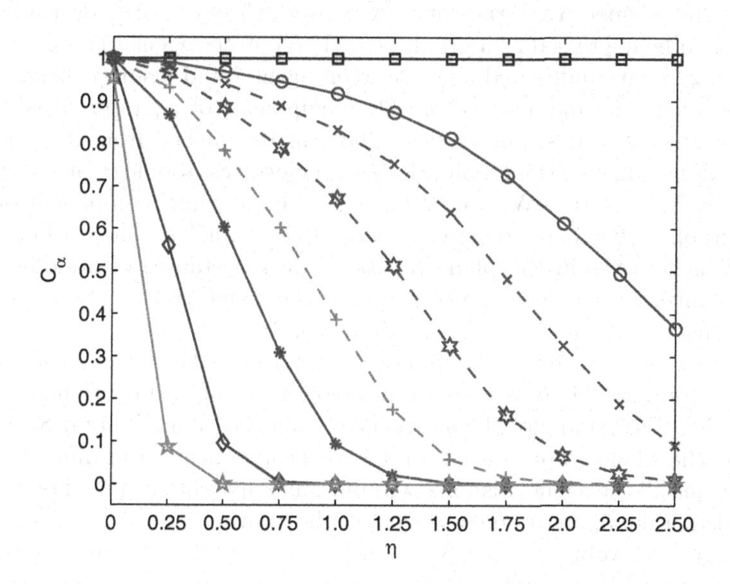

Fig. 10.11. The average speed coefficient C_α decreases to zero as the noise amplitude η increases. All estimates are the results of averaging over 200 realizations.

given level of local direction consensus, one may wonder if the transient time may be longer even though the convergence probability is higher. However, as can be seen from Fig. 10.6, the value of α does not have a significant influence on the transient time. Denote τ as the time step required for the average speed of all agents to reach 98% of maximum speed v_0. We find that τ obeys a simple log scaling law of the form (Fig. 10.7):

$$\tau \approx 4 + \beta \left[\log_{10} \alpha\right], \quad \alpha \geq 1, \tag{10.12}$$

where $\beta \approx 4.67$. Therefore, even for a high value of $\alpha = 1000$, most agents will move with nearly the maximum speed in just less than 18 steps. This behavior looks somewhat like the applause phenomenon which turns suddenly into synchronized clapping [20].

Why is the convergence probability enhanced as the exponent α increases in the adaptive velocity model? This is because the adaptive velocity strategy with large value of α tends to hold the local agents together to form large cluster. When in the approximate isotropy region, $\phi \approx 0$ which implies agents move in scattered directions, the speeds of agents are relatively small according to adaptive velocity strategy with positive value of α. Even for $0 \ll \phi < 1$, the speeds of agents are still small for large values of α. Thus transformations of those agents' positions are indistinctive, so neighbors tend to be also neighbors in the next step or even later, and communications between them continue to be held, which are beneficial to directional consensus.

From the perspective of the complex network theory [1, 21], swarm topology can also be expressed as a graph $G = (V, E)$: every agent i is represented by a vertex v_i; an undirected edge between agent i and agent j means that they are neighbors and *vice versa*. The component of a graph to which a vertex belongs is that set of vertices that can be reached from it by paths running along edges of the graph [21]. As time evolves, topology of the graph $G(k) = (V, E(k))$ varies. We are interested in the maximal component of the swarm graph (MCSG) in the steady state. Recent analysis shows that for a swarm which moves in the plane instead of in a rectangle of periodic conditions, convergence or emergence is due to the connectivity between agents [16, 19], instead of long-range interactions [30, 31].

Denote S as the ratio of the number of vertices in MCSG in steady state versus the total number of vertices in the whole graph of the underlying swarm. Clearly, $0 < S \leq 1$ and global convergence is achieved if and only if $S = 1$. In this case, the whole graph consists of only one component (swarming cluster). $S \approx 0$ means all the agents disperse without any apparent clusters. For $S \gg 0$, there exists a dominate or giant cluster in the swarm.

For any given value of α, MCSG is understandably a decreasing function of the maximum speed v_0, and it decreases much more slowly for larger values of α (Fig. 10.8); while for any given value of v_0, MCSG is an increasing function of the exponent α and smaller values of v_0 result in higher values of MCSG (Fig. 10.9). Thus, in the case of a large maximum speed v_0, although it is quite difficult or even impossible to form a giant cluster in the constant speed Vicsek model which corresponds to $\alpha = 0$, it is much easier to form a giant cluster for the adaptive velocity model if α is large enough. This also indicates that attractive actions between agents is not a necessity for swarm aggregations.

Figs. 10.10 and 10.11 show the influence of uniformly distributed noise added to the moving direction of each agent with noise amplitude η based on the global order parameter Φ_θ defined in (10.3) and the average speed coefficient C_α defined as:

$$C_\alpha \triangleq \frac{1}{N} \sum_{i=1}^{N} c_\alpha(\phi_i) = \frac{1}{N} \sum_{i=1}^{N} \phi_i^\alpha. \tag{10.13}$$

We can see from Fig. 10.10 that for large noise amplitude η and large exponent α, the global order parameter Φ_θ decreases along the same straight line, which deserves further investigation.

Comparing Figs. 10.10 and 10.11, one finds that the more robust the speed, the less emergence the swarm in the exposure of noise. The value $\alpha = 0$ corresponds to constant speed and it is the least anti-noise case of the swarm.

10.5 Conclusions

We propose an adaptive velocity model in which each agent not only adjusts its moving direction but also adjusts its speed based on the local degree of

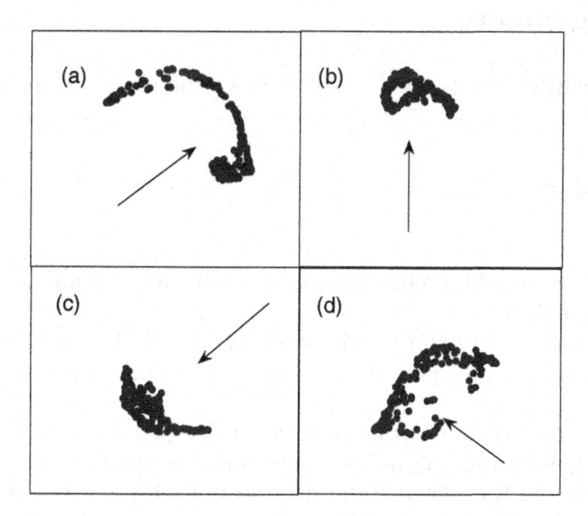

Fig. 10.12. Some interesting shapes that swarms take on. These are all coherent moving cases. The arrows denote the coherent moving direction of swarms. The parameters here are $N = 200, R = 1.2, v_0 = 0.4, \alpha = 0$.

direction consensus among its neighbors at every time step. Each agent takes its moving direction as the average angle of its local order parameter with its speed proportional to the power function of the magnitude of its local complex-valued order parameter at each step. The adaptive velocity model reduces to the constant speed Vicsek model when the power-law exponent $\alpha = 0$. A larger value of α implies that an agent will move with a slower speed in the face of a given level of non-complete local direction consensus, which results in higher convergence probability and larger swarm clusters.

Some difficult yet important problems about the adaptive velocity model remain to be further investigated. For example, under what conditions can we guarantee the existence of a critical value of α such that above the value, a given convergence probability or average MCSG can be guaranteed? Furthermore, stability analysis about the linearized Vicsek's model has been focused on the topology of swarms in the process of evolution [16], but the question of what initial distribution condition of the underlying swarm can guarantee this topology restriction remains unsolved. More practical stability analysis for the adaptive velocity model needs to be explored.

The properties of evolutional graphs of swarms over time may serve as a promising topic for further research. Unlike regular (or quasi-regular) geometric shape the attraction–repulsion models [5, 14, 22] take on (see the figures in the reference papers), what shapes (for example, see Fig. 10.12) the non-attraction-repulsion models, such as adaptive velocity model, will take in the coherent moving state also remains elusive. These questions remain interesting and challenging for further investigation.

Acknowledgements

This work was supported by the NSF of P. R. China under Grant No. 70431002 and 60674045.

References

1. Albert R, Barabasi AL (2002) Statistical mechanics of complex networks. Rev Mod Phys 74:47–97
2. Aldana M, Huepe C (2003) Phase transitions in self-driven many-particle systems and related non-equilibrium models: a network approach. J Stat Phys 112:135–153
3. Buhl J, Sumpter DJT, Couzin ID, Hale JJ, Despland E, Miller ER, Simpson SJ (2006) From disorder to order in marching locusts. Science 312:1402–1406
4. Chen HY, Leung KT (2006) Rotating states of self-propelling particles in two dimensions. Phys Rev E 73:056107–056109
5. Couzin ID, Krause J, James R, Ruxton GD, Franks NR (2002) Collective memory and spatial sorting in animal groups. J Theor Biol 218:1–11
6. Couzin ID, Krause J, Franks NR, Levin SA (2005) Effective leadership and decision-making in animal groups on the move. Nature 433:513–516
7. Czirok A, Stanley HE, Vicsek T (1997) Spontaneously ordered motion of self-propelled particles. J Phys A 30:1375–1385
8. Czirok A, Vicsek T (2000) Collective behavior of interacting self-propelled particles. Physica A 281:17–29
9. D'Orsogna MR, Chuang YL, Bertozzi AL, Chayes LS (2006) Self-propelled particles with soft-core interactions: patterns, stability, and collapse. Phys Rev Lett 96:104302–104304
10. Erdmann U, Ebeling W, Mikhailov AS (2005) Noise-induced transition from translational to rotational motion of swarms. Phys Rev E 71:051904–051907
11. Gazi V, Passino KM (2003) Stability analysis of swarms. IEEE Trans Automatic Control 48:692–697
12. Gazi V, Passino KM (2004) Stability analysis of social foraging swarms. IEEE Trans Syst Man Cybern-B 34:539–557
13. Gregoire G, Chate H (2004) Onset of collective and cohesive motion. Phys Rev Lett 92:025702–025704
14. Gregoire G, Chate H, Tu YH (2003) Moving and staying together without a leader. Physica D 181:157–170
15. Huepe C, Aldana M (2004) Intermittency and clustering in a system of self-driven particles. Phys Rev Lett 92:168701–168704
16. Jadbabaie A, Lin J, Morse AS (2003) Coordination of groups of mobile autonomous agents using nearest neighbor rules. IEEE Trans Automatic Control 48:988–1001
17. Levine H, Rappel WJ, Cohen I (2000) Self-organization in systems of self-propelled particles. Phys Rev E 63:017101–017104
18. Mogilner A, Edelstein-Keshet L (1999) A non-local model for a swarm. J Math Biol 38:534–570
19. Moreau L (2005) Stability of multiagent systems with time-dependent communication links. IEEE Trans Automatic Control 50:169–182

20. Neda Z, Ravasz E, Brechet Y, Vicsek T, Barabasi AL (2000) The sound of many hands clapping - tumultuous applause can transform itself into waves of synchronized clapping. Nature 403:849–850
21. Newman MEJ (2003) The structure and function of complex networks. SIAM Review 45:167–256
22. Olfati-Saber R (2006) Flocking for multi-agent dynamic systems: algorithms and theory. IEEE Trans Automatic Control 51:401–420
23. Parrish JK (1999) Complexity, pattern, and evolutionary trade-offs in animal aggregation. Science 284:99–101
24. Parrish JK, Viscido SV, Grunbaum D (2002) Self-organized fish schools: an examination of emergent properties. Biol Bull 202:296–305
25. Reynolds CW (1987) Flocks, herds, and schools: a distributed behavioral model. Computer Graphics (ACM) 21:25–34
26. Strogatz SH (2000) From Kuramoto to Crawford: exploring the onset of synchronization in populations of coupled oscillators. Physica D 143:1–20
27. Toner J, Tu Y (1995) Long-range order in a two-dimensional dynamical XY model: how birds fly together. Phys Rev Lett 23:4326–4329
28. Toner J, Tu Y (1998) Flocks, herds, and schools: a quantitative theory of flocking. Phys Rev E 58:4828–4858
29. Topaz CM, Bertozzi AL, Lewis MA (2006) A nonlocal continuum model for biological aggregation. Bull Math Biol 68:1601–1623
30. Vicsek T, Czirok A, Ben-Jacob E, Cohen I, Shochet O (1995) Novel type of phase transition in a system of self-driven particles. Phys Rev Lett 75:1226–1229
31. Vicsek T (2001) A question of scale. Nature 411:421

11

Design of Robust Strictly Positive Real Transfer Functions

Wensheng Yu[1] and Long Wang[2]

[1] Institute of Automation, Chinese Academy of Sciences
Beijing 100080, P. R. China wensheng.yu@ia.ac.cn
[2] Peking University, Beijing 100871, P. R. China longwang@pku.edu.cn

Abstract. This chapter studies the robust synthesis problem for strictly positive real (SPR) transfer functions. The concepts of SPR regions and weak SPR regions are introduced. By using the complete discrimination system (CDS) for polynomials, complete characterization of the (weak) SPR regions for transfer functions in coefficient space is given. It is shown that the weak monic SPR region associated with a fixed polynomial is bounded and the intersection of several weak monic SPR regions associated with different polynomials cannot be a single point. Furthermore, we show how to construct a point in the SPR region from a point in the weak SPR region. Based on these theoretical development, we propose an algorithm for robust design of SPR transfer functions. This algorithm works well for both low-order and high-order polynomial families. Especially, the derived conditions are necessary and sufficient for robust SPR design of polynomial segment or low-order ($n \leq 4$) interval polynomials. Illustrative examples are provided to show the effectiveness of this algorithm.

Keywords. Robustness, strictly positive realness (SPR), synthesis method, transfer functions, weak strictly positive realness (WSPR), WSPR regions.

11.1 Introduction

The strict positive realness (SPR) of transfer functions is an important performance specification which plays a critical role in various fields such as absolute stability/hyperstability theory [29, 36], network realizability theory/passivity analysis [8, 21], quadratic optimal control [7] and adaptive system theory [31]. Since there always exist uncertainties in real systems, it is imperative to study the robust SPR. In recent years, stimulated by the parametrization method in robust stability analysis [1, 9, 13], the study of robust SPR systems has received much attention, and great progress has been made. However, most results belong to the category of robust strictly positive real analysis. Much work remains to be done in robust strictly positive real synthesis.

Generally speaking, the synthesis problem is mathematically more difficult than the analysis problem, since the proof is usually constructive, i.e., the proof not only proves the existence of the solution, but also provides a constructive procedure to find it. The synthesis problem is of more practical significance from the engineering application viewpoint.

The basic statement of the robust SPR synthesis problem is as follows: Given an nth-order robustly stable polynomial set F, does there exist, and how to construct a (fixed) polynomial $b(s)$ such that, $\forall a(s) \in F$, $b(s)/a(s)$ is strictly positive real? If such a polynomial $b(s)$ exists, then we say that F is synthesizable.

In this chapter, we summarize some of our recently-obtained results on the design of robust SPR transfer functions. We first introduce the concepts of SPR regions and weak SPR regions and give a complete characterization of them. We show that the monic SPR region associated with a fixed polynomial is unbounded, whereas the weak monic SPR region is bounded. We then prove that the intersection of several weak monic SPR regions associated with different polynomials cannot be a single point. Furthermore, we show how to construct a point in the SPR region from a point in the weak SPR region. Based on these theoretical development, we propose an algorithm for robust design of SPR transfer functions. This algorithm works well for both low-order and high-order polynomial families. The derived conditions are necessary and sufficient for robust SPR design of polynomial segment or low-order ($n \leq 4$) interval polynomials. Illustrative examples are provided to show the effectiveness of this algorithm.

11.2 Definitions and Notation

The concept of strict positive realness stems from different area such as control systems, network analysis, etc. There are some slightly different definitions in the literature [55]. In this chapter, we will employ the following definitions.

Denote $\mathbb{R}^n$ as an n dimensional real field, P^n as the set of all nth-order polynomials of s with real coefficients. Denote $H^n \subset P^n$ as the set of all nth-order Hurwitz stable polynomials (all roots lie within the left half of the complex plane).

In the following definitions, $b(\cdot) \in P^m, a(\cdot) \in P^n$, and $p(s) = b(s)/a(s)$ is a rational function.

Definition 11.1. $p(s)$ is said to be strictly positive real (SPR), denoted by $p(s) \in$ SPR, if $b(s) \in P^n$, $a(s) \in H^n$, and $\forall \omega \in \mathbb{R}$, $\mathrm{Re}[p(j\omega)] > 0$. □

Definition 11.2. $p(s)$ is said to be weak strictly positive real (WSPR), denoted by $p(s) \in$ WSPR, if $b(s) \in P^{n-1}$, $a(s) \in H^n$, and $\forall \omega \in \mathbb{R}$, $\mathrm{Re}[p(j\omega)] > 0$. □

Definition 11.3. Given $a(s) \in H^n$, the set of the coefficients (in $\mathbb{R}^{n+1}$) of all the $b(s)$'s in P^n such that $p(s) := \dfrac{b(s)}{a(s)} \in$ SPR is said to be the SPR region associated with $a(s)$, denoted by Ω_a. □

Definition 11.4. Given $a(s) \in H^n$, the set of the coefficients (in $\mathbb{R}^n$) of all the $b(s) \in P^{n-1}$ such that $p(s) := \dfrac{b(s)}{a(s)} \in$ WSPR is said to be the WSPR region associated with $a(s)$, denoted by Ω_a^W. □

For notational convenience, Ω_a (Ω_a^W) sometimes also stands for the set of all the polynomials $b(s)$ in P^n (P^{n-1}) such that $p(s) := \dfrac{b(s)}{a(s)} \in$ SPR (WSPR).

From the definitions above, it is easy to get the following properties:

Proposition 11.1 ([5, 27]). If $p(s) \in$ SPR (WSPR), then

$$|\arg(b(j\omega)) - \arg(a(j\omega))| < \frac{\pi}{2},$$

$\forall \omega \in \mathbb{R}$, where $\arg(\cdot)$ stands for the argument of the complex number, and the difference of two arguments can differ by an integer number of 2π. □

Proposition 11.2 ([27]). Given $a(s) \in H^n$ and Ω_a is a non-empty, open, convex cone in $\mathbb{R}^{n+1}$. □

Proposition 11.3 ([17]). Given $a(s) \in H^n$, we have $\Omega_a \subset H^n$ and $\Omega_a^W \subset H^{n-1}$. □

Proposition 11.4 ([5, 36, 55]). Given $a(s) \in H^n$ and $b(s) \in P^m$, if $\forall \omega \in \mathbb{R}$, $\mathrm{Re}[p(j\omega)] > 0$, then $|m - n| \leq 1$. □

The problem we are interested in is: Given a family of Hurwitz stable polynomials, how can we find a fixed polynomial such that their ratios will be SPR-invariant? In what follows, will first give some characterization of SPR (WSPR) regions, and then propose an efficient synthesis procedure for this problem.

11.3 Some Properties of SPR (WSPR) Regions

By definition, an SPR (WSPR) transfer function multiplied by a positive integer is still SPR (WSPR). Thus, without loss of generality, let

$$a(s) = s^n + a_1 s^{n-1} + \cdots + a_n \in H^n.$$

Denote as Ω_{1a} the set of the coefficients of all the $b(s) = s^n + x_1 s^{n-1} + \cdots + x_n \in P^n$, i.e., $(x_1, x_2, \ldots, x_n) \in \mathbb{R}^n$, such that $p(s) = \dfrac{b(s)}{a(s)} \in$ SPR; and denote as Ω_{1a}^W the set of the coefficients of all the $b(s) = s^{n-1} + x_1 s^{n-2} + \cdots + x_{n-1} \in P^{n-1}$, i.e., $(x_1, x_2, \ldots, x_{n-1}) \in \mathbb{R}^{n-1}$, such that $p(s) = \dfrac{b(s)}{a(s)} \in$ WSPR. Obviously, we have

$$\{1\} \times \Omega_{1a} = \left\{ \left(1, \frac{b_1}{b_0}, \frac{b_2}{b_0}, \ldots, \frac{b_n}{b_0}\right) \middle| \ \forall (b_0, b_1, b_2, \ldots, b_n) \in \Omega_a \right\},$$

$$\{1\} \times \Omega_{1a}^W = \left\{ \left(1, \frac{b_1}{b_0}, \frac{b_2}{b_0}, \ldots, \frac{b_{n-1}}{b_0}\right) \middle| \ \forall (b_0, b_1, b_2, \ldots, b_{n-1}) \in \Omega_a^W \right\}.$$

For notational convenience, Ω_{1a} (Ω_{1a}^W) sometimes also stands for the corresponding polynomial set.

As we know [26, 27], Ω_a is a non-empty, open, convex cone in $\mathbb{R}^{n+1}$. Thus, Ω_a is an unbounded set in $\mathbb{R}^{n+1}$. In what follows, we will show that Ω_{1a} is also an unbounded set in $\mathbb{R}^n$.

Theorem 11.1. Given $a(s) \in H^n$, Ω_{1a} is a non-empty, open, unbounded convex set in $\mathbb{R}^n$.

Proof. Obviously, we have $a(s) \in \Omega_a$. If the leading coefficient of $a(s)$ is a_0, then

$$\left(\frac{a_1}{a_0}, \frac{a_2}{a_0}, \ldots, \frac{a_n}{a_0}\right) \in \Omega_{1a}.$$

Thus, Ω_{1a} is not empty.

Moreover, $\forall (x_1, x_2, \ldots, x_n) \in \Omega_{1a}$, then $(1, x_1, x_2, \ldots, x_n) \in \Omega_a$. By Proposition 11.2, Ω_a is open. Thus, there exists $\delta > 0$, such that, when

$$\sqrt{(1-y_0)^2 + (x_1 - y_1)^2 + \cdots + (x_n - y_n)^2} < \delta,$$

we have $(y_0, y_1, y_2, \ldots, y_n) \in \Omega_a$.

For this δ, if $(z_1, z_2, \ldots, z_n) \in \mathbb{R}^n$ satisfying

$$\sqrt{(x_1 - z_1)^2 + \cdots + (x_n - z_n)^2} < \delta,$$

then, obviously

$$\sqrt{(1-1)^2 + (x_1 - z_1)^2 + \cdots + (x_n - z_n)^2} < \delta,$$

thus $(1, z_1, z_2, \ldots, z_n) \in \Omega_a$. Furthermore $(z_1, z_2, \ldots, z_n) \in \Omega_{1a}$. Hence, Ω_{1a} is open.

The convexity of Ω_{1a} is a direct consequence of the definition.

In what follows, we will prove that Ω_{1a} is unbounded. For this purpose, we first introduce some notation, which will be used also in other proofs later. Let

$$a(s) = s^n + a_1 s^{n-1} + \cdots + a_n \in H^n,$$
$$b(s) = x_0 s^n + x_1 s^{n-1} + \cdots + x_n \in P^n \cup P^{n-1}.$$

Then, $\forall \omega \in \mathbb{R}$, we have

$$\begin{aligned}
\mathrm{Re}\left[\frac{b(j\omega)}{a(j\omega)}\right] &= \frac{1}{|a(j\omega)|^2}\mathrm{Re}[b(j\omega)a(-j\omega)] \\
&= \frac{1}{|a(j\omega)|^2}\sum_{l=0}^{n}\left(\sum_{k=0}^{n} a_k x_{2l-k}(-1)^{l+k}\right)\omega^{2(n-l)} \\
&= \frac{1}{|a(j\omega)|^2}\sum_{l=0}^{n} c_l \omega^{2(n-l)},
\end{aligned}$$

where $c_l := \sum_{k=0}^{n} a_k x_{2l-k}(-1)^{l+k}$, and $a_0 = 1$; and let $a_i = 0, x_i = 0$, when $i < 0$ or $i > n$, $l = 0, 1, \ldots, n$.

Define the matrices

$$H_a := \begin{bmatrix} a_1 & 1 & 0 & 0 & 0 & \cdots & 0 \\ a_3 & a_2 & a_1 & 1 & 0 & \cdots & 0 \\ a_5 & a_4 & a_3 & a_2 & a_1 & \cdots & 0 \\ \vdots & \vdots & \vdots & \vdots & \vdots & \ddots & \vdots \\ a_{2n-1} & a_{2n-2} & a_{2n-3} & a_{2n-4} & a_{2n-5} & \cdots & a_n \end{bmatrix},$$

$$E_n := \begin{bmatrix} 1 & & & & \\ & -1 & & & \\ & & 1 & & \\ & & & -1 & \\ & & & & \ddots \end{bmatrix},$$

$$A := \begin{bmatrix} 1 & 0 & \cdots & 0 \\ -a_2 & & & \\ a_4 & & E_n H_a E_n & \\ \vdots & & & \end{bmatrix}, \quad b := \begin{bmatrix} x_0 \\ x_1 \\ \vdots \\ x_n \end{bmatrix}, \quad c := \begin{bmatrix} c_0 \\ c_1 \\ \vdots \\ c_n \end{bmatrix},$$

where $a_i = 0$ when $i > n$. Then it is easy to verify that

$$c = Ab. \tag{11.1}$$

Divide the matrices in (11.1) as follows

$$\left[\begin{array}{c|ccc} 1 & 0 & \cdots & 0 \\ \hline -a_2 & & & \\ a_4 & & E_n H_a E_n & \\ \vdots & & & \end{array}\right] \left[\begin{array}{c} x_0 \\ \hline x_1 \\ \vdots \\ x_n \end{array}\right] = \left[\begin{array}{c} c_0 \\ \hline c_1 \\ \vdots \\ c_n \end{array}\right].$$

Since $a(s) \in H^n$, we know that $E_n H_a E_n$ is invertible. Now take any $d = [d_1, d_2, \ldots, d_n]^T \in \mathbb{R}^n$, such that all elements of d are positive. Denote $\bar{a} = [-a_2, a_4, -a_6, a_8, \ldots, (-1)^n a_{2n}]^T$, where $a_i = 0$ if $i > n$.

Let $\bar{b} = (E_n H_a E_n)^{-1}(d - \bar{a}) := [b_1, b_2, \ldots, b_n]^T$. Then obviously, we have $[1, d_1, d_2, \ldots, d_n]^T = A[1, b_1, b_2, \ldots, b_n]^T$, and $b_n = \dfrac{d_n}{a_n}$. By (11.1), we have $(b_1, b_2, \ldots b_n) \in \Omega_{1a}$.

On the other hand, due to the arbitrariness of d, d_n can be taken arbitrarily large. Thus b_n can also be arbitrarily large. Therefore, Ω_{1a} is unbounded. This completes the proof. □

Since Ω_a and Ω_{1a} are both unbounded sets, when considering the robust SPR synthesis problem, we must check if two (or more) SPR regions intersect or not, which is hardly tractable when operating on unbounded sets. This is the reason that we introduce the concept of WSPR regions, which are bounded as shown below.

Theorem 11.2. Given $a(s) \in H^n$, Ω_{1a}^W is a non-empty, bounded convex set in $\mathbb{R}^{n-1}$.

Proof. Let H_a, E_n and A be the same as in the proof of Theorem 11.1. Divide the matrices in (11.1) as follows:

$$\left[\begin{array}{cc|ccc} 1 & 0 & 0 & \cdots & 0 \\ \hline -a_2 & a_1 & & & \\ a_4 & -a_3 & & B & \\ \vdots & \vdots & & & \\ \hline 0 & 0 & 0 & \cdots & a_n \end{array}\right] \left[\begin{array}{c} x_0 \\ x_1 \\ \hline x_2 \\ \vdots \\ x_{n-1} \\ \hline x_n \end{array}\right] = \left[\begin{array}{c} c_0 \\ \hline c_1 \\ \vdots \\ c_{n-1} \\ \hline c_n \end{array}\right],$$

where B is an $(n-1) \times (n-1)$ matrix formed by the first $n-1$ row and last $n-1$ column of the matrix $E_n H_a E_n$. Obviously, B is also invertible. Denote $\bar{a} := [a_1, -a_3, a_5, -a_7, \ldots, (-1)^{n-1} a_{2(n-1)+1}]^T$, and let $a_i = 0$ when $i > n$.

Let $\bar{b} := -B^{-1}\bar{a} = [b_1, b_2, \ldots, b_{n-1}]^T$. Since $a(s) \in H^n$, it is easy to verify that $b_{n-1} > 0$. Denote $b = [0, 1, b_1, b_2, \ldots, b_{n-1}]^T$. Take $c_0 = c_1 = \cdots = c_{n-1} = 0$, $c_n = a_n b_{n-1}$ in $c := [c_0, c_1, \ldots, c_n]$, it is easy to see that (11.1) is true. Thus we have $(b_1, b_2, \ldots, b_{n-1}) \in \Omega_{1a}^W$, namely, Ω_{1a}^W is not empty.

The convexity of Ω_{1a}^W is a direct consequence of the definition.

In what follows, we will prove that Ω_{1a}^W is bounded.

Take any $(x_1, x_2, \ldots, x_{n-1}) \in \Omega_{1a}^W$. Then we have

$$\frac{s^{n-1} + x_1 s^{n-2} + \cdots + x_{n-1}}{s^n + a_1 s^{n-1} + \cdots + a_n} \in \text{WSPR}.$$

By Proposition 11.3, we have $s^{n-1} + x_1 s^{n-2} + \cdots + x_{n-1} \in H^{n-1}$. Moreover, $\forall \omega \in \mathbb{R}$, we have

$$\operatorname{Re}\left(\left. \frac{s^{n-1} + x_1 s^{n-2} + \cdots + x_{n-1}}{s^n + a_1 s^{n-1} + \cdots + a_n} \right|_{s=j\omega} \right) > 0.$$

Hence,

$$\mathrm{Re}\left(\frac{s^n + a_1 s^{n-1} + \cdots + a_n}{s^{n-1} + x_1 s^{n-2} + \cdots + x_{n-1}}\bigg|_{s=j\omega}\right) > 0, \quad \forall \omega \in \mathbb{R}.$$

Obviously

$$\begin{aligned}&\frac{s^n + a_1 s^{n-1} + \cdots + a_n}{s^{n-1} + x_1 s^{n-2} + \cdots + x_{n-1}} \\ &\quad = s + \frac{(a_1 - x_1)s^{n-1} + (a_2 - x_2)s^{n-2} + \cdots + (a_{n-1} - x_{n-1})s + a_n}{s^{n-1} + x_1 s^{n-2} + \cdots + x_{n-1}}.\end{aligned}$$

Thus

$$\begin{aligned}&\mathrm{Re}\left(\frac{(a_1 - x_1)s^{n-1} + (a_2 - x_2)s^{n-2} + \cdots + (a_{n-1} - x_{n-1})s + a_n}{s^{n-1} + x_1 s^{n-2} + \cdots + x_{n-1}}\bigg|_{s=j\omega}\right) \\ &\quad = \mathrm{Re}\left(\frac{s^n + a_1 s^{n-1} + \cdots + a_n}{s^{n-1} + x_1 s^{n-2} + \cdots + x_{n-1}}\bigg|_{s=j\omega}\right) - \mathrm{Re}(j\omega) \\ &\quad = \mathrm{Re}\left(\frac{s^n + a_1 s^{n-1} + \cdots + a_n}{s^{n-1} + x_1 s^{n-2} + \cdots + x_{n-1}}\bigg|_{s=j\omega}\right) > 0, \quad \forall \omega \in \mathbb{R}.\end{aligned}$$

By Proposition 11.4, we have

$$\begin{aligned}&\frac{(a_1 - x_1)s^{n-1} + (a_2 - x_2)s^{n-2} + \cdots + (a_{n-1} - x_{n-1})s + a_n}{s^{n-1} + x_1 s^{n-2} + \cdots + x_{n-1}} \\ &\quad \in \{\mathrm{SPR}\} \cup \{\mathrm{WSPR}\}.\end{aligned}$$

Again, by Proposition 11.3, we have

$$\begin{aligned}&(a_1 - x_1)s^{n-1} + (a_2 - x_2)s^{n-2} + \cdots + (a_{n-1} - x_{n-1})s + a_n \\ &\quad \in H^{n-1} \cup H^{n-2}.\end{aligned}$$

Hence,

$$0 < x_1 \le a_1,\ 0 < x_2 < a_2, \ldots, 0 < x_{n-1} < a_{n-1}.$$

Namely,

$$\begin{aligned}\Omega_{1a}^W &\subset \left\{(x_1, x_2, \ldots, x_{n-1}) \middle| \alpha(s) := \sum_{i=1}^{n}(a_i - x_i)s^{n-i} \in H^{n-1} \cup H^{n-2}, x_n = 0\right\} \\ &\subset \{(x_1, x_2, \ldots, x_{n-1}) | 0 < x_1 \le a_1, 0 < x_2 < a_2, \ldots, 0 < x_{n-1} < a_{n-1}\}.\end{aligned}$$

Thus, Ω_{1a}^W is bounded. This completes the proof. □

It should be pointed out that Ω_{1a}^W is not an open set in $\mathbb{R}^{n-1}$. In fact, from the proof of Theorem 11.2, we know that Ω_{1a}^W is tangent to the hyperplane $x_1 = a_1$ in $\mathbb{R}^{n-1}$. And there exist some points of Ω_{1a}^W in this hyperplane. Thus, Ω_{1a}^W cannot be an open set. Obviously, Ω_a^W is a non-empty, convex cone in $\mathbb{R}^{n-1}$. Thus, Ω_a^W is also unbounded. It is easy to know that Ω_a^W is not an open set either.

Though Ω_{1a}^W is not an open set, the following theorem guarantees such a fact: when the intersection of two or more WSPR regions is not empty, then the intersection must be a region, not a single point.

Theorem 11.3. Given $a(s) \in H^n$, if $(x_1, x_2, \ldots, x_{n-1}) \in \Omega_{1a}^W$, then for sufficiently small $\varepsilon > 0$, we have $(x_1 - \varepsilon, x_2 - \varepsilon, \ldots, x_{n-1} - \varepsilon) \in \Omega_{1a}^W$.

Proof. $\forall (x_1, x_2, \ldots, x_{n-1}) \in \Omega_{1a}^W$, and $\forall \omega \in \mathbb{R}$, we have

$$\mathrm{Re}\left(\frac{s^{n-1} + x_1 s^{n-2} + \cdots + x_{n-1}}{s^n + a_1 s^{n-1} + \cdots + a_n}\bigg|_{s=j\omega}\right) > 0,$$

$\forall \varepsilon > 0$, since

$$\begin{aligned}
&\mathrm{Re}\left(\frac{s^{n-1} + (x_1 - \varepsilon)s^{n-2} + \cdots + (x_{n-1} - \varepsilon)}{s^n + a_1 s^{n-1} + \cdots + a_n}\bigg|_{s=j\omega}\right) \\
&= \mathrm{Re}\left[\left(\frac{s^{n-1} + x_1 s^{n-2} + \cdots + x_{n-1}}{s^n + a_1 s^{n-1} + \cdots + a_n} + \frac{(-\varepsilon)(s^{n-2} + s^{n-3} + \cdots + 1)}{s^n + a_1 s^{n-1} + \cdots + a_n}\right)\bigg|_{s=j\omega}\right] \\
&= \mathrm{Re}\left(\frac{s^{n-1} + x_1 s^{n-2} + \cdots + x_{n-1}}{s^n + a_1 s^{n-1} + \cdots + a_n}\bigg|_{s=j\omega}\right) + \frac{(-\varepsilon)}{|a(j\omega)|^2}(-\omega^{2(n-1)} + \tilde{c}(\omega)),
\end{aligned}$$

where $\tilde{c}(\omega)$ is a real polynomial with order not greater than $2(n-2)$. Thus, when $|\omega|$ is sufficiently large, the sign of $(-\varepsilon)(-\omega^{2(n-1)} + \tilde{c}(\omega))$ is positive. Namely, there exists $\omega_1 > 0$ such that, for all$|\omega| \geq \omega_1$,

$$\mathrm{Re}\left(\frac{s^{n-1} + (x_1 - \varepsilon)s^{n-2} + \cdots + (x_{n-1} - \varepsilon)}{s^n + a_1 s^{n-1} + \cdots + a_n}\bigg|_{s=j\omega}\right) > 0.$$

Denote

$$M_1 = \inf_{|\omega| \leq \omega_1} \mathrm{Re}\left(\frac{s^{n-1} + x_1 s^{n-2} + \cdots + x_{n-1}}{s^n + a_1 s^{n-1} + \cdots + a_n}\bigg|_{s=j\omega}\right),$$

$$N_1 = \sup_{|\omega| \leq \omega_1}\left|\mathrm{Re}\left(\frac{1}{|a(j\omega)|^2}(\omega^{2(n-1)} - \tilde{c}(\omega))\right)\right|.$$

Then $M_1 > 0$ and $N_1 > 0$. Choosing $0 < \varepsilon < \dfrac{M_1}{N_1}$, then it is easy to see that

$$\mathrm{Re}\left(\frac{s^{n-1} + (x_1 - \varepsilon)s^{n-2} + \cdots + (x_{n-1} - \varepsilon)}{s^n + a_1 s^{n-1} + \cdots + a_n}\bigg|_{s=j\omega}\right) > 0, \quad \forall \omega \in \mathbb{R}.$$

Therefore, $\dfrac{s^{n-1}+(x_1-\varepsilon)s^{n-2}+\cdots+(x_{n-1}-\varepsilon)}{s^n+a_1s^{n-1}+\cdots+a_n} \in$ WSPR, namely,

$$(x_1-\varepsilon, x_2-\varepsilon, \ldots, x_{n-1}-\varepsilon) \in \Omega_{1a}^W.$$

This completes the proof. □

The following theorem shows the relationship between Ω_{1a}^W and Ω_a, and plays an important role in robust SPR synthesis.

Theorem 11.4. Given $a(s) \in H^n$, if $(x_1, x_2, \ldots, x_{n-1}) \in \Omega_{1a}^W$, then $\forall(1, \alpha_1, \alpha_2, \ldots, \alpha_n) \in \mathbb{R}^{n+1}$, we can take sufficiently small $\varepsilon > 0$ such that

$$(0, 1, x_1, x_2, \ldots, x_{n-1}) + \varepsilon(1, \alpha_1, \alpha_2, \ldots, \alpha_n) \in \Omega_a.$$

Proof. Denote $b(s) = s^{n-1}+x_1s^{n-2}+\cdots+x_{n-1}$, $\alpha(s) = s^n+\alpha_1s^{n-1}+\cdots+\alpha_n$, and $\tilde{b}(s) = b(s)+\varepsilon\alpha(s)$. Since $(x_1, x_2, \ldots, x_{n-1}) \in \Omega_{1a}^W$, we have

$$\mathrm{Re}\left[\frac{b(j\omega)}{a(j\omega)}\right] > 0, \quad \forall\omega \in \mathbb{R}.$$

We only need to show that, for sufficiently small $\varepsilon > 0$,

$$\mathrm{Re}\left[\frac{\tilde{b}(j\omega)}{a(j\omega)}\right] > 0, \quad \forall\omega \in \mathbb{R}.$$

Obviously, $\tilde{b}(s)$ and $a(s)$ have the same order n. Thus, there exists $\omega_2 > 0$ such that, for all $|\omega| \geq \omega_2$, we have $\mathrm{Re}\left[\dfrac{\tilde{b}(j\omega)}{a(j\omega)}\right] > 0$.

Denote

$$M_2 = \inf_{|\omega|\leq\omega_2} \mathrm{Re}\left[\frac{b(j\omega)}{a(j\omega)}\right],$$

$$N_2 = \sup_{|\omega|\leq\omega_2} \left|\mathrm{Re}\frac{\alpha(j\omega)}{a(j\omega)}\right|.$$

Then $M_2 > 0$ and $N_2 > 0$. Choosing $0 < \varepsilon < \dfrac{M_2}{N_2}$, by simple computation, we have

$$\mathrm{Re}\left[\frac{\tilde{b}(j\omega)}{a(j\omega)}\right] > 0, \quad \forall\omega \in \mathbb{R}.$$

This completes the proof. □

11.4 Characterization of SPR (WSPR) Regions

In this section, we will consider the following problem.

Problem 11.1. Given $a(s) \in H^n$, how can we find all $b(s)$, such that $p(s) = \frac{b(s)}{a(s)} \in \text{SPR (WSPR)}$? □

This problem is important in robust SPR analysis and synthesis. Expressing $b(s)$ explicitly based on the coefficients of $a(s)$ is an unsolved problem proposed by Huang, Hollot and Xu [27] in 1990.

Verification of SPR (WSPR) of transfer functions can, in principle, be transformed into checking the real roots of polynomials. The classical Sturm method can be used to check the distribution of real roots of polynomials [22, 52], but it is not efficient for polynomials with symbolic coefficients [50, 52].

Recently, Yang *et al.* established the complete discrimination system (CDS) for polynomials [22, 52], which can express the roots distribution explicitly based on the coefficients of polynomials. Using the CDS, we can give a complete characterization of SPR (WSPR) regions.

Given

$$f(x) = a_0 x^n + a_1 x^{n-1} + \cdots + a_n \in P^n.$$

The Sylvester matrix of $f(x)$ and its derivative $f'(x)$ [50, 52]

$$\begin{bmatrix} a_0 & a_1 & a_2 & \cdots & a_n & & & & \\ 0 & na_0 & (n-1)a_1 & \cdots & a_{n-1} & & & & \\ & a_0 & a_1 & \cdots & a_{n-1} & a_n & & & \\ & 0 & na_0 & \cdots & 2a_{n-1} & a_{n-1} & & & \\ & & & \cdots & \cdots & & & & \\ & & & \cdots & \cdots & & & & \\ & & & & a_0 & a_1 & \cdots & a_n \\ & & & & 0 & na_0 & \cdots & a_{n-1} \end{bmatrix}$$

is called the discrimination matrix of $f(x)$, denoted by Discr(f).

The n-tuple

$$[D_1(f), D_2(f), \ldots, D_n(f)],$$

which are the determinants of the first n even-order main submatrices of Discr(f), formed by the first $2k$ row and first $2k$ column, $k = 1, 2, \ldots, n$, are called the discriminant sequence of $f(x)$.

Furthermore

$$[\text{sign}(D_1), \text{sign}(D_2), \ldots, \text{sign}(D_n)]$$

is called the sign list of the discriminant sequence $[D_1, D_2, \ldots, D_n]$, where sign$(\cdot)$ is the sign function, namely

$$\text{sign}(x) = \begin{cases} 1, & \text{if } x > 0, \\ 0, & \text{if } x = 0, \\ -1, & \text{if } x < 0. \end{cases}$$

Given a sign list $[s_1, s_2, \ldots, s_n]$, we can construct a revised sign list

$$[\varepsilon_1, \varepsilon_2, \ldots, \varepsilon_n]$$

as follows:

(i) If $[s_i, s_{i+1}, \ldots, s_{i+j}]$ is a section of the given list, where $s_i \neq 0; s_{i+1} = s_{i+2} = \cdots = s_{i+j-1} = 0; s_{i+j} \neq 0$, then, we replace the subsection

$$[s_{i+1}, s_{i+2}, \ldots, s_{i+j-1}]$$

by

$$[-s_i, -s_i, s_i, s_i, -s_i, -s_i, s_i, s_i, -s_i, \ldots],$$

namely, $\varepsilon_{i+r} = (-1)^{\left[\frac{r+1}{2}\right]} s_i$, $r = 1, 2, \ldots, j-1$, where the notation $\left[\dfrac{r+1}{2}\right]$ stands for the largest integer that is smaller than or equal to the real number $\dfrac{r+1}{2}$.

(ii) Otherwise, let $\varepsilon_k = s_k$, i.e., no changes for other terms.

Lemma 11.1 ([50, 52]). *Given a real polynomial $f(x) = a_0 x^n + a_1 x^{n-1} + \cdots + a_n \in P^n$, if the number of sign changes in the revised sign list of its discriminant sequence is ν, and the number of non-zero elements in the revised sign list of its discriminant sequence is μ, then the number of distinct real roots of $f(x)$ is $\mu - 2\nu$.* □

The discriminant sequence of the polynomial $f(x)$ can also be constructed by the main submatrices of the Bezout matrix of $f(x)$ and $f'(x)$ [50, 52]; the number of distinct real roots of the polynomial $f(x)$ can also be determined by the sign differences of the Bezout matrix of $f(x)$ and $f'(x)$ [50, 52].

The original complete discrimination system of polynomials [50, 52] is more general than Lemma 11.1, which can also be used to determine the number of complex roots and the multiplicities of repeated roots [50, 52].

We are now in a position to give a complete characterization of SPR (WSPR) regions. As in the last section, let

$$a(s) = s^n + a_1 s^{n-1} + \cdots + a_n \in H^n,$$

$$b(s) = x_0 s^n + x_1 s^{n-1} + \cdots + x_n \in \cup P^n \cup P^{n-1}.$$

Consider the polynomial

$$f(\omega) = \sum_{l=0}^{n} c_l \omega^{2(n-l)}$$

where $c_l := \sum_{k=0}^{n} a_k x_{2l-k} (-1)^{l+k}$, $a_0 = 1$, and let $a_i = 0, x_i = 0$, when $i < 0$ or $i > n$, $l = 0, 1, \ldots, n$. Then, we have the following theorem.

Theorem 11.5. Given $a(s) = s^n + a_1 s^{n-1} + \cdots + a_n \in H^n$, suppose that $b(s) = x_0 s^n + x_1 s^{n-1} + \cdots + x_n \in P^n \cup P^{n-1}$, then $p(s) := \dfrac{b(s)}{a(s)} \in \{\text{SPR}\} \cup \{\text{WSPR}\}$ if and only if the number of sign changes ν in the revised sign list of the discriminant sequence of $f(\omega) = \sum_{l=0}^{n} c_l \omega^{2(n-l)}$ and its number of non-zero elements μ satisfy $\mu = 2\nu$ when $c_0 = c_1 = \cdots = c_i = 0$, $c_{i+1} > 0$ $(i = 0, 1, \ldots, n-1)$.

Proof. From the proof of Theorem 11.1, we know that

$$\text{Re}\left[\frac{b(j\omega)}{a(j\omega)}\right] = \frac{1}{|a(j\omega)|^2} \sum_{l=0}^{n} c_l \omega^{2(n-l)}, \quad \forall \omega \in \mathbb{R},$$

where $c_l := \sum_{k=0}^{n} a_k x_{2l-k} (-1)^{l+k}$, where $a_0 = 1$, when $i < 0$ or $i > n$. Let $a_i = 0$, $x_i = 0$, $l = 0, 1, \ldots, n$.

By introducing E_n, H_a, A, b and c as before, it is easy to verify that

$$c = Ab.$$

By the definition of SPR (WSPR), in order to have $p(s) := \dfrac{b(s)}{a(s)} \in \{\text{SPR}\} \cup \{\text{WSPR}\}$, we must have

$$f(\omega) = \sum_{l=0}^{n} c_l \omega^{2(n-l)} > 0, \quad \forall \omega \in \mathbb{R},$$

which is equivalent to the following condition: when $c_0 = c_1 = \cdots = c_i = 0$, $c_{i+1} > 0$ $(i = 0, 1, \ldots, n-1)$, $f(\omega)$ has no real roots. (Notice that we are considering a sequence of polynomials with positive leading coefficients.) Thus, by Lemma 11.1, we complete the proof. □

Necessary and sufficient conditions are obtained in Theorem 11.5. By using a computer, it is easy to get the sign list of the discriminant sequence for a polynomial with symbolic coefficients [50, 52]. Thus, Theorem 11.5 provides an efficient on-line algorithm for robust SPR synthesis.

Recently, Yang and Xia obtain a new criterion for the distribution of positive roots or negative roots of a polynomial [51]. It is similar to Lemma 11.1, but is more efficient in computation.

The conditions in Theorem 11.5 are very complicated when expressed explicitly. Since $a(s)$ and $b(s)$ are both Hurwitz stable, by some simplification, we can get simpler expressions. In what follows, we will present some simplified expressions for SPR (WSPR) regions corresponding to third- and fourth-order Hurwitz polynomials.

Corollary 11.1. Given $a(s) = s^3 + a_1 s^2 + a_2 s + a_3 \in H^3$, suppose that $b(s) = s^2 + xs + y, (x, y) \in \mathbb{R}^2$. Then $p(s) := \dfrac{b(s)}{a(s)} \in$ WSPR if and only if

$$\begin{aligned}(x,y) \in \Omega_{1a}^{W} := \{(x,y)|\ & a_2^2 x^2 + 2(2a_3 - a_1 a_2)xy + a_1^2 y^2 \\ & - 2a_2 a_3 x - 2a_1 a_3 y + a_3^2 < 0\} \\ & \cup \{(x,y)|x \le a_1, a_2 x - a_1 y - a_3 \ge 0, y > 0\}.\end{aligned}$$

□

Corollary 11.2. Given $a(s) = s^3 + a_1 s^2 + a_2 s + a_3 \in H^3$, suppose that $b(s) = s^3 + xs^2 + ys + z, (x, y, z) \in \mathbb{R}^3$. Then $p(s) := \dfrac{b(s)}{a(s)} \in$ SPR if and only if

$$\begin{aligned}(x,y,z) \in \Omega_{1a} := & \{(x,y,z)|z > 0, \Delta_3 < 0\} \\ & \cup \{(x,y,z)|z > 0, p > 0, \Delta_1 > 0, \Delta_2 > 0, \Delta_3 \ge 0\} \\ & \cup \{(x,y,z)|p > 0, \Delta_1 = 0, \Delta_2 = 0, \Delta_3 = 0\},\end{aligned}$$

where

$$p = a_1 x - y - a_2,\ r = a_2 y - a_3 x - a_1 z,\ t = a_3 z,$$

$$\Delta_1 = p^2 - 3r,\ \Delta_2 = rp^2 + 3tp - 4r^2,$$

$$\Delta_3 = -4r^3 + 18rtp + p^2 r^2 - 4p^3 t - 27t^2,$$

$$[D_1, D_2, D_3, D_4, D_5, D_6] = [1, -p, -p\Delta_1, \Delta_1\Delta_2, \Delta_2\Delta_3, -t\Delta_3^2].$$ □

Corollary 11.3. Given $a(s) = s^4 + a_1 s^3 + a_2 s^2 + a_3 s + a_4 \in H^4$, suppose that $b(s) = s^3 + xs^2 + ys + z, (x, y, z) \in \mathbb{R}^3$. Then $p(s) := \dfrac{b(s)}{a(s)} \in$ WSPR if and only if

$$\begin{aligned}(x,y,z) \in \Omega_{1a}^{W} := & \{(x,y,z)|\ x < a_1, z > 0, \Delta_3 < 0\} \\ & \cup \{(x,y,z)|\ x < a_1, z > 0, p > 0, \Delta_1 > 0, \Delta_2 > 0, \Delta_3 \ge 0\} \\ & \cup \{(x,y,z)|\ x < a_1, p > 0, \Delta_1 = 0, \Delta_2 = 0, \Delta_3 = 0\} \\ & \cup \{(x,y,z)|\ x = a_1, a_1 a_2 - a_3 - a_1 y + z > 0, z > 0, \Delta' < 0\} \\ & \cup \{(x,y,z)|\ x = a_1, a_1 a_2 - a_3 - a_1 y + z > 0, \\ & \qquad z > 0,\ p' > 0,\ \Delta' \ge 0\} \\ & \cup \{(x,y,z)|\ x = a_1, a_1 a_2 - a_3 - a_1 y + z = 0, \\ & \qquad z > 0,\ a_3 y - a_2 z - a_1 a_4 \ge 0\}.\end{aligned}$$

where

$$p=\frac{a_2x-a_1y-a_3+z}{a_1-x},\ r=\frac{a_3y-a_4x-a_2z}{a_1-x},\ t=\frac{a_4z}{a_1-x},$$

$$\Delta_1=p^2-3r,\ \Delta_2=rp^2+3tp-4r^2,$$

$$\Delta_3=-4r^3+18rtp+p^2r^2-4p^3t-27t^2,$$

$$[D_1,D_2,D_3,D_4,D_5,D_6]=[1,-p,-p\Delta_1,\Delta_1\Delta_2,\Delta_2\Delta_3,-t\Delta_3^2],$$

$$p'=\frac{a_3y-a_2z-a_1a_4}{a_1a_2-a_3-a_1y+z},\ r'=\frac{a_4z}{a_1a_2-a_3-a_1y+z},\ \Delta'=p'^2-4r'.$$

$$[D_1',D_2',D_3',D_4']=[1,-p',-p'\Delta',r'\Delta'^2].$$

□

The expressions in the corollaries above are complete for high-order systems. In principle, we can derive explicit expressions for higher-order systems. But the resulting expressions become more and more complicated, and the number of expressions also increase quickly. In the following corollary, the so-called "lazy strategy" is employed to express the conditions [50, 52].

Corollary 11.4. Given $a(s)=s^4+a_1s^3+a_2s^2+a_3s+a_4\in H^4$, suppose that $b(s)=s^4+xs^3+ys^2+zs+u,(x,y,z,u)\in\mathbb{R}^4$. Then $p(s):=\dfrac{b(s)}{a(s)}\in$ SPR if and only if the number of sign changes ν in the revised sign list of $[D_1,D_2,D_3,D_4,D_5,D_6,D_7,D_8]$ and the number of non-zero elements μ satisfy $\mu=2\nu$, where

$$p=a_1x-a_2-y,\ q=a_4+u+a_2y-a_3x-a_1z,$$

$$r=a_3z-a_2u-a_4y,\ t=a_4u,$$

$$\Delta_1=-8q+3p^2,\ \Delta_2=3rp+qp^2-4q^2,$$

$$\Delta_3=28qrp-8q^3+32tq-6p^3r+2p^2q^2-12p^2t-36r^2,$$

$$\begin{aligned}\Delta_4&=18qpr^2-7rp^2t+q^2p^2r+3qp^3t-16t^2p-4p^3r^2\\&\quad-4rq^3-12q^2pt+48rtq-27r^3,\end{aligned}$$

$$\begin{aligned}\Delta_5&=144qt^2p^2+144qtr^2-192rt^2p+18qr^3p-6p^2tr^2-4p^2q^3t\\&\quad+p^2q^2r^2+18p^3rtq-80rptq^2-128t^2q^2-4p^3r^3-27p^4t^2\\&\quad-27r^4-4q^3r^2+16q^4t+256t^3,\end{aligned}$$

$$\begin{aligned}&[D_1,D_2,D_3,D_4,D_5,D_6,D_7,D_8]\\&\quad=[1,-p,-p\Delta_1,\Delta_1\Delta_2,\Delta_2\Delta_3,-\Delta_3\Delta_4,-\Delta_4\Delta_5,t\Delta_5^2].\end{aligned}$$

□

11.5 Robust SPR Synthesis: Intersection of WSPR Regions

Generally speaking, the design problem is more difficult than the analysis problem, since it is usually constructive, i.e., it not only shows the existence of the solution, but also provides a constructive procedure to find it. In this section, we will propose an algorithm for robust design of SPR transfer functions. This algorithm works well for both low-order and high-order polynomial families. Illustrative examples are provided to show the effectiveness of this algorithm.

Suppose that

$$F = \{a_i(s) = s^n + \sum_{l=1}^{n} a_l^{(i)} s^{n-l}, \quad i = 1, 2, \ldots, m\}.$$

How do we find a polynomial $b(s)$, such that

$$p_i(s) := \frac{b(s)}{a_i(s)} \in \text{SPR}, \quad i = 1, 2, \ldots, m?$$

As observed earlier, the existence of such a polynomial $b(s)$ boils down to the condition that the intersection of the SPR regions associated with $a_i(s)$ is not empty. From the results in the previous section, we know that SPR regions are unbounded, whereas monic WSPR regions are bounded. Thus, from computational considerations, we first consider the intersection of monic WSPR regions, and then construct a polynomial $b(s)$ by using the technique presented in the previous section.

Since SPR (WSPR) transfer functions with fixed numerator (or denominator) enjoy convexity property, namely, if there exists a polynomial $c(s)$, such that $\frac{c(s)}{a(s)}$ and $\frac{c(s)}{b(s)}$ are both SPR (WSPR), then, it is easy to verify that, for any $\alpha \geq 0$, $\beta \geq 0$ and $(\alpha, \beta) \neq (0, 0)$, we have $\frac{c(s)}{\alpha a(s) + \beta b(s)} \in$ SPR (WSPR). Therefore, the assumptions made on F do not lose any generality. Actually, the method proposed in this chapter also applies to convex combinations of polynomials, interval polynomials, and more generally, polytopic polynomials and multilinearly perturbed polynomials.

By the results presented in the last two sections, the problem above can be transformed into checking first whether $\bigcap_{i=1}^{m} \Omega_{1a_i}^{W}$ is empty or not. If $\bigcap_{i=1}^{m} \Omega_{1a_i}^{W} \neq \phi$, then by Theorem 11.4, we can find a $b(s)$. If $\bigcap_{i=1}^{m} \Omega_{1a_i}^{W} \neq \phi$, it is easy to see that $\bigcap_{i=1}^{m} \Omega_{1a_i}^{W}$ must be a bounded convex set, and it cannot be an isolated point (see Theorems 11.2 and 11.3). Moreover, $\Omega_{1a_i}^{W}$, $i = 1, 2, \ldots, m$, can be characterized by Theorem 11.5.

In fact, by the discussions in the last two sections, we can get some more information about $\bigcap_{i=1}^{m} \Omega_{1a_i}^{W}$. For example,

(i) $\bigcap_{i=1}^{m} \Omega_{1a_i}^{W} \subset \{(x_1, x_2, \dots, x_{n-1}) | \, 0 < x_1 \le \alpha_1, \ 0 < x_2 < \alpha_2, \dots, 0 < x_{n-1} < \alpha_{n-1}\}$, where $\alpha_l = \min\{a_l^{(i)}, i = 1, 2, \dots, m\}$, $l = 1, 2, \dots, n-1$ (by Theorem 11.2 and its proof).

(ii) If $b := (b_1, b_2, \dots, b_{n-1}) \in \bigcap_{i=1}^{m} \Omega_{1a_i}^{W}$, then the $(n-1)$st-order polynomial with coefficients $(1, b_1, b_2, \dots, b_{n-1})$ is in H^{n-1} (by Proposition 11.3).

(iii) If $b := (b_1, b_2, \dots, b_{n-1}) \in \bigcap_{i=1}^{m} \Omega_{1a_i}^{W}$, then for $i = 1, 2, \dots, m$, the polynomials with coefficients $(a_1^{(i)} - b_1, a_2^{(i)} - b_2, \dots, a_{n-1}^{(i)} - b_{n-1}, a_n^{(i)})$, respectively, are in $H^{n-1} \cup H^{n-2}$ (by Theorem 11.2 and its proof).

The three points above are very useful in checking if $\bigcap_{i=1}^{m} \Omega_{1a_i}^{W}$ is empty or not, and in effectively finding the elements of $\bigcap_{i=1}^{m} \Omega_{1a_i}^{W}$.

By the results presented in the previous section, we propose the following design procedure.

Step 1. Test the robust stability of the convex hull of F, i.e., $\overline{F}$. If $\overline{F}$ is robustly stable, then go to Step 2; otherwise, print "there does not exist such a $b(s)$" (by Definition 11.1 and Definition 11.2).

Step 2. Let $\alpha_l = \min\{a_l^{(i)}, i = 1, 2, \dots, m\}$, $l = 1, 2, \dots, n-1$. Grid the hyperrectangle

$$D := \{(x_1, x_2, \dots, x_{n-1}) \mid 0 < x_l < \alpha_l, \ l = 1, 2, \dots, n-1\}$$

according to the precision required (by Theorem 11.2 and its proof).

Step 3. Take $b := (b_1, b_2, \dots, b_{n-1})$ at each gridding point. Test whether b belongs to $\bigcap_{i=1}^{m} \Omega_{1a_i}^{W}$ by the following steps:

(i) Test if the $(n-1)$st-order polynomial with coefficients $(1, b_1, b_2, \dots, b_{n-1})$ belongs to H^{n-1} (by Proposition 11.3).

(ii) For $i = 1, 2, \dots, m$, test if the polynomial with coefficients $(a_1^{(i)} - b_1, a_2^{(i)} - b_2, \dots, a_{n-1}^{(i)} - b_{n-1}, a_n^{(i)})$ belongs to $H^{n-1} \cup H^{n-2}$, respectively (by Theorem 11.2 and its proof).

(iii) Test if b belongs to $\Omega_{1a_i}^{W}$, $i = 1, 2, \dots, m$.

If all three points above are satisfied, go to the next step; otherwise, move to the next gridding point and test the three points again (If all gridding points have been tested, then print "there does not exist such a b in $\bigcap_{i=1}^{m} \Omega_{1a_i}^{W}$ with the given precision").

Step 4. Take a sufficiently small $\varepsilon > 0$ such that

$$(\varepsilon, 1, b_1, b_2, \dots, b_{n-1}) \in \bigcap_{i=1}^{m} \Omega_{a_i}.$$

Hence, the nth-order polynomial with coefficients $(\varepsilon, 1, b_1, b_2, \dots, b_{n-1})$ satisfies the design requirement (by Theorem 11.4).

In the next two sections, we will show that for the low-order stable interval polynomial family ($n \le 4$) or arbitrary order stable convex combination, existence of the solution to the design problem is always guaranteed. Given adequate precision, our method will surely find a polynomial that satisfies the design requirement. As shown by numerous examples below, our method is also effective for higher-order polynomial families.

Example 11.1. Suppose that $F = \{a_1(s) = s^4 + 11s^3 + 56s^2 + 88s + 1,\ a_2(s) = s^4 + 11s^3 + 56s^2 + 88s + 50,\ a_3(s) = s^4 + 89s^3 + 56s^2 + 88s + 1,\ a_4(s) = s^4 + 89s^3 + 56s^2 + 88s + 50\}$. It is easy to see that the convex hull $\overline{F}$ of F is robustly stable [30]. Using our method, it is easy to get $b(s) = s^3 + 3.3s^2 + 2.24s + 1.76 \in \bigcap_{i=1}^4 \Omega_{1a_i}^W$ (Note also that, in this example, $b(s)$ is not unique. Using our method, we can get all such $b(s)$'s with given precision. This is also true for the examples below.) Thus, let $c(s) := \varepsilon s^4 + b(s)$, $\varepsilon > 0$, ε sufficiently small, e.g., take $\varepsilon \leq 0.3$ (ε is determined by Theorem 11.4), then the design requirement has been met. □

In this example, if we take $F = \{a_1(s) = s^4 + 11s^3 + 56s^2 + 88s + 50,\ a_2(s) = s^4 + 89s^3 + 56s^2 + 88s + 50\}$, then it is just the counterexample given in [12]. It can be verified that the sufficient conditions presented in [5, 18] are not satisfied. But we can use the methods in [26, 27, 35] to do SPR synthesis. When F is enlarged to be the set of four polynomials as in this example, the methods in [26, 27, 35] do not work either. Using our method, it is easy to get the design done.

In what follows, we will provide some more examples for higher-order polynomial families.

Example 11.2. Suppose that $F = \{a_1(s) = s^6 + 3.5s^5 + 26.5s^4 + 60.5s^3 + 61s^2 + 27.5s + 4.5,\ a_2(s) = s^6 + 8.5s^5 + 33.5s^4 + 59.5s^3 + 59s^2 + 32.5s + 7.5,\ a_3(s) = s^6 + 8.5s^5 + 26.5s^4 + 59.5s^3 + 61s^2 + 32.5s + 4.5,\ a_4(s) = s^6 + 3.5s^5 + 33.5s^4 + 60.5s^3 + 59s^2 + 27.5s + 7.5\}$. It is easy to see that the convex hull $\overline{F}$ of F is robustly stable [30]. Using our method, it is easy to get $b(s) = s^5 + 3.15s^4 + 13s^3 + 19.83s^2 + 14.75s + 5.5 \in \bigcap_{i=1}^4 \Omega_{1a_i}^W$. Thus, let $c(s) := \varepsilon s^6 + b(s)$, $\varepsilon > 0$, ε sufficiently small, e.g., take $\varepsilon \leq 0.003$, then the design requirement has been met. □

Example 11.3. Suppose that $F = \{a_1(s) = s^6 + 12s^5 + 70s^4 + 300s^3 + 500s^2 + 600s + 300,\ a_2(s) = s^6 + 14s^5 + 70s^4 + 240s^3 + 500s^2 + 700s + 300,\ a_3(s) = s^6 + 12s^5 + 80s^4 + 300s^3 + 450s^2 + 600s + 400,\ a_4(s) = s^6 + 14s^5 + 80s^4 + 240s^3 + 500s^2 + 700s + 400\}$. It is easy to see that the convex hull $\overline{F}$ of F is robustly stable [30]. Using our method, it is easy to get $b(s) = s^5 + 6s^4 + 35s^3 + 80s^2 + 120s + 120 \in \bigcap_{i=1}^4 \Omega_{1a_i}^W$. Thus, let $c(s) := \varepsilon s^6 + b(s)$, $\varepsilon > 0$, ε sufficiently small, e.g., take $\varepsilon \leq 0.03$, then the design requirement has been met. □

Example 11.4. Suppose that $F = \{a_1(s) = s^7 + 9s^6 + 31s^5 + 71.5s^4 + 111.5s^3 + 109s^2 + 76s + 12.5,\ a_2(s) = s^7 + 9.5s^6 + 31s^5 + 71s^4 + 111.5s^3 + 109.5s^2 + 76s + 12,\ a_3(s) = s^7 + 9s^6 + 31.5s^5 + 71.5s^4 + 111s^3 + 109s^2 + 76.5s + 12.5,\ a_4(s) = s^7 + 9.5s^6 + 31.5s^5 + 71s^4 + 111s^3 + 109.5s^2 + 76.5s + 12\}$. It is easy to see that the convex hull $\overline{F}$ of F is robustly stable [30]. Using our method, it is easy to get $b(s) = s^6 + 7.2s^5 + 18.6s^4 + 42.6s^3 + 44.4s^2 + 43.6s + 15.2 \in \bigcap_{i=1}^4 \Omega_{1a_i}^W$. Then let $c(s) := \varepsilon s^7 + b(s)$, where $\varepsilon > 0$ is sufficiently small, e.g., let $\varepsilon \leq 0.1$, it is easy to check that the design requirement has been met. □

The three examples above are taken from [9], where only Hurwitz stability was discussed. Robust SPR synthesis was not discussed there.

Example 11.5. Let $F = \{a_1(s) = s^9 + 11s^8 + 52s^7 + 145s^6 + 266s^5 + 331s^4 + 280s^3 + 155s^2 + 49s + 6,\ a_2(s) = s^9 + 11s^8 + 52s^7 + 146s^6 + 265.5s^5 + 332s^4 + 278.5s^3 + 151s^2 + 48s + 2\}$. It can be verified that $a_2(s) - a_1(s) = s^6 - 0.5s^5 + s^4 - 1.5s^3 - 4s^2 - s - 4$ satisfies the extended alternating Hurwitz minor condition [10, 37], namely, it is a convex direction for Hurwitz stability [10, 37]. Moreover, it is easy to see that $a_1(s)$ and $a_2(s)$ are both Hurwitz stable polynomials. Thus, the convex hull $\overline{F}$ of F is robustly stable [10, 37]. Using our method, it is easy to get $b(s) = s^8 + 8.8s^7 + 41.6s^6 + 87s^5 + 159.3s^4 + 132.4s^3 + 111.4s^2 + 30.2s + 9.6 \in \Omega_{1a_1}^W \cap \Omega_{1a_2}^W$. Thus, let $c(s) := \varepsilon s^9 + b(s)$, $\varepsilon > 0$, ε sufficiently small, e.g., take $\varepsilon \leq 0.07$, then the design requirement has been met. □

Note that our design method is also effective when dealing with discrete time systems. Note also that, in Examples 11.1–11.5, $b(s)$ is not unique. Using our method, we can get all such $b(s)$s with given precision.

It should be pointed out that there is hardly any example with order higher than 4 in the literature. In 1998, a sixth-order example of interval family was given in [32] as follows. Unfortunately, that example is incorrect.

Example 11.6. Suppose that $F = \{a_1(s) = s^6 + 0.8s^5 + 58.06s^4 + 50.9s^3 + 1028.5s^2 + 163.82s + 1042.5,\ a_2(s) = s^6 + 1.5s^5 + 58.06s^4 + 28.3s^3 + 1028.5s^2 + 376.36s + 1042.5,\ a_3(s) = s^6 + 0.8s^5 + 68.62s^4 + 50.9s^3 + 755.47s^2 + 163.82s + 3286.7,\ a_4(s) = s^6 + 1.5s^5 + 68.62s^4 + 28.3s^3 + 755.47s^2 + 376.36s + 3286.7\}$.
Find a polynomial $b(s)$, such that $p_i(s) := \dfrac{b(s)}{a_i(s)} \in \mathrm{SPR},\ i = 1, 2, 3, 4$.

By Definitions 11.1 and 11.2 and Proposition 11.3, a prerequisite of the robust SPR design problem is that the convex hull $\overline{F}$ of F is robustly stable. But it is easy to check that $\overline{F}$ is not robustly stable. In fact, $(1.0446 \pm 5.8969i)$ are roots of $a_1(s)$ with positive real part; $(1.037 \pm 4.9835i)$ are roots of $a_2(s)$ with positive real part; $(0.03291 \pm 7.5026i)$ and $(0.68089 \pm 2.4933i)$ are roots of $a_3(s)$ with positive real part; $(0.87123 \pm 2.867i)$ are roots of $a_4(s)$ with positive real part. Thus, it does not make sense to consider robust SPR design in this case. □

11.6 Applications to Robust SPR Synthesis for Low-order Systems

In this section, it is shown that the derived conditions presented in the previous sections are necessary and sufficient for robust SPR design of low-order

($n \leq 4$) polynomial segment or interval polynomials. For low-order ($n \leq 4$) robustly stable polynomial segments and interval polynomial families, it is verified that there always exists a fixed polynomial such that their ratio is SPR-invariant. A rigorous proof is given for Anderson's claim [5, 12] on SPR synthesis of the fourth-order stable interval polynomials. Moreover, the close relationship between SPR synthesis for low-order polynomial segment and SPR synthesis for low-order interval polynomials is also discussed.

Denote that $F = \{a_i(s) = s^n + \sum_{l=1}^{n} a_l^{(i)} s^{n-l}, i = 1, 2\}$ as the two endpoint polynomials of a polynomial segment $\overline{F}$ (convex combination), it is easy to prove the following lemma.

Lemma 11.2 ([5]). If $F = \{a_i(s) = s^n + \sum_{l=1}^{n} a_l^{(i)} s^{n-l}, i = 1, 2\}$ is the set of the two endpoint polynomials of a segment of polynomials $\overline{F}$ (convex combination), then there exists a fixed polynomial $b(s)$ such that $\forall a(s) \in \overline{F}$, $\dfrac{b(s)}{a(s)}$ is strictly positive real, if and only if $\dfrac{b(s)}{a_i(s)}$, $i = 1, 2$, are strictly positive real. □

Consider an interval polynomial family

$$K = \{a(s) = s^n + \sum_{i=1}^{n} a_i s^{n-i}, \quad a_i \in [a_i^-, a_i^+], \quad i = 1, 2, \ldots, n\}.$$

Denote

$$a_1(s) = s^n + \cdots + a_{n-3}^+ s^3 + a_{n-2}^+ s^2 + a_{n-1}^- s + a_n^-,$$
$$a_2(s) = s^n + \cdots + a_{n-3}^- s^3 + a_{n-2}^- s^2 + a_{n-1}^+ s + a_n^+,$$
$$a_3(s) = s^n + \cdots + a_{n-3}^+ s^3 + a_{n-2}^- s^2 + a_{n-1}^- s + a_n^+,$$
$$a_4(s) = s^n + \cdots + a_{n-3}^- s^3 + a_{n-2}^+ s^2 + a_{n-1}^+ s + a_n^-,$$

and $F = \{a_i(s) = s^n + \sum_{l=1}^{n} a_l^{(i)} s^{n-l}, i = 1, 2, 3, 4\}$ as the set of the four Kharitonov vertex polynomials of the interval polynomial family K [30].

Lemma 11.3 ([30]). If $F = \{a_i(s) = s^n + \sum_{l=1}^{n} a_l^{(i)} s^{n-l}, i = 1, 2, 3, 4\}$ is the set of the four Kharitonov vertex polynomials of the interval polynomial family K, then K is robustly stable if and only if $a_i(s) \in H^n, i = 1, 2, 3, 4$. □

The following result was proved by Dasgupta and Bhagwat [18].

Lemma 11.4 ([18]). If $F = \{a_i(s) = s^n + \sum_{l=1}^{n} a_l^{(i)} s^{n-l}, i = 1, 2, 3, 4\}$ is the set of the four Kharitonov vertex polynomials of the interval polynomial family K, then there exists a fixed polynomial $b(s)$ such that $\forall a(s) \in K$, $\dfrac{b(s)}{a(s)}$ is strictly positive real, if and only if $\dfrac{b(s)}{a_i(s)}$, $i = 1, 2, 3, 4$, are strictly positive real. □

11.6.1 The Third-order SPR Synthesis

First, for a low-order ($n \leq 3$) stable convex combination of polynomials, we have the following lemmas.

Lemma 11.5. Let $a(s) = s^3 + a_1 s^2 + a_2 s + a_3 \in H^3$. Then the following quadratic curve

$$a_2^2 x^2 + 2(2a_3 - a_1 a_2)xy + a_1^2 y^2 - 2a_2 a_3 x - 2a_1 a_3 y + a_3^2 = 0$$

is an ellipse lying in the first quadrant of the x-y plane. Moreover, this elliptic curve is tangent to lines $x = 0$, $y = 0$, $x = a_1$, and $y = a_2$ at $\left(0, \frac{a_3}{a_1}\right)$, $\left(\frac{a_3}{a_2}, 0\right)$, $\left(a_1, a_2 - \frac{a_3}{a_1}\right)$, and $\left(a_1 - \frac{a_3}{a_2}, a_2\right)$, respectively.

Proof. Since $a(s) = s^3 + a_1 s^2 + a_2 s + a_3 \in H^3$, it is straightforward to verify the above conclusions. □

Lemma 11.6. Assume that $a(s) = s^3 + a_1 s^2 + a_2 s + a_3 \in H^3$. Let $c(s) = s^2 + xs + y$, $(x, y) \in \mathbb{R}^2$. Then $\forall \omega \in \mathbb{R}$, $\mathrm{Re}\left[\frac{c(j\omega)}{a(j\omega)}\right] > 0$ if and only if

$$\begin{aligned}
&(x,y) \in \Omega \\
&\quad = \{(x, y)|\ a_2^2 x^2 + 2(2a_3 - a_1 a_2)xy + a_1^2 y^2 - 2a_2 a_3 x - 2a_1 a_3 y + a_3^2 < 0\} \\
&\qquad \cup \{(x, y)|x \leq a_1,\ a_2 x - a_1 y - a_3 \geq 0,\ y > 0\}.
\end{aligned}$$

Proof. Since $a(s) = s^3 + a_1 s^2 + a_2 s + a_3 \in H^3$ and $c(s) = s^2 + xs + y$, $(x, y) \in \mathbb{R}^2$, for all $\omega \in \mathbb{R}$,

$$\mathrm{Re}\left[\frac{c(j\omega)}{a(j\omega)}\right] > 0$$

$$\Leftrightarrow \mathrm{Re}\left[\frac{c(j\omega)}{a(j\omega)}\right] = \frac{1}{|a(j\omega)|^2}[(a_1 - x)\omega^4 + (a_2 x - a_1 y - a_3)\omega^2 + a_3 y] > 0$$

$\Leftrightarrow$ (i) $a_1 - x > 0$, $a_3 y > 0$, $[a_2 x - a_1 y - a_3]^2 - 4(a_1 - x)a_3 y < 0$, or (ii) $a_1 - x \geq 0$, $a_2 x - a_1 y - a_3 \geq 0$, $y > 0$

$\Leftrightarrow (x, y) \in \Omega$ (see Lemma 11.5). □

Lemma 11.6 is consistent with Corollary 11.1.

Denote

$$\Omega_e^a := \{(x, y)|\ a_2^2 x^2 + 2(2a_3 - a_1 a_2)xy + a_1^2 y^2 - 2a_2 a_3 x - 2a_1 a_3 y + a_3^2 < 0\},$$

$$\Omega_t^a := \left\{(x, y)\middle|\ x \leq a_1,\ a_2 x - a_1 y - a_3 \geq 0,\ 0 < y \leq a_2 - \frac{a_3}{a_1}\right\},$$

$$\Omega_e^b := \{(x, y)|\ b_2^2 x^2 + 2(2b_3 - b_1 b_2)xy + b_1^2 y^2 - 2b_2 b_3 x - 2b_1 b_3 y + b_3^2 < 0\},$$

$$\Omega_t^b := \left\{(x, y)\middle|\ x \leq b_1,\ b_2 x - b_1 y - b_3 \geq 0,\ 0 < y \leq b_2 - \frac{b_3}{b_1}\right\}.$$

Lemma 11.7. Let $a(s) = s^3 + a_1 s^2 + a_2 s + a_3 \in H^3$ and $b(s) = s^3 + b_1 s^2 + b_2 s + b_3 \in H^3$. Then $\Omega_t^a \cap \Omega_t^b \neq \phi \Rightarrow \Omega_e^a \cap \Omega_e^b \neq \phi$; $\Omega_t^a \cap \Omega_e^b \neq \phi \Rightarrow \Omega_e^a \cap \Omega_e^b \neq \phi$; $\Omega_e^a \cap \Omega_t^b \neq \phi \Rightarrow \Omega_e^a \cap \Omega_e^b \neq \phi$.

Proof. Without loss of generality, let $b_1 \geq a_1$.

If $\Omega_t^a \cap \Omega_t^b \neq \phi$, then $\frac{b_3}{b_2} \leq a_1$. When $\frac{b_3}{b_2} \geq \frac{a_3}{a_2}$, the open straight line segment $\overline{B_1 B_2}$ linking the points $B_1 \left(\frac{b_3}{b_2}, 0\right)$ and $B_2 \left(0, \frac{b_3}{b_1}\right)$ must intersect with Ω_e^a, since $\overline{B_1 B_2} \subset \Omega_e^b$ (see Lemma 11.5). We conclude that $\Omega_e^a \cap \Omega_e^b \neq \phi$. When $\frac{b_3}{b_2} \leq \frac{a_3}{a_2}$, the open straight line segment $\overline{B_1 B_2'}$ linking the points $B_1 \left(\frac{b_3}{b_2}, 0\right)$ and $B_2' \left(b_1, b_2 - \frac{b_3}{b_1}\right)$ must intersect with Ω_e^a, since $\overline{B_1 B_2'} \subset \Omega_e^b$ (see Lemma 11.5). We conclude that $\Omega_e^a \cap \Omega_e^b \neq \phi$.

If $\Omega_t^a \cap \Omega_e^b \neq \phi$, by the definitions of Ω_t^a and Ω_e^b and the propositions of Ω_e^a and Ω_e^b (see Lemma 11.5), the open straight line segment $\overline{A_1 A_2}$ linking the points $A_1 \left(\frac{a_3}{a_2}, 0\right)$ and $A_2 \left(a_1, a_2 - \frac{a_3}{a_1}\right)$ must intersect with Ω_e^b, since $\overline{A_1 A_2} \subset \Omega_e^a$. We conclude that $\Omega_e^a \cap \Omega_e^b \neq \phi$. Similarly, we can prove that $\Omega_e^a \cap \Omega_t^b \neq \phi \Rightarrow \Omega_e^a \cap \Omega_e^b \neq \phi$. □

Lemma 11.8. Assume that $a(s) = s^3 + a_1 s^2 + a_2 s + a_3 \in H^3$ and $b(s) = s^3 + b_1 s^2 + b_2 s + b_3 \in H^3$. Let $c(s) = s^2 + xs + y$. Then $\forall \omega \in \mathbb{R}$, $\mathrm{Re}\left[\frac{c(j\omega)}{a(j\omega)}\right] > 0$ and $\mathrm{Re}\left[\frac{c(j\omega)}{b(j\omega)}\right] > 0$ if and only if $\Omega_e^a \cap \Omega_e^b \neq \phi$.

Proof. Denote $\Omega^a := \Omega_e^a \cup \Omega_t^a$ and $\Omega^b := \Omega_e^b \cup \Omega_t^b$. Since $a(s) = s^3 + a_1 s^2 + a_2 s + a_3 \in H^3$, $b(s) = s^3 + b_1 s^2 + b_2 s + b_3 \in H^3$, and $c(s) = s^2 + xs + y$, for all $\omega \in \mathbb{R}$,

$$
\begin{aligned}
&\mathrm{Re}\left[\frac{c(j\omega)}{a(j\omega)}\right] > 0 \text{ and } \mathrm{Re}\left[\frac{c(j\omega)}{b(j\omega)}\right] > 0 \\
&\Leftrightarrow \Omega^a \cap \Omega^b \neq \phi \text{ (see Lemma 11.6)} \\
&\Leftrightarrow \Omega^a \cap \Omega^b = (\Omega_e^a \cup \Omega_t^a) \cap (\Omega_e^b \cup \Omega_t^b) \neq \phi \\
&\Leftrightarrow (\Omega_e^a \cup \Omega_t^a) \cap (\Omega_e^b \cup \Omega_t^b) = (\Omega_e^a \cap \Omega_e^b) \cup (\Omega_t^a \cap \Omega_e^b) \cup (\Omega_e^a \cap \Omega_t^b) \cup (\Omega_t^a \cup \Omega_t^b) \neq \phi \\
&\Leftrightarrow \Omega_e^a \cap \Omega_e^b \neq \phi \text{ (see Lemma 11.7).}
\end{aligned}
$$

□

Lemma 11.9. If $a(s) = s^3 + a_1 s^2 + a_2 s + a_3 \in H^3$, $b(s) = s^3 + b_1 s^2 + b_2 s + b_3 \in H^3$, $c(s) = s^2 + xs + y, \forall \omega \in \mathbb{R}$, $\mathrm{Re}\left[\frac{c(j\omega)}{a(j\omega)}\right] > 0$ and $\mathrm{Re}\left[\frac{c(j\omega)}{b(j\omega)}\right] > 0$, let

$$\tilde{c}(s) := c(s) + r \cdot a(s), \quad r > 0, \quad r \text{ small enough, or}$$

$$\tilde{c}(s) := c(s) + r \cdot b(s), \quad r > 0, \quad r \text{ small enough.}$$

Then both $\frac{\tilde{c}(s)}{a(s)}$ and $\frac{\tilde{c}(s)}{b(s)}$ are SPR.

Proof. See the proof of Theorem 11.4.

Lemma 11.10. Let $a(s) = s^3 + a_1 s^2 + a_2 s + a_3 \in H^3$ and $b(s) = s^3 + b_1 s^2 + b_2 s + b_3 \in H^3$. Then $\Omega_e^a \cap \Omega_e^b \neq \phi$ if and only if $\lambda b(s) + (1-\lambda)a(s) \in H^3, \lambda \in [0,1]$.

Proof. Necessity. If $\Omega_e^a \cap \Omega_e^b \neq \phi$, by Lemma 11.8, there exists a polynomial $c(s) = s^2 + xs + y$ such that $\forall \omega \in \mathbb{R}$, $\mathrm{Re}\left[\frac{c(j\omega)}{a(j\omega)}\right] > 0$ and $\mathrm{Re}\left[\frac{c(j\omega)}{b(j\omega)}\right] > 0$. By Lemma 11.9, we can obtain a polynomial $\tilde{c}(s)$ such that both $\frac{\tilde{c}(s)}{a(s)}$ and $\frac{\tilde{c}(s)}{b(s)}$ are SPR. It is easy to verify that $\lambda \in [0,1]$, $\frac{\tilde{c}(s)}{\lambda b(s) + (1-\lambda)a(s)}$ is also SPR. From the Proposition 11.3, we can conclude that $\lambda b(s) + (1-\lambda)a(s) \in H^3$, $\lambda \in [0,1]$.

Sufficiency. $\forall \lambda \in [0,1]$, $\lambda b(s) + (1-\lambda)a(s) \in H^3$, this implies $a_i > 0$, $b_i > 0$, $i = 1,2,3$, and for all $\lambda \in [0,1]$

$$[(b_1 - a_1)(b_2 - a_2)]\lambda^2 + [a_1(b_2 - a_2) + a_2(b_1 - a_1) - (b_3 - a_3)]\lambda + [a_1 a_2 - a_3] > 0. \quad (11.2)$$

Assume that $\Omega_e^a \cap \Omega_e^b = \phi$. Without loss of generality, let $b_1 > \frac{b_3}{b_2} > a_1 > \frac{a_3}{a_2}$, $b_2 > \frac{b_3}{b_1} > a_2 > \frac{a_3}{a_1}$. Therefore, $\exists u \in [a_1, b_1]$, $v \in [a_2, b_2]$ such that the following straight line l

$$l\colon\ \frac{x}{u} + \frac{y}{v} = 1$$

is tangent to Ω_e^a and Ω_e^b simultaneously.

Since l is tangent to Ω_e^a, consider the following equations

$$\begin{cases} \frac{x}{u} + \frac{y}{v} = 1, \\ a_2^2 x^2 + 2(2a_3 - a_1 a_2)xy + a_1^2 y^2 - 2a_2 a_3 x - 2a_1 a_3 y + a_3^2 = 0. \end{cases} \quad (11.3)$$

Notice that the Hurwitzness of the polynomial $a(s)$, via a lengthy but straightforward computation, can be verified so that the following equation holds:

$$uv - a_2 u + a_3 - a_1 v = 0. \quad (11.4)$$

Since l is tangent to Ω_e^b, for the same reasons, we have

$$uv - b_2 u + b_3 - b_1 v = 0. \quad (11.5)$$

Combining (11.4) and (11.5), we take $u_1 = u - a_1, v_1 = v - a_2$, and obtain

$$\begin{cases} u_1v_1 = a_1a_2 - a_3, \\ u_1v_1 + [a_2 - b_2]u_1 + [a_1 - b_1]v_1 + [a_1a_2 - a_1b_2 - a_2b_1 + b_3] = 0, \end{cases} \tag{11.6}$$

which yields

$$(b_2-a_2)\frac{u_1^2}{(b_1-a_1)}+[a_1(b_2-a_2)+a_2(b_1-a_1)-(b_3-a_3)]\frac{u_1}{(b_1-a_1)}+[a_1a_2-a_3]=0.$$

Let $\tilde{\lambda} = \dfrac{u_1}{(b_1 - a_1)}$. Obviously, $\tilde{\lambda} \in [0, 1]$. It follows that

$$[(b_1-a_1)(b_2-a_2)]\tilde{\lambda}^2+[a_1(b_2-a_2)+a_2(b_1-a_1)-(b_3-a_3)]\tilde{\lambda}+[a_1a_2-a_3]=0. \tag{11.7}$$

This equation contradicts (11.2), which completes the proof.

Then we can obtain the following theorem.

Theorem 11.6. Let $a(s) = s^3 + a_1s^2 + a_2s + a_3 \in H^3$ and $b(s) = s^3 + b_1s^2 + b_2s + b_3 \in H^3$. A necessary and sufficient condition for the existence of $c(s)$ such that both $c(s)/a(s)$ and $c(s)/b(s)$ are SPR is

$$\lambda b(s) + (1 - \lambda)a(s) \in H^3, \lambda \in [0, 1].$$

Proof. Combining Lemmas 11.5–11.10, one can easily complete the proof of the theorem. □

The cases of $n < 3$ are trivial. The conditions of Lemma 11.8 can easily be checked by plotting of Ω_e^a and Ω_e^b in the x-y plane with a computer.

Given an interval stable polynomial of nth degree, $n \leq 3$, using the above method, one can easily find a fixed polynomial which SPR stabilizes the whole interval polynomials. In fact, we have the following theorem.

Theorem 11.7. If

$$F = \{a_i(s) = s^n + \sum_{l=1}^{n} a_l^{(i)} s^{n-l}, \quad i = 1, 2, 3, 4\}$$

is the set of the four Kharitonov vertex polynomials of a low-order ($n \leq 3$) stable interval polynomial family, then we have $\bigcap_{i=1}^4 \Omega_{1a_i}^W \neq \phi$.

Proof. The statement is obviously true for the cases when $n = 1$ or $n = 2$. We will prove it for the case when $n = 3$.

Suppose that F is the set of the four Kharitonov vertex polynomials of the 3rd-order interval polynomial family $s^3 + \alpha s^2 + \beta s + \gamma$, $\alpha \in [\underline{a}_1, \overline{a}_1]$, $\beta \in [\underline{a}_2, \overline{a}_2]$, $\gamma \in [\underline{a}_3, \overline{a}_3]$, and $F \subset H^3$. Obviously, we have $\underline{a}_1\underline{a}_2 - \overline{a}_3 > 0$. Now take x, y in such a way that $y > 0$ is sufficiently small, $0 < x < \underline{a}_1$, and x is sufficiently close to $\underline{a}_1$, such that $\underline{a}_2x - \overline{a}_3 - \overline{a}_1y > 0$. It is easy to verify that $(x, y) \in \bigcap_{i=1}^4 \Omega_{1a_i}^W$. This completes the proof. □

11.6.2 The Fourth-order SPR Synthesis

For a fourth-order stable interval polynomial family, or stable convex combination of two polynomials, does there exist a polynomial such that their ratios are SPR-invariant? This section gives this problem a positive answer.

Consider the fourth-order interval polynomials given by

$$K = \{a(s) = s^4 + a_1 s^3 + a_2 s^2 + a_3 s + a_4,\ a_i \in [a_i^-, a_i^+],\ i = 1,2,3,4\}.$$

Denote

$$\begin{aligned} a_1(s) &= s^4 + a_1^+ s^3 + a_2^+ s^2 + a_3^- s + a_4^-, \\ a_2(s) &= s^4 + a_1^- s^3 + a_2^- s^2 + a_3^+ s + a_4^+, \\ a_3(s) &= s^4 + a_1^+ s^3 + a_2^- s^2 + a_3^- s + a_4^+, \\ a_4(s) &= s^4 + a_1^- s^3 + a_2^+ s^2 + a_3^+ s + a_4^-, \end{aligned}$$

as the four Kharitonov vertex polynomials of K [30].

Lemma 11.11. Suppose that $a(s) = s^4 + a_1 s^3 + a_2 s^2 + a_3 s + a_4 \in H^4$. Then the following quadratic curve is an ellipse in the first quadrant of the x-y plane:

$$\begin{aligned} &(a_2^2 - 4a_4)x^2 + 2(2a_3 - a_1 a_2)xy + a_1^2 y^2 \\ &\quad -2(a_2 a_3 - 2a_1 a_4)x - 2a_1 a_3 y + a_3^2 = 0 \end{aligned}$$

and this ellipse is tangent to the y axis at $(0, \frac{a_3}{a_1})$, tangent to the lines $x = a_1$ and $a_3 y - a_4 x = 0$ at $\left(a_1, a_2 - \frac{a_3}{a_1}\right)$ and $\left(\frac{a_3^2}{a_2 a_3 - a_1 a_4}, \frac{a_3 a_4}{a_2 a_3 - a_1 a_4}\right)$, respectively.

Proof. Since $a(s)$ is Hurwitz stable, Lemma 11.11 is proved by a direct calculation. □

Let $a(s) = s^4 + a_1 s^3 + a_2 s^2 + a_3 s + a_4 \in H^4$. For notational simplicity, denote

$$\begin{aligned} \Omega_e^a &:= \{(x,y) |\ (a_2^2 - 4a_4)x^2 + 2(2a_3 - a_1 a_2)xy + a_1^2 y^2 \\ &\qquad -2(a_2 a_3 - 2a_1 a_4)x - 2a_1 a_3 y + a_3^2 < 0\} \\ \Omega_t^a &:= \{(x,y) |\ a_1 - x \geq 0, a_2 x - a_1 y - a_3 \geq 0, \\ &\qquad a_3 y - a_4 x \geq 0\} \\ \Omega^a &:= \Omega_e^a \cup \Omega_t^a. \end{aligned}$$

Apparently, Ω^a is a bounded convex set in the x-y plane.

Lemma 11.12. Suppose that $a(s) = s^4 + a_1 s^3 + a_2 s^2 + a_3 s + a_4 \in H^4$ and $(x,y) \in \Omega^a$. Let $c(s) := s^3 + xs^2 + ys + \varepsilon$, where ε is positive and sufficiently small, then $\forall \omega \in \mathbb{R}$, $\mathrm{Re}\left[\frac{c(j\omega)}{a(j\omega)}\right] > 0$, namely, $(x, y, \varepsilon) \in \Omega_{1a}^W$.

Proof. Suppose that $(x,y) \in \Omega^a$. Let $c(s) := s^3 + xs^2 + ys + \varepsilon$, where $\varepsilon > 0$ and ε is sufficiently small.

$\forall \omega \in \mathbb{R}$, consider

$$\begin{aligned}\mathrm{Re}\left[\frac{c(j\omega)}{a(j\omega)}\right] &= \frac{1}{|a(j\omega)|^2}[(a_1-x)\omega^6 \\ &\quad + (a_2x - a_1y - a_3)\omega^4 + (a_3y - a_4x)\omega^2 \\ &\quad + \varepsilon(\omega^4 - a_2\omega^2 + a_4)].\end{aligned}$$

In order to prove that $\forall \omega \in \mathbb{R}$, $\mathrm{Re}\left[\frac{c(j\omega)}{a(j\omega)}\right] > 0$, let $t = \omega^2$. We only need to prove that, for any $\varepsilon > 0$, ε sufficiently small,

$$\begin{aligned}f(t) := t[(a_1-x)t^2 + (a_2x - a_1y - a_3)t + (a_3y - a_4x)] \\ +\varepsilon(t^2 - a_2t + a_4) > 0, \ \forall t \in [0, +\infty).\end{aligned}$$

Since $(x,y) \in \Omega^a$, by definition of Ω^a and Lemma 11.11, (x,y) satisfies $a_1 - x > 0$, $a_3y - a_4x > 0$, and

$$[a_2x - a_1y - a_3]^2 - 4(a_1-x)(a_3y - a_4x) < 0,$$

or

$$a_1 - x \geq 0, \quad a_2x - a_1y - a_3 \geq 0, a_3y - a_4x \geq 0.$$

Since $a(s) \in H^4$, $a_1 - x$, $a_2x - a_1y - a_3$ and $a_3y - a_4x$ cannot be 0 simultaneously. Thus, $\forall t \in (0, +\infty)$

$$(a_1-x)t^2 + (a_2x - a_1y - a_3)t + (a_3y - a_4x) > 0.$$

On the other hand, we have $f(0) > 0$, and for any $\varepsilon > 0$, if t is a sufficiently large or a sufficiently small positive number, we have $f(t) > 0$. Namely, there exist $0 < t_1 < t_2$ such that, for all $\varepsilon > 0$, $t \in [0, t_1] \cup [t_2, +\infty)$, $f(t) > 0$.

Denote

$$M = \inf_{t \in [t_1, t_2]} t[(a_1-x)t^2 + (a_2x - a_1y - a_3)t + (a_3y - a_4x)],$$

$$N = \sup_{t \in [t_1, t_2]} |t^2 - a_2t + a_4|.$$

Then $M > 0$ and $N > 0$. Choosing $0 < \varepsilon < \frac{M}{N}$, by a direct calculation, we have

$$\begin{aligned}f(t) &= t[(a_1-x)t^2 + (a_2x - a_1y - a_3)t + (a_3y - a_4x)] \\ &\quad + \varepsilon(t^2 - a_2t + a_4) > 0, \quad \forall t \in [0, +\infty).\end{aligned}$$

Namely

$$\forall \omega \in \mathbb{R}, \quad \mathrm{Re}\left[\frac{b(j\omega)}{a(j\omega)}\right] > 0.$$

This completes the proof. □

Lemma 11.13. Suppose that $a(s) = s^4 + a_1 s^3 + a_2 s^2 + a_3 s + a_4 \in H^4$, $b(s) = s^4 + b_1 s^3 + b_2 s^2 + b_3 s + a_4 \in H^4$, if $\lambda b(s) + (1-\lambda)a(s) \in H^4$, $\lambda \in [0,1]$, then $\Omega_e^a \bigcap \Omega_e^b \neq \phi$.

Proof. If $\forall \lambda \in [0,1]$, $\lambda b(s) + (1-\lambda)a(s) \in H^4$, by Lemma 11.11, for any $\lambda \in [0,1]$,

$$\begin{aligned}\Omega_e^{a_\lambda} := \{(x,y) | (a_{\lambda 2}^2 - 4a_{\lambda 4})x^2 + 2(2a_{\lambda 3} - a_{\lambda 1}a_{\lambda 2})xy + a_{\lambda 1}^2 y^2 \\ - 2(a_{\lambda 2}a_{\lambda 3} - 2a_{\lambda 1}a_{\lambda 4})x - 2a_{\lambda 1}a_{\lambda 3}y + a_{\lambda 3}^2 < 0\}\end{aligned}$$

is also an elliptic region in the first quadrant of the x-y plane, where $a_{\lambda i} := a_i + \lambda(b_i - a_i)$, $i = 1,2,3,4$. Apparently, when λ changes continuously from 0 to 1, $\Omega_e^{a_\lambda}$ will change continuously from Ω_e^a to Ω_e^b.

Now assume $\Omega_e^a \cap \Omega_e^b = \phi$, by Lemma 11.11 (without loss of generality, suppose that $b_3/b_1 > a_3/a_1$), $\exists v \in \left[\frac{a_3}{a_1}, \frac{b_3}{b_1}\right]$ and $u \neq 0$, such that the following line l

$$l:\ \frac{x}{u} + \frac{y}{v} = 1$$

is tangent to Ω_e^a and Ω_e^b simultaneously.

Since l is tangent to Ω_e^a, consider

$$\begin{cases} \dfrac{x}{u} + \dfrac{y}{v} = 1, \\ (a_2^2 - 4a_4)x^2 + 2(2a_3 - a_1 a_2)xy + a_1^2 y^2 \\ \qquad -2(a_2 a_3 - 2a_1 a_4)x - 2a_1 a_3 y + a_3^2 = 0. \end{cases} \tag{11.8}$$

Since $a(s)$ is Hurwitz stable and $u \neq 0$, by a direct calculation, we know that the necessary and sufficient condition for l being tangent to Ω_e^a is

$$uv^2 - a_1 v^2 - a_2 uv + a_3 v + a_4 u = 0. \tag{11.9}$$

Since l is tangent to Ω_e^b, for the same reason, we have:

$$uv^2 - b_1 v^2 - b_2 uv + b_3 v + b_4 u = 0. \tag{11.10}$$

From (11.9) and (11.10), we obviously have $\forall \lambda \in [0,1]$,

$$uv^2 - a_{\lambda 1}v^2 - a_{\lambda 2}uv + a_{\lambda 3}v + a_{\lambda 4}u = 0. \tag{11.11}$$

Equation (11.11) shows that l is also tangent to $\Omega_e^{a_\lambda}$ ($\forall \lambda \in [0,1]$), but l separates Ω_e^a and Ω_e^b, and when λ changes continuously from 0 to 1, $\Omega_e^{a_\lambda}$ will change continuously from Ω_e^a to Ω_e^b, which is obviously impossible. This completes the proof. □

Lemma 11.14. If $F = \{a_i(s), i = 1,2,3,4\}$ is the set of the four Kharitonov vertex polynomials of a fourth-order stable interval polynomial family, then $\Omega^{a_2} \subset \Omega^{a_4}$ and $\Omega^{a_3} \subset \Omega^{a_1}$.

Proof. By the definition of Ω^a, it is easy to see that

$$\Omega^{a_1} = \{(x,y)|(a_1^+ - x)t^2 + (a_2^+ x - a_1^+ y - a_3^-)t + (a_3^- y - a_4^- x) > 0, \forall t \in [0,\infty)\},$$

$$\Omega^{a_2} = \{(x,y)|(a_1^- - x)t^2 + (a_2^- x - a_1^- y - a_3^+)t + (a_3^+ y - a_4^+ x) > 0, \forall t \in [0,\infty)\},$$

$$\Omega^{a_3} = \{(x,y)|(a_1^+ - x)t^2 + (a_2^- x - a_1^+ y - a_3^-)t + (a_3^- y - a_4^+ x) > 0, \forall t \in [0,\infty)\},$$

$$\Omega^{a_4} = \{(x,y)|(a_1^- - x)t^2 + (a_2^+ x - a_1^- y - a_3^+)t + (a_3^+ y - a_4^- x) > 0, \forall t \in [0,\infty)\}.$$

Obviously, we have $\Omega^{a_2} \subset \Omega^{a_4}$ and $\Omega^{a_3} \subset \Omega^{a_1}$. This completes the proof. □

Lemma 11.15. If $F = \{a_i(s), i = 1,2,3,4\}$ is the set of the four Kharitonov vertex polynomials of a fourth-order stable interval polynomial family, then $\bigcap_{i=1}^4 \Omega^{a_i} \neq \phi$. □

Lemma 11.15 plays an important role in proving Anderson's claim on robust SPR synthesis for the fourth-order stable interval polynomial family. For a complete understanding of it, we give three different proofs in the sequel.

Proof. Method 1. By Lemma 11.14, we only need to prove that $\Omega^{a_2} \cap \Omega^{a_3} \neq \phi$, By Lemma 11.13, we know that $\Omega_e^{a_2} \cap \Omega_e^{a_3} \neq \phi$, but $\Omega^{a_2} = \Omega_e^{a_2} \cup \Omega_t^{a_2}$ and $\Omega^{a_3} = \Omega_e^{a_3} \cup \Omega_t^{a_3}$. Thus, $\Omega^{a_2} \cap \Omega^{a_3} \neq \phi$. This completes the proof. □

Proof. Method 2. Since F is the set of the four Kharitonov vertex polynomials of a fourth-order stable interval polynomial family, by Lemma 11.11, in the x-y plane, $\Omega_e^{a_2}$ and $\Omega_e^{a_4}$ are both tangent to $x = 0$ at $\left(0, \frac{a_3^+}{a_1^-}\right)$ (denote this tangent point as A_{24}); $\Omega_e^{a_1}$ and $\Omega_e^{a_3}$ are both tangent to $x = 0$ at $\left(0, \frac{a_3^-}{a_1^+}\right)$ (denote this tangent point as A_{13}). Denote the tangent point of $\Omega_e^{a_2}$ ($\Omega_e^{a_4}$) and $x = a_1^-$ as $A_2(a_1^-, a_2^- - a_3^+/a_1^-)$ ($A_4(a_1^-, a_2^+ - a_3^+/a_1^-)$); and denote the tangent point of $\Omega_e^{a_1}$ ($\Omega_e^{a_3}$) and $x = a_1^+$ as $A_1(a_1^+, a_2^+ - a_3^-/a_1^+)$ ($A_3(a_1^+, a_2^- - a_3^-/a_1^+)$). Furthermore, denote the intersection points of $x = a_1^-$ and the straight lines $a_3^+ y - a_4^+ x = 0$, $a_3^+ y - a_4^- x = 0$ as $B_2\left(a_1^-, \frac{a_1^- a_4^+}{a_3^+}\right)$, $B_4\left(a_1^-, \frac{a_1^- a_4^-}{a_3^+}\right)$, respectively; and denote the intersection points of $x = a_1^+$ and the straight lines $a_3^- y - a_4^- x = 0$, $a_3^- y - a_4^+ x = 0$ as $B_1\left(a_1^+, \frac{a_1^+ a_4^-}{a_3^-}\right)$, $B_3\left(a_1^+, \frac{a_1^+ a_4^+}{a_3^-}\right)$, respectively.

In what follows, $[A, B]$ stands for the set of points in the line segment connecting the point A and the point B, including the endpoints A and B,

$(A, B]$ stands for the set of points in the line segment connecting the point A and the point B, including the endpoint B, but not A. Then it is easy to see that $[A_2, B_2] \subset \Omega^{a_2}$, $[A_2, B_2] \subset [A_4, B_4] \subset \Omega^{a_4}$, $[A_3, B_3] \subset \Omega^{a_3}$, $[A_3, B_3] \subset [A_1, B_1] \subset \Omega^{a_1}$, and $(A_{24}, A_2] \subset \Omega^{a_2}$, $(A_{24}, A_2] \subset \Omega^{a_4}$, $(A_{13}, A_3] \subset \Omega^{a_3}$, $(A_{13}, A_3] \subset \Omega^{a_1}$.

Denote $A_3^\star$ as $\left(a_1^-, \left(\frac{a_2^-}{a_1^+} - 2\frac{a_3^-}{a_1^{+2}}\right) a_1^- + \frac{a_3^-}{a_1^+}\right)$. Then, $A_3^\star \in (A_{13}, A_3]$.

If $\frac{a_3^+}{a_1^-} = \frac{a_3^-}{a_1^+}$, i.e., $a_1^- = a_1^+$ and $a_3^- = a_3^+$. Then, take $\delta > 0$, δ sufficiently small. By Lemma 11.11, it is easy to verify that $\left(\delta, \frac{a_3^+}{a_1^-}\right) \in \bigcap_{i=1}^4 \Omega_e^{a_i}$. Thus, $\bigcap_{i=1}^4 \Omega^{a_i} \neq \phi$.

Now, suppose that $\frac{a_3^+}{a_1^-} > \frac{a_3^-}{a_1^+}$ and

$$a_2^- - \frac{a_3^+}{a_1^-} \geq \left(\frac{a_2^-}{a_1^+} - 2\frac{a_3^-}{a_1^{+2}}\right) a_1^- + \frac{a_3^-}{a_1^+}.$$

It is easy to verify that

$$\left(\frac{a_2^-}{a_1^+} - 2\frac{a_3^-}{a_1^{+2}}\right) a_1^- + \frac{a_3^-}{a_1^+} > \frac{a_1^- a_4^+}{a_3^+}.$$

Thus, we have $A_3^\star \in [A_2, B_2]$. Hence $A_3^\star \in [A_2, B_2] \bigcap (A_{13}, A_3]$. Therefore $A_3^\star \in \bigcap_{i=1}^4 \Omega^{a_i}$. Thus $\bigcap_{i=1}^4 \Omega^{a_i} \neq \phi$.

Finally, with $\frac{a_3^+}{a_1^-} > \frac{a_3^-}{a_1^+}$, if

$$a_2^- - \frac{a_3^+}{a_1^-} < \left(\frac{a_2^-}{a_1^+} - 2\frac{a_3^-}{a_1^{+2}}\right) a_1^- + \frac{a_3^-}{a_1^+},$$

then it is easy to see that $(A_{13}, A_3] \cap (A_{24}, A_2] \neq \phi$ and $(A_{13}, A_3] \cap (A_{24}, A_2] \subset \bigcap_{i=1}^4 \Omega^{a_i}$. Thus, we also have $\bigcap_{i=1}^4 \Omega^{a_i} \neq \phi$. This completes the proof. □

Proof. Method 3. $A_{13}\left(0, \frac{a_3^-}{a_1^+}\right)$, $A_{24}\left(0, \frac{a_3^+}{a_1^-}\right)$, $B_2\left(a_1^-, \frac{a_1^- a_4^+}{a_3^+}\right)$, and $B_3\left(a_1^+, \frac{a_1^+ a_4^+}{a_3^-}\right)$ are defined identically as in the Method 2 above. (A, B) stands for the set of points in the line segment connecting the point A and the point B, but not including the endpoints A and B.

If $\frac{a_3^+}{a_1^-} = \frac{a_3^-}{a_1^+}$, i.e., $a_1^- = a_1^+$ and $a_3^- = a_3^+$. Then, take $\delta > 0$, δ sufficiently small. By Lemma 11.11, it is easy to verify that $\left(\delta, \frac{a_3^+}{a_1^-}\right) \in \bigcap_{i=1}^4 \Omega_e^{a_i}$. Thus, $\bigcap_{i=1}^4 \Omega^{a_i} \neq \phi$.

Now, suppose that $\frac{a_3^+}{a_1^-} > \frac{a_3^-}{a_1^+}$. Then, it is easy to see that $(A_{13}, B_3) \cap (A_{24}, B_2) \neq \phi$ and $(A_{13}, B_3) \cap (A_{24}, B_2) \subset \bigcap_{i=1}^4 \Omega^{a_i}$. Thus, we also have $\bigcap_{i=1}^4 \Omega^{a_i} \neq \phi$. This completes the proof. □

Lemma 11.16. If $a(s) = s^4 + a_1 s^3 + a_2 s^2 + a_3 s + a_4 \in H^4$, $b(s) = s^3 + xs^2 + ys + z$, and $\forall \omega \in \mathbb{R}$, $\mathrm{Re}\left[\frac{b(j\omega)}{a(j\omega)}\right] > 0$, take

$$\tilde{b}(s) := b(s) + rc(s), \quad r > 0, \quad r \text{ small enough},$$

where $c(s)$ is an arbitrarily fixed fourth-order monic polynomial. Then $\frac{\tilde{b}(s)}{a(s)}$ is SPR.

Proof. See the proof of Theorem 11.4. □

Now we are in a position to establish the following theorem.

Theorem 11.8. If $F = \{a_i(s), i = 1, 2\}$ is the set of the two endpoint polynomials of a fourth-order stable segment of polynomials (convex combination), then there always exists a fixed polynomial $b(s)$ such that $\frac{b(s)}{a_1(s)}$ and $\frac{b(s)}{a_2(s)}$ are strictly positive real.

Proof. The theorem is proved by simply combining Lemmas 11.11–11.13 and 11.16. □

Theorem 11.9. If $F = \{a_i(s), i = 1, 2, 3, 4\}$ is the set of the four Kharitonov vertex polynomials of a fourth-order stable interval polynomial family K, then there always exists a fixed polynomial $b(s)$ such that $\forall a(s) \in K$, $\frac{b(s)}{a(s)}$ is strictly positive real.

Proof. The theorem is proved by simply combining Lemmas 11.11–11.12 and Lemmas 11.15–11.16. □

Note that in Theorem 11.8, $\frac{b(s)}{a_1(s)}$ and $\frac{b(s)}{a_2(s)}$ being strictly positive implies $\forall \lambda \in [0, 1]$, $\frac{b(s)}{\lambda a_1(s) + (1 - \lambda) a_2(s)}$ being strictly positive real (by Lemma 11.2); similarly, in Theorem 11.9, $\forall a(s) \in F$, $\frac{b(s)}{a(s)}$ being strictly positive real implies $\forall a(s) \in K$, $\frac{b(s)}{a(s)}$ being strictly positive real (by Lemma 11.4).

Robust stability of a polynomial segment can be checked by many efficient methods, e.g., eigenvalue method, root locus method, value set method [1, 9,

13]. Robust stability of K in Theorem 11.9 can be ascertained by checking only two Kharitonov vertex polynomials [6].

From the proofs of Lemma 11.15, we can establish the relationship between SPR synthesis for the fourth-order polynomial segments and SPR synthesis for the fourth-order interval polynomials. In fact, it is easy to see that Theorem 11.8 implies Theorem 11.9. Similarly, Theorem 11.6 implies Theorem 11.7. However, we do not know whether similar results is true for higher-order ($n \geq 5$) systems. This subject is currently under investigation.

By Theorems 11.8 and 11.9, for the low-order stable interval polynomial family or low-order stable convex combination, existence of the solution to the synthesis problem is always guaranteed. As shown by numerous examples, our method is very effective.

Let us consider Example 11.1 again.

Example 11.7. Suppose that $F = \{a_1(s) = s^4 + 11s^3 + 56s^2 + 88s + 1, a_2(s) = s^4+11s^3+56s^2+88s+50, a_3(s) = s^4+89s^3+56s^2+88s+1, a_4(s) = s^4+89s^3+56s^2+88s+50\}$. It is easy to see that the convex hull $\overline{F}$ of F is robustly stable. By our method as in the constructive proof of Theorem 11.9, it is easy to get $(11, 7.6657) \in \bigcap_{i=1}^{4} \Omega^{a_i}$. Thus, choosing $b(s) = s^3 + 11s^2 + 7.76657s + \varepsilon$, where ε is a sufficiently small positive number (ε is determined by Lemma 11.12, in this example, $0 < \varepsilon \leq 3$), and taking $\varepsilon = 2$, by Lemma 11.12, $\forall \omega \in \mathbb{R}$, $\mathrm{Re}\left[\frac{b(j\omega)}{a_i(j\omega)}\right] > 0$, $i = 1,2,3,4$. Finally, let $\tilde{b}(s) := b(s) + rs^4$, where $r > 0$, r sufficiently small (r is determined by Lemma 11.16, in this example, $0 < r \leq 0.5$). It is easy to check that $\frac{\tilde{b}(s)}{a_i(s)}$, $i = 1,2,3,4$, are strictly positive real (note that $b(s)$ and $\tilde{b}(s)$ are not unique). □

In what follows, we will provide some more examples for fourth-order interval polynomial families.

Example 11.8. Suppose that $a_1(s) = s^4+5s^3+6s^2+4s+0.5$, $a_2(s) = s^4+2s^3+6s^2+6s+1$, $a_3(s) = s^4+5s^3+6s^2+4s+1$, $a_4(s) = s^4+2s^3+6s^2+6s+0.5$ are the four Kharitonov vertex polynomials of a fourth-order interval polynomial set K. It is easy to check using Kharitonov's Theorem (Lemma 11.3) that K is robustly stable. By our method as in the constructive proof of Theorem 11.9, it is easy to get $(2, 2.56) \in \bigcap_{i=1}^{4} \Omega^{a_i}$. Thus, choosing $b(s) = s^3+2s^2+2.56s+\varepsilon$, where ε is a sufficiently small positive number (in this example, $0 < \varepsilon \leq 1$), and taking $\varepsilon = 0.5$, by Lemma 11.12, $\forall \omega \in \mathbb{R}$, $\mathrm{Re}\left[\frac{b(j\omega)}{a_i(j\omega)}\right] > 0$, $i = 1,2,3,4$. Finally, let $\tilde{b}(s) := b(s)+rs^4$, where $r > 0$, r sufficiently small (in this example, $0 < r \leq 0.5$). It is easy to check that $\frac{\tilde{b}(s)}{a_i(s)}$, $i = 1,2,3,4$, are strictly positive real. □

Example 11.9. Suppose that $a_1(s) = s^4 + 2.5s^3 + 6s^2 + 4s + 0.5$, $a_2(s) = s^4 + 2s^3 + 5s^2 + 6s + 5$, $a_3(s) = s^4 + 2.5s^3 + 5s^2 + 4s + 5$, $a_4(s) = s^4 + 2s^3 + 6s^2 + 6s + 0.5$ are the four Kharitonov vertex polynomials of a fourth-order interval polynomial set K, it is easy to check using Kharitonov's Theorem (Lemma 11.3) that K is robustly stable. By our method as in the constructive proof of Theorem 11.9, it is easy to get $(1.1475, 2.4262) \in \bigcap_{i=1}^{4} \Omega^{a_i}$. Thus, choosing $b(s) = s^3 + 1.1475s^2 + 2.4262s + \varepsilon$, where ε is a sufficiently small positive number (in this example, $0 < \varepsilon \leq 1$), and taking $\varepsilon = 0.5$, by Lemma 11.12, $\forall \omega \in \mathbb{R}$, $\mathrm{Re}\left[\dfrac{b(j\omega)}{a_i(j\omega)}\right] > 0$, $i = 1, 2, 3, 4$. Finally, let $\tilde{b}(s) := b(s) + rs^4$, where $r > 0$, r sufficiently small (in this example, $0 < r \leq 0.2$). It is easy to check that $\dfrac{\tilde{b}(s)}{a_i(s)}$, $i = 1, 2, 3, 4$, are strictly positive real. □

Finally, it should also be pointed out that, for the vertex set $F = \{a_i(s) = s^n + \sum_{l=1}^{n} a_l^{(i)} s^{n-l}, i = 1, 2, \ldots, m\}$ of a general polytopic polynomial family $\overline{F}$, even if $\overline{F}$ is robustly stable, it is still possible that there does not exist a polynomial $c(s) \in H^{n-1}$ such that $c(s)/a(s) \in$ WSPR, for all $a(s) \in \overline{F}$. Therefore, $\bigcap_{i=1}^{m} \Omega_{1a_i}^{W} = \phi$.

To see this, let us look at an example of a third-order triangle polynomial family.

Example 11.10. Let $F = \{a_1(s) = s^3 + 2.6s^2 + 37s + 64, a_2(s) = s^3 + 17s^2 + 83s + 978, a_3(s) = s^3 + 15s^2 + 28s + 415\}$. It is easy to verify that $a_i(s)$, $i = 1, 2, 3$, are Hurwitz stable. Moreover, all edges of $\overline{F}$, i.e., $\lambda a_i(s) + (1 - \lambda) a_j(s)$, $\lambda \in [0, 1]$, $i, j = 1, 2, 3$, are also Hurwitz stable. Therefore, by the Edge Theorem [1, 9, 13, 11], $\overline{F}$ is robustly stable. On the other hand, by a direct computation using Corollary 11.1, we can easily see that $\Omega_{1a_1}^{W} \cap \Omega_{1a_2}^{W} \cap \Omega_{1a_3}^{W} = \phi$. Henceforth, there does not exist a polynomial $c(s) \in H^2$, such that $c(s)/a_i(s) \in$ WSPR, $i = 1, 2, 3$ (although $\Omega_{1a_1}^{W} \cap \Omega_{1a_2}^{W} \neq \phi$, $\Omega_{1a_1}^{W} \cap \Omega_{1a_3}^{W} \neq \phi$, and $\Omega_{1a_2}^{W} \cap \Omega_{1a_3}^{W} \neq \phi$).

Note that, in this example, though we have $\Omega_{1a_1}^{W} \cap \Omega_{1a_2}^{W} \cap \Omega_{1a_3}^{W} = \phi$, but it is easy to check $(6, 73, 68) \in \Omega_{1a_1} \cap \Omega_{1a_2} \cap \Omega_{1a_3}$. Let $\tilde{c}(s) := s^3 + 6s^2 + 73s + 68$. It is easy to check that $\dfrac{\tilde{c}(s)}{a_i(s)}$, $i = 1, 2, 3$, are strictly positive real. □

The above Example 11.10 shows some intrinsic differences between the SPR synthesis of interval polynomial families and the SPR synthesis of polytopic polynomial families. This problem deserves further investigation.

In the next section, we will show that there always exists a polynomial such that their ratios are SPR-invariant for an arbitrary order stable convex combination of two polynomials. But for a higher-order stable interval polynomial family, does there exist a polynomial such that their ratios are SPR-invariant? This is still an open problem. From our numerous examples, given a stable interval polynomial family, it seems that such a polynomial can always be found. No counterexample has been found. Thus, we conjecture that this problem has a positive answer.

11.7 Robust SPR Synthesis for Polynomial Segment of Arbitrary Order

This section presents a constructive proof of the following statement: for any two nth-order polynomials $a(s)$ and $b(s)$, the Hurwitz stability of their convex combination is necessary and sufficient for the existence of a polynomial $c(s)$ such that $c(s)/a(s)$ and $c(s)/b(s)$ are both SPR.

11.7.1 Main Results

The following theorem is the main result of this section.

Theorem 11.10. Suppose that $a(s) = s^n + a_1 s^{n-1} + \cdots + a_n \in H^n$ and $b(s) = s^n + b_1 s^{n-1} + \cdots + b_n \in H^n$. The necessary and sufficient condition for the existence of a polynomial $c(s)$ such that $c(s)/a(s)$ and $c(s)/b(s)$ are both strictly positive real, is

$$\lambda b(s) + (1-\lambda)a(s) \in H^n, \quad \lambda \in [0,1]. \qquad \square$$

We first introduce some lemmas.

Lemma 11.17. Suppose that $a(s) = s^n + a_1 s^{n-1} + \cdots + a_n \in H^n$. Then, for every $k \in \{1, 2, \ldots, n-2\}$, the following quadratic curve is an ellipse in the first quadrant (i.e., $x_i \geq 0$, $i = 1, 2, \ldots, n-1$) of the $\mathbb{R}^{n-1}$ space $(x_1, x_2, \ldots, x_{n-1})$[3]:

[3] When $n = 3$, the ellipse equation is:

$$(a_2x_1 - a_1x_2 - a_3)^2 - 4(a_1 - x_1)a_3x_2 = 0.$$

When $n = 4$, the two ellipse equations are:

$$\begin{cases} (a_2x_1 + x_3 - a_1x_2 - a_3)^2 - 4(a_1 - x_1)(a_3x_2 - a_2x_3 - a_4x_1) = 0, \\ a_4x_3 = 0, \end{cases}$$

$$\begin{cases} (a_3x_2 - a_2x_3 - a_4x_1)^2 - 4(a_2x_1 + x_3 - a_1x_2 - a_3)a_4x_3 = 0, \\ a_1 - x_1 = 0. \end{cases}$$

When $n = 5$, the three ellipse equations are:

$$\begin{cases} (a_2x_1 + x_3 - a_1x_2 - a_3)^2 - 4(a_1 - x_1)(a_5 + a_3x_2 + a_1x_4 - a_2x_3 - a_4x_1) = 0, \\ a_4x_3 - a_3x_4 - a_5x_2 = 0, \ a_5x_4 = 0, \end{cases}$$

$$\begin{cases} (a_5 + a_3x_2 + a_1x_4 - a_2x_3 - a_4x_1)^2 - 4(a_2x_1 + x_3 - a_1x_2 \\ \quad - a_3)(a_4x_3 - a_3x_4 - a_5x_2) = 0, \\ a_1 - x_1 = 0, \ a_5x_4 = 0, \end{cases}$$

$$\begin{cases} (a_4x_3 - a_3x_4 - a_5x_2)^2 - 4(a_5 + a_3x_2 + a_1x_4 - a_2x_3 - a_4x_1)a_5x_4 = 0, \\ a_1 - x_1 = 0, \ a_2x_1 + x_3 - a_1x_2 - a_3 = 0. \end{cases}$$

When $n = 6$, the four ellipse equations are:

$$\begin{cases} (a_2x_1 + x_3 - a_1x_2 - a_3)^2 - 4(a_1 - x_1)(a_5 + a_3x_2 + a_1x_4 - x_5 - a_2x_3 \\ \quad - a_4x_1) = 0, \\ a_6x_1 + a_4x_3 + a_2x_5 - a_3x_4 - a_5x_2 = 0, \ a_5x_4 - a_4x_5 - a_6x_3 = 0, \ a_6x_5 = 0, \end{cases}$$

$$\begin{cases} c_{k+1}^2 - 4c_k c_{k+2} = 0, \\ c_l = 0, \\ l \in \{1, 2, \ldots, n\}, \ l \neq k, k+1, k+2, \end{cases}$$

and this ellipse is tangent to the line

$$\begin{cases} c_l = 0, \\ l \in \{1, 2, \ldots, n\}, \ l \neq k+1, k+2, \end{cases}$$

and the line

$$\begin{cases} c_l = 0, \\ l \in \{1, 2, \ldots, n\}, \ l \neq k, k+1, \end{cases}$$

respectively, where $c_l := \sum_{j=0}^{n} (-1)^{l+j} a_j x_{2l-j-1}$, $l = 1, 2, \ldots, n$, $a_0 = 1$, $x_0 = 1$, $a_i = 0$ if $i < 0$ or $i > n$, and $x_i = 0$ if $i < 0$ or $i > n-1$.

Proof. Since $a(s)$ is Hurwitz stable, by using mathematical induction, the lemma is proved by a direct calculation (see the appendix for details).

For notational simplicity, for $a(s) = s^n + a_1 s^{n-1} + \cdots + a_n \in H^n$, $b(s) = s^n + b_1 s^{n-1} + \cdots + b_n \in H^n$, $\forall k \in \{1, 2, \ldots, n-2\}$, denote

$$\begin{aligned} \Omega_{ek}^a := \{(x_1, x_2, \ldots, x_{n-1}) | \ & c_{k+1}^2 - 4c_k c_{k+2} < 0, \\ & c_l = 0, \ l \in \{1, 2, \ldots, n\}, \ l \neq k, k+1, k+2\}, \end{aligned}$$

and

$$\begin{aligned} \Omega_{ek}^b := \{(x_1, x_2, \ldots, x_{n-1}) | \ & d_{k+1}^2 - 4d_k d_{k+2} < 0, \\ & d_l = 0, \ l \in \{1, 2, \ldots, n\}, \ l \neq k, k+1, k+2\}, \end{aligned}$$

where $c_l := \sum_{j=0}^{n} (-1)^{l+j} a_j x_{2l-j-1}$, $d_l := \sum_{j=0}^{n} (-1)^{l+j} b_j x_{2l-j-1}$, $l = 1, 2, \ldots, n$, $a_0 = 1$, $b_0 = 1$, $x_0 = 1$, $a_i = 0$ and $b_i = 0$ if $i < 0$ or $i > n$, and $x_i = 0$ if $i < 0$ or $i > n-1$.

In what follows, (A, B) stands for the set of points in the line segment connecting the point A and the point B in the $\mathbb{R}^{n-1}$ space $(x_1, x_2, \ldots, x_{n-1})$, not including the endpoints A and B. Denote

---

$$\begin{cases} (a_5 + a_3 x_2 + a_1 x_4 - x_5 - a_2 x_3 - a_4 x_1)^2 \\ \quad -4(a_2 x_1 + x_3 - a_1 x_2 - a_3)(a_6 x_1 + a_4 x_3 + a_2 x_5 - a_3 x_4 - a_5 x_2) = 0, \\ a_1 - x_1 = 0, \ a_5 x_4 - a_4 x_5 - a_6 x_3 = 0, \ a_6 x_5 = 0, \end{cases}$$

$$\begin{cases} (a_6 x_1 + a_4 x_3 + a_2 x_5 - a_3 x_4 - a_5 x_2)^2 \\ \quad -4(a_5 + a_3 x_2 + a_1 x_4 - x_5 - a_2 x_3 - a_4 x_1)(a_5 x_4 - a_4 x_5 - a_6 x_3) = 0, \\ a_1 - x_1 = 0, \ a_2 x_1 + x_3 - a_1 x_2 - a_3 = 0, \ a_6 x_5 = 0, \end{cases}$$

$$\begin{cases} (a_5 x_4 - a_4 x_5 - a_6 x_3)^2 - 4(a_6 x_1 + a_4 x_3 + a_2 x_5 - a_3 x_4 - a_5 x_2) a_6 x_5 = 0, \\ a_1 - x_1 = 0, \ a_2 x_1 + x_3 - a_1 x_2 - a_3 = 0, \ a_5 + a_3 x_2 + a_1 x_4 \\ \quad -x_5 - a_2 x_3 - a_4 x_1 = 0. \end{cases}$$

$$\Omega^a := \{(x_1, x_2, \ldots, x_{n-1}) |\ (x_1, x_2, \ldots, x_{n-1}) \in \bigcup_{i=1, i<j\leq n-2}^{n-3} (A_i, A_j), \\ \forall A_i \in \Omega^a_{ei},\ i \in \{1, 2, \ldots, n-2\}\}$$

and

$$\Omega^b := \{(x_1, x_2, \ldots, x_{n-1}) |\ (x_1, x_2, \ldots, x_{n-1}) \in \bigcup_{i=1, i<j\leq n-2}^{n-3} (B_i, B_j), \\ \forall B_i \in \Omega^b_{ei},\ i \in \{1, 2, \ldots, n-2\}\}.$$

Lemma 11.18. Suppose that $a(s) = s^n + a_1 s^{n-1} + \cdots + a_n \in H^n$, $b(s) = s^n + b_1 s^{n-1} + \cdots + b_n \in H^n$. If $\Omega^a \cap \Omega^b \neq \phi$, take $(x_1, x_2, \ldots, x_{n-1}) \in \Omega^a \cap \Omega^b$, and let $c(s) := s^{n-1} + (x_1 - \varepsilon)s^{n-2} + x_2 s^{n-3} + \cdots + x_{n-2}s + (x_{n-1} + \varepsilon)$ (ε is a sufficiently small positive number). Then, for $\dfrac{c(s)}{a(s)}$ and $\dfrac{c(s)}{b(s)}$, we have $\forall \omega \in \mathbb{R}, \mathrm{Re}\left[\dfrac{c(j\omega)}{a(j\omega)}\right] > 0$ and $\mathrm{Re}\left[\dfrac{c(j\omega)}{b(j\omega)}\right] > 0$.

Proof. Suppose that $(x_1, x_2, \ldots, x_{n-1}) \in \Omega^a \bigcap \Omega^b$, and let $c(s) := s^{n-1} + (x_1 - \varepsilon)s^{n-2} + x_2 s^{n-3} + \cdots + x_{n-2}s + (x_{n-1} + \varepsilon)$, $\varepsilon > 0$, ε sufficiently small.

$\forall \omega \in \mathbb{R}$, consider

$$\begin{aligned}
\mathrm{Re}\left[\frac{c(j\omega)}{a(j\omega)}\right] &= \frac{1}{|a(j\omega)|^2}[c_1\omega^{2(n-1)} + c_2\omega^{2(n-2)} + \cdots + c_{n-1}\omega^2 + c_n] \\
&\quad + \mathrm{Re}\left[\frac{-\varepsilon(j\omega)^{n-2} + \varepsilon}{a(j\omega)}\right] \\
&= \frac{1}{|a(j\omega)|^2}[c_1\omega^{2(n-1)} + c_2\omega^{2(n-2)} + \cdots + c_{n-1}\omega^2 + c_n] \\
&\quad + \frac{(-\varepsilon)}{|a(j\omega)|^2}\left(-\omega^{2(n-1)} + \tilde{c}(\omega^2)\right),
\end{aligned}$$

where $c_l := \sum_{j=0}^{n} (-1)^{l+j} a_j x_{2l-j-1}$, $l = 1, 2, \ldots, n$, $a_0 = 1$, $x_0 = 1$, $a_i = 0$ if $i < 0$ or $i > n$, and $x_i = 0$ if $i < 0$ or $i > n-1$, and $\tilde{c}(\omega^2)$ is a real polynomial with order not greater than $2(n-2)$.

In order to prove that $\forall \omega \in \mathbb{R}, \mathrm{Re}\left[\dfrac{c(j\omega)}{a(j\omega)}\right] > 0$, let $t = \omega^2$. We only need to prove that, for any $\varepsilon > 0$, ε sufficiently small, the following polynomial $f_1(t)$ satisfies

$$f_1(t) := c_1 t^{n-1} + c_2 t^{n-2} + \cdots + c_{n-1}t + c_n \\ + \varepsilon(t^{n-1} - \tilde{c}(t)) > 0, \quad \forall t \in [0, +\infty).$$

Since $(x_1, x_2, \ldots, x_{n-1}) \in \Omega^a$, by the definition of Ω^a, it is easy to know that

$$g_1(t) := c_1 t^{n-1} + c_2 t^{n-2} + \cdots + c_{n-1}t + c_n > 0, \quad \forall t \in (0, +\infty).$$

Moreover, we obviously have $f_1(0) > 0$, and for any $\varepsilon > 0$, when t is a sufficiently large or sufficiently small positive number, we have $f_1(t) > 0$;

namely, there exist $0 < t_1 < t_2$ such that, for all $\varepsilon > 0$, $t \in [0, t_1] \cup [t_2, +\infty)$, we have $f_1(t) > 0$.

Denote

$$M_1 = \inf_{t\in[t_1,t_2]} g_1(t),$$

$$N_1 = \sup_{t\in[t_1,t_2]} |t^{n-1} - \tilde{c}(t))|.$$

Then $M_1 > 0$ and $N_1 > 0$. Choosing $0 < \varepsilon < \dfrac{M_1}{N_1}$, by a direct calculation, we have

$$\begin{aligned} f_1(t) &:= c_1 t^{n-1} + c_2 t^{n-2} + \cdots + c_{n-1} t + c_n \\ &\quad + \varepsilon(t^{n-1} - \tilde{c}(t)) > 0, \quad \forall t \in [0, +\infty). \end{aligned}$$

Therefore,

$$\forall \omega \in \mathbb{R}, \quad \mathrm{Re}\left[\frac{c(j\omega)}{a(j\omega)}\right] > 0.$$

Similarly, since $(x_1, x_2, \ldots, x_{n-1}) \in \Omega^b$, there exist $0 < t_3 < t_4$ such that, for all $\varepsilon > 0$, $t \in [0, t_3] \bigcup [t_4, +\infty)$, we have $f_2(t) > 0$,where

$$\begin{aligned} f_2(t) &:= d_1 t^{n-1} + d_2 t^{n-2} + \cdots + d_{n-1} t + d_n \\ &\quad + \varepsilon(t^{n-1} - \tilde{d}(t)), \end{aligned}$$

$$d_l := \sum_{j=0}^{n} (-1)^{l+j} b_j x_{2l-j-1}, \quad l = 1, 2, \ldots, n,$$

$b_0 = 1$, $x_0 = 1$, $b_i = 0$ if $i < 0$ or $i > n$, and $x_i = 0$ if $i < 0$ or $i > n - 1$, and $\tilde{d}(\omega^2)$ is a real polynomial with order not greater than $2(n-2)$, which is determined by the following equation:

$$\mathrm{Re}\left[\frac{-\varepsilon(j\omega)^{n-2} + \varepsilon}{b(j\omega)}\right] = \frac{(-\varepsilon)}{|b(j\omega)|^2}(-\omega^{2(n-1)} + \tilde{d}(\omega^2)).$$

Denote

$$g_2(t) := d_1 t^{n-1} + d_2 t^{n-2} + \cdots + d_{n-1} t + d_n,$$

$$M_2 = \inf_{t\in[t_3,t_4]} g_2(t),$$

$$N_2 = \sup_{t\in[t_3,t_4]} |t^{n-1} - \tilde{d}(t))|.$$

Then $M_2 > 0$ and $N_2 > 0$. Choosing $0 < \varepsilon < \dfrac{M_2}{N_2}$, we have

$$\forall \omega \in \mathbb{R}, \quad \mathrm{Re}\left[\frac{c(j\omega)}{b(j\omega)}\right] > 0.$$

Thus, by choosing $0 < \varepsilon < \min\left\{\dfrac{M_1}{N_1}, \dfrac{M_2}{N_2}\right\}$, the lemma is proved. □

Lemma 11.19. Suppose that $a(s) = s^n + a_1 s^{n-1} + \cdots + a_n \in H^n$, $b(s) = s^n + b_1 s^{n-1} + \cdots + b_n \in H^n$. If $\lambda b(s) + (1-\lambda)a(s) \in H^n$, $\lambda \in [0,1]$, then $\Omega^a \cap \Omega^b \neq \phi$.

Proof. If $\forall \lambda \in [0,1], a_\lambda(s) := \lambda b(s) + (1-\lambda)a(s) \in H^n$, by Lemma 11.17, for any $\lambda \in [0,1]$, $\Omega_{ek}^{a_\lambda}$, $k = 1, 2, \ldots, n-2$, are $n-2$ ellipses in the first quadrant of the $\mathbb{R}^{n-1}$ space $(x_1, x_2, \ldots, x_{n-1})$.

$\forall \lambda \in [0,1]$, denote

$$\Omega^{a_\lambda} := \{(x_1, x_2, \ldots, x_{n-1}) | \ (x_1, x_2, \ldots, x_{n-1}) \in \bigcup_{i=1, i<j\le n-2}^{n-3} (A_{\lambda i}, A_{\lambda j}), \\ \forall A_{\lambda i} \in \Omega_{ei}^{a_\lambda}, \ i \in \{1, 2, \ldots, n-2\}\}.$$

Apparently, when λ changes continuously from 0 to 1, Ω^{a_λ} will change continuously from Ω^a to Ω^b, and $\Omega_{ek}^{a_\lambda}$ will change continuously from Ω_{ek}^a to Ω_{ek}^b, $k = 1, 2, \ldots, n-2$.

Now assume that $\Omega^a \cap \Omega^b = \phi$. By the definitions of Ω^a and Ω^b and Lemma 11.7.1, $\exists u_1 > 0$, $u_2 > 0, u_1 \neq a_1, u_1 \neq b_1$, and $\exists \tilde{k} \in \{1, 2, \ldots, n-2\}$, such that the following hyperplane L in the $\mathbb{R}^{n-1}$ space $(x_1, x_2, \ldots, x_{n-1})$

$$L\colon \ \frac{x_1}{u_1} + \frac{x_2}{u_2} + \cdots + \frac{x_{n-1}}{u_{n-1}} = 1$$

separates Ω^a and Ω^b. Meanwhile, L is tangent to $\Omega_{e1}^a, \Omega_{e2}^a, \ldots, \Omega_{e(n-2)}^a$ and $\Omega_{e\tilde{k}}^b$ simultaneously (or tangent to $\Omega_{e1}^b, \Omega_{e2}^b, \ldots, \Omega_{e(n-2)}^b$ and $\Omega_{e\tilde{k}}^a$ simultaneously).

Without loss of generality, suppose that L is tangent to $\Omega_{e1}^a, \Omega_{e2}^a, \ldots, \Omega_{e(n-2)}^a$ and $\Omega_{e\tilde{k}}^b$ simultaneously.

In what follows, the notation $[x]$ stands for the largest integer that is smaller than or equal to the real number x, and $\langle y \rangle_z$ stands for the remainder of the nonnegative integer y divided by the positive integer z.[4]

Since L is tangent to $\Omega_{e1}^a, \Omega_{e2}^a, \ldots, \Omega_{e(n-2)}^a$ and $\Omega_{e\tilde{k}}^b$ simultaneously, note that $a(s)$ is Hurwitz stable and $u_1 > 0, u_1 \neq a_1, u_2 > 0$, using mathematical induction, by a lengthy calculation, we know that the necessary and sufficient condition for L being tangent to $\Omega_{e1}^a, \Omega_{e2}^a, \ldots, \Omega_{e(n-2)}^a$ simultaneously is[5]

---

[4] For example, $[1.5] = 1$, $[0.5] = 0$, $[-1.5] = -2$, and $\langle 0 \rangle_2 = 0$, $\langle 1 \rangle_2 = 1$, $\langle 11 \rangle_3 = 2$.

[5] When $n = 3$, we have:

$$u_1 u_2 - a_1 u_2 - a_2 u_1 + a_3 = 0.$$

When $n = 4$, we have:

$$\begin{cases} u_1 u_2^2 - a_1 u_2^2 - a_2 u_1 u_2 + a_3 u_2 + a_4 u_1 = 0, \\ u_3 = -u_1 u_2. \end{cases}$$

When $n = 5$, we have:

$$\begin{cases} u_1 u_2^2 - a_1 u_2^2 - a_2 u_1 u_2 + a_3 u_2 + a_4 u_1 - a_5 = 0, \\ u_3 = -u_1 u_2, \ u_4 = -u_2^2. \end{cases}$$

$$\sum_{i=0}^{n}(-1)^{\left[\frac{i+1}{2}\right]}a_i u_1^{\langle i+1\rangle_2}u_2^{\left[\frac{n}{2}\right]-\left[\frac{i}{2}\right]}=0, \tag{11.12}$$

and

$$u_j=(-1)^{\left[\frac{j-1}{2}\right]}u_1^{\langle j\rangle_2}u_2^{\left[\frac{j}{2}\right]},\quad j=3,4,\ldots,n-1, \tag{11.13}$$

where $a_0=1$.

Since $u_j=(-1)^{\left[\frac{j-1}{2}\right]}u_1^{\langle j\rangle_2}u_2^{\left[\frac{j}{2}\right]}$, $j=3,4,\ldots,n-1$, L is tangent to $\Omega^b_{e\tilde{k}}$, by a direct calculation. We have

$$\sum_{i=0}^{n}(-1)^{\left[\frac{i+1}{2}\right]}b_i u_1^{\langle i+1\rangle_2}u_2^{\left[\frac{n}{2}\right]-\left[\frac{i}{2}\right]}=0, \tag{11.14}$$

where $b_0=1$.

From (11.12) and (11.14), we obviously have $\forall\lambda\in[0,1]$,

$$\sum_{i=0}^{n}(-1)^{\left[\frac{i+1}{2}\right]}a_{\lambda i} u_1^{\langle i+1\rangle_2}u_2^{\left[\frac{n}{2}\right]-\left[\frac{i}{2}\right]}=0, \tag{11.15}$$

where $a_{\lambda i}:=a_i+\lambda(b_i-a_i)$, $a_0=1$, $b_0=1$, $i=0,1,2,\ldots,n$. Equations (11.15) and (11.13) show that L is also tangent to $\Omega^{a_\lambda}_{e\tilde{k}}$ ($\forall\lambda\in[0,1]$), but L separates $\Omega^a_{e\tilde{k}}$ and $\Omega^b_{e\tilde{k}}$, and when λ changes continuously from 0 to 1, $\Omega^{a_\lambda}_{e\tilde{k}}$ will change continuously from $\Omega^a_{e\tilde{k}}$ to $\Omega^b_{e\tilde{k}}$, which is obviously impossible. This completes the proof. □

Lemma 11.20. Suppose that $a(s)=s^n+a_1s^{n-1}+\cdots+a_n\in H^n$, $b(s)=s^n+b_1s^{n-1}+\cdots+b_n\in H^n$, $c(s)=s^{n-1}+x_1s^{n-2}+\cdots+x_{n-1}$. If $\forall\omega\in\mathbb{R}$, $\mathrm{Re}\left[\frac{c(j\omega)}{a(j\omega)}\right]>0$ and $\mathrm{Re}\left[\frac{c(j\omega)}{b(j\omega)}\right]>0$, take

$$\tilde{c}(s):=c(s)+\delta\cdot h(s),\quad \delta>0,\ \delta\ \text{sufficiently small},$$

where $h(s)$ is an arbitrarily given monic nth order polynomial, then $\frac{\tilde{c}(s)}{a(s)}$ and $\frac{\tilde{c}(s)}{b(s)}$ are both strictly positive real.

Proof. See the proof of Theorem 11.4.

When $n=6$, we have:

$$\begin{cases} u_1u_2^3-a_1u_2^3-a_2u_1u_2^2+a_3u_2^2+a_4u_1u_2-a_5u_2-a_6u_1=0,\\ u_3=-u_1u_2,\ u_4=-u_2^2,\ u_5=u_1u_2^2. \end{cases}$$

Proof of Theorem 11.10. The statement is obviously true for the cases when $n = 1$ or $n = 2$. We will prove it for the case when $n \geq 3$.

Since SPR transfer functions enjoy convexity property, we get the necessary part of the theorem.

The sufficiency of Theorem 11.10 is now proved by simply combining Lemmas 11.17– 11.20. □

Remark 11.1. From the proof of Theorem 11.10, we can see that this section not only proves the existence, but also provides a design method. In fact, based on the main idea, we have developed a geometric algorithm with order reduction for robust SPR synthesis which is very efficient for high-order polynomial segments [47]. □

Remark 11.2. The method provided in this section is constructive, and is insightful and helpful in solving more general robust SPR synthesis problems for polynomial polytopes, multilinear families, etc. □

Remark 11.3. Our main results in this section can also be extended to discrete-time case. In fact, by using the bilinear transformation, we can transform the unit circle into the left half plane. Hence, Theorem 11.10 can be generalized to the discrete-time case. Moreover, in the discrete-time case, the order of the polynomial obtained by our method is bounded by the order of the given polynomial segment [53, 55]. □

Remark 11.4. If $\dfrac{c(s)}{a(s)}$ and $\dfrac{c(s)}{b(s)}$ are both SPR, it is easy to show that $\forall \lambda \in [0, 1]$,

$$\frac{c(s)}{\lambda a(s) + (1 - \lambda) b(s)}$$

is also SPR. □

Remark 11.5. The stability of a polynomial segment can be checked by many efficient methods, e.g., eigenvalue method, root locus method, value set method, etc. [1, 9, 13]. □

11.7.2 Design Procedure and Some Examples

As observed earlier, the method provided in this section is constructive. By our theoretical analysis, we propose the following design procedure.

Step 1. Test the robust Hurwitz stability of the convex combination of the two polynomials $a(s)$ and $b(s)$. If the convex combination is robustly stable, then go to the next step. Otherwise, print "there does not exist such a $c(s)$" (by Definition 11.1 and Proposition 11.1).

Step 2. Construct Ω^a_{ek}, Ω^b_{ek}, $k = 1, 2, \ldots, n-2$, and find a point

$$(x_1, x_2, \ldots, x_{n-1} \in \left(\bigcup_{k=1}^{n-2} \Omega_{ek}^a\right) \bigcup \left(\bigcup_{k=1}^{n-2} \Omega_{ek}^b\right)$$

such as $\bar{c}(s) := s^{n-1} + x_1 s^{n-2} + x_2 s^{n-3} + \cdots + x_{n-2}s + x_{n-1}$ and $\forall \omega \in \mathbb{R}\backslash\{0\}$, $\mathrm{Re}\left[\frac{\bar{c}(j\omega)}{a(j\omega)}\right] > 0$ and $\mathrm{Re}\left[\frac{\bar{c}(j\omega)}{b(j\omega)}\right] > 0$ (by the Lemmas 11.18 and 11.19).

Step 3. Let $c(s) := s^{n-1} + (x_1 - \varepsilon)s^{n-2} + x_2 s^{n-3} + \cdots + x_{n-2}s + (x_{n-1} + \varepsilon)$, ε is a sufficiently small positive number, then for $\frac{c(s)}{a(s)}$ and $\frac{c(s)}{b(s)}$, we have $\forall \omega \in \mathbb{R}, \mathrm{Re}\left[\frac{c(j\omega)}{a(j\omega)}\right] > 0$ and $\mathrm{Re}\left[\frac{c(j\omega)}{b(j\omega)}\right] > 0$ (by the Lemmas 11.18 and 11.19).

Step 4. Take $\tilde{c}(s) := \delta s^n + c(s)$, $\delta > 0$, δ sufficiently small, then this $\tilde{c}(s)$ satisfies the design requirement (by Lemma 11.20). □

There is hardly any example with order higher than 6 in the literature. In what follows, let us consider Examples 11.4 and 11.5 again. As shown by the examples below, our method is very effective.

Example 11.11. Suppose that $a(s) = s^7 + 9s^6 + 31s^5 + 71.5s^4 + 111.5s^3 + 109s^2 + 76s + 12.5$, $b(s) = s^7 + 9.4s^6 + 31.2s^5 + 71.3s^4 + 111s^3 + 109.2s^2 + 76.4s + 12$. It is easy to see that the convex combination of the two polynomials $a(s)$ and $b(s)$ is robustly Hurwitz stable. Using our method, we can get $(9, 14.9409, 34.08, 26.5088, 4.36, 0) \in \Omega_{e2}^a = \{(x_1, x_2, x_3, x_4, x_5, x_6) | \; c_1 = c_5 = c_6 = c_7 = 0,\; c_3^2 - c_2 c_4 = 5.4505 - 2.704x_3 + 18.501x_5 + 0.99999x_3^2 - 15.295x_3x_5 + 58.723x_5^2 < 0\}$ such as $\bar{c}(s) := s^6 + 9s^5 + 14.9409s^4 + 34.08s^3 + 26.5088s^2 + 4.36s$ and $\forall \omega \in \mathbb{R}\backslash\{0\}, \mathrm{Re}\left[\frac{\bar{c}(j\omega)}{a(j\omega)}\right] > 0$ and $\mathrm{Re}\left[\frac{\bar{c}(j\omega)}{b(j\omega)}\right] > 0$. Let $c(s) = s^6 + (9 - \varepsilon)s^5 + 14.9409s^4 + 34.08s^3 + 26.5088s^2 + 4.36s + \varepsilon$, ε is a sufficiently small positive number (in this example, $0 < \varepsilon \leq 1$), and take $\varepsilon = 0.1$. Then for $\frac{c(s)}{a(s)}$ and $\frac{c(s)}{b(s)}$, we have $\forall \omega \in \mathbb{R}$, $\mathrm{Re}\left[\frac{c(j\omega)}{a(j\omega)}\right] > 0$ and $\mathrm{Re}\left[\frac{c(j\omega)}{b(j\omega)}\right] > 0$. Thus, let $\tilde{c}(s) := c(s) + \delta s^6$, $\delta > 0$, δ sufficiently small, e.g., $\delta \leq 0.4$. Then, the design requirement has been met. □

Example 11.12. Suppose that $a(s) = s^9 + 11s^8 + 52s^7 + 145s^6 + 266s^5 + 331s^4 + 280s^3 + 155s^2 + 49s + 6$, $b(s) = s^9 + 11s^8 + 52s^7 + 146s^6 + 265.5s^5 + 332s^4 + 278.5s^3 + 151s^2 + 48s + 2$. It is easy to see that the convex combination of the two polynomials $a(s)$ and $b(s)$ is robustly Hurwitz stable. Using our method, we can get $(11, 24.2122, 70.5046, 87.3862, 56.27, 18.4975, 2.265, 0) \in \Omega_{e2}^a = \{(x_1, x_2, x_3,\; x_4,\; x_5, x_6, x_7, x_8) | \; c_1 = c_5 = c_6 = c_7 = c_8 = c_9 = 0,\; c_3^2 - c_2c_4 = (-1982.9 + 2713.2x_5 - 67345.x_7)^2 - 4(378.97 - 267.89x_5 + 6590.0x_7)(1607.3 - 3682.0x_5 + 92407.x_7) < 0\}$ such as $\bar{c}(s) := s^8 + 11s^7 + 24.2122s^6 + 70.5046s^5 + 87.3862s^4 + 56.27s^3 + 18.4975s^2 + 2.265s$ and $\forall \omega \in \mathbb{R}\backslash\{0\}$, $\mathrm{Re}\left[\frac{\bar{c}(j\omega)}{a(j\omega)}\right] > 0$

and $\mathrm{Re}\left[\frac{\bar{c}(j\omega)}{b(j\omega)}\right] > 0$. Let $c(s) = s^8 + (11 - \varepsilon)s^7 + 24.2122s^6 + 70.5046s^5 + 87.3862s^4 + 56.27s^3 + 18.4975s^2 + 2.265s + \varepsilon$, ε is a sufficiently small positive number (in this example, $0 < \varepsilon \leq 0.001$), and take $\varepsilon = 0.0005$. Then for $\frac{c(s)}{a(s)}$ and $\frac{c(s)}{b(s)}$, we have $\forall \omega \in \mathbb{R}$, $\mathrm{Re}\left[\frac{c(j\omega)}{a(j\omega)}\right] > 0$ and $\mathrm{Re}\left[\frac{c(j\omega)}{b(j\omega)}\right] > 0$. Thus, letting $\tilde{c}(s) := c(s) + \delta s^6$, $\delta > 0$, δ sufficiently small, e.g., taking $\delta \leq 0.3$, the design requirement has been met. □

Note that, in our Examples 11.11 and 11.12, $c(s)$ is not unique.

11.7.3 Appendix: Proof of Lemma 11.17

We only need consider the case when $n \geq 3$.

In what follows, the notation $\langle y \rangle_z$ stands for the remainder of the nonnegative integer y divided by the positive integer z.

To prove Lemma 11.17, for convenience, we also introduce some lemmas.

Lemma 11.21. Suppose that $a(s) = s^n + a_1 s^{n-1} + \cdots + a_n \in H^n$. Then, $\alpha(s) = s^{n-1} + \alpha_1 s^{n-2} + \cdots + \alpha_{n-1} \in H^{n-1}$, where[6]

$$\alpha_i = a_i^{\langle n-i \rangle_2} \left(\frac{a_i a_{n-1} - a_{i-1} a_n}{a_{n-1}} \right)^{\langle n-1-i \rangle_2},$$

$i = 1, 2, \ldots, n-1$, $a_0 = 1$, $a_l = 0$ if $l < 0$ or $l > n$.

Proof. The lemma is proved by direct use of the Hurwitz criterion. □

---

[6] For example, when $n = 3$, we have:

$$\alpha_1 = \frac{a_1 a_2 - a_3}{a_2}, \ \alpha_2 = a_2.$$

When $n = 4$, we have:

$$\alpha_1 = a_1, \ \alpha_2 = \frac{a_2 a_3 - a_1 a_4}{a_3}, \ \alpha_3 = a_3.$$

When $n = 5$, we have:

$$\alpha_1 = \frac{a_1 a_4 - a_5}{a_4}, \ \alpha_2 = a_2, \ \alpha_3 = \frac{a_3 a_4 - a_2 a_5}{a_4}, \ \alpha_4 = a_4.$$

When $n = 6$, we have:

$$\alpha_1 = a_1, \ \alpha_2 = \frac{a_2 a_5 - a_1 a_6}{a_5}, \ \alpha_3 = a_3, \ \alpha_4 = \frac{a_4 a_5 - a_3 a_6}{a_5}, \ \alpha_5 = a_5.$$

Lemma 11.22. Suppose that $a(s) = s^n + a_1 s^{n-1} + \cdots + a_n \in H^n$. Then, $\beta(s) = s^{n-1} + \beta_1 s^{n-2} + \cdots + \beta_{n-1} \in H^{n-1}$, where[7]

$$\beta_i = (a_1 a_{i+1} - a_{i+2})^{\langle i \rangle_2} \left(\frac{a_{i+1}}{a_1} \right)^{\langle i+1 \rangle_2},$$

$i = 1, 2, \ldots, n-1$, $a_0 = 1$, $a_l = 0$ if $l < 0$ or $l > n$.

Proof. The lemma is proved by direct use of the Hurwitz criterion. □

Lemma 11.23. Suppose that $a(s) = s^3 + a_1 s^2 + a_2 s + a_3 \in H^3$. Then, the following quadratic curve is an ellipse in the first quadrant (i.e., $x_i \geq 0$, $i = 1, 2$) of the $\mathbb{R}^2$ space (x_1, x_2):

$$(a_2 x_1 - a_1 x_2 - a_3)^2 - 4(a_1 - x_1) a_3 x_2 = 0,$$

and this ellipse is tangent to the line

$$a_1 - x_1 = 0,$$

and the line

$$a_3 x_2 = 0,$$

respectively.

Proof. Since $a(s)$ is Hurwitz stable, the lemma is proved by direct calculation (see Corollary 11.1 and Lemma 11.5).

We are now in a position to prove Lemma 11.17 using mathematical induction.

Proof of Lemma 11.17. First, the conclusion of Lemma 11.17 is true when $n = 3$ (Lemma 11.23).

[7] For example, when $n = 3$, we have:

$$\beta_1 = a_1 a_2 - a_3, \quad \beta_2 = \frac{a_3}{a_1}.$$

When $n = 4$, we have:

$$\beta_1 = a_1 a_2 - a_3, \quad \beta_2 = \frac{a_3}{a_1}, \quad \beta_3 = a_1 a_4.$$

When $n = 5$, we have:

$$\beta_1 = a_1 a_2 - a_3, \quad \beta_2 = \frac{a_3}{a_1}, \quad \beta_3 = a_1 a_4 - a_5, \quad \beta_4 = \frac{a_5}{a_1}.$$

When $n = 6$, we have:

$$\beta_1 = a_1 a_2 - a_3, \quad \beta_2 = \frac{a_3}{a_1}, \quad \beta_3 = a_1 a_4 - a_5, \quad \beta_4 = \frac{a_5}{a_1}, \quad \beta_5 = a_1 a_6.$$

Second, we suppose that Lemma 11.17 holds for the nth-order polynomial case, and we show that Lemma 11.17 is true for $(n+1)$st-order polynomial case.

Suppose that $a(s) = s^{n+1} + a_1 s^n + \cdots + a_n s + a_{n+1} \in H^{n+1}$, by Lemma 11.21 and Lemma 11.22, we have

$$\alpha(s) = s^n + \alpha_1 s^{n-1} + \cdots + \alpha_n \in H^n,$$

and

$$\beta(s) = s^n + \beta_1 s^{n-1} + \cdots + \beta_n \in H^n,$$

where

$$\alpha_i = a_i^{\langle n+1-i \rangle_2} \left(\frac{a_i a_n - a_{i-1} a_{n+1}}{a_n} \right)^{\langle n-i \rangle_2},$$

and

$$\beta_i = (a_1 a_{i+1} - a_{i+2})^{\langle i \rangle_2} \left(\frac{a_{i+1}}{a_1} \right)^{\langle i+1 \rangle_2},$$

$i = 1, 2, \ldots, n$, $a_0 = 1$, $a_l = 0$ if $l < 0$ or $l > n+1$.

Since we suppose that Lemma 11.17 holds for the nth-order polynomial case, for every $k \in \{1, 2, \ldots, n-2\}$, the following quadratic curve is an ellipse in the first quadrant (i.e., $y_i \geq 0$, $i = 1, 2, \ldots, n-1$) of the $\mathbb{R}^{n-1}$ space $(y_1, y_2, \ldots, y_{n-1})$:

$$\begin{cases} c_{\alpha,k+1}^2 - 4c_{\alpha,k} c_{\alpha,k+2} = 0, \\ c_{\alpha,l} = 0, \\ l \in \{1, 2, \ldots, n\},\ l \neq k, k+1, k+2, \end{cases} \tag{11.16}$$

and this ellipse is tangent to the line

$$\begin{cases} c_{\alpha,l} = 0, \\ l \in \{1, 2, \ldots, n\},\ l \neq k+1, k+2, \end{cases}$$

and the line

$$\begin{cases} c_{\alpha,l} = 0, \\ l \in \{1, 2, \ldots, n\},\ l \neq k, k+1, \end{cases}$$

respectively, where $c_{\alpha,l} := \sum_{j=0}^{n} (-1)^{l+j} \alpha_j y_{2l-j-1}$, $l = 1, 2, \ldots, n$, $\alpha_0 = 1$, $y_0 = 1$, $\alpha_i = 0$ if $i < 0$ or $i > n$, and $y_i = 0$ if $i < 0$ or $i > n-1$.

Meanwhile, for every $k \in \{1, 2, \ldots, n-2\}$, the following quadratic curve is an ellipse in the first quadrant (i.e., $z_i \geq 0$, $i = 1, 2, \ldots, n-1$) of the $\mathbb{R}^{n-1}$ space $(z_1, z_2, \ldots, z_{n-1})$:

$$\begin{cases} c_{\beta,k+1}^2 - 4c_{\beta,k} c_{\beta,k+2} = 0, \\ c_{\beta,l} = 0, \\ l \in \{1, 2, \ldots, n\},\ l \neq k, k+1, k+2, \end{cases} \tag{11.17}$$

and this ellipse is tangent to the line

$$\begin{cases} c_{\beta,l} = 0, \\ l \in \{1, 2, \ldots, n\},\ l \neq k+1, k+2, \end{cases}$$

and the line

$$\begin{cases} c_{\beta,l} = 0, \\ l \in \{1, 2, \ldots, n\},\ l \neq k, k+1, \end{cases}$$

respectively, where $c_{\beta,l} := \sum_{j=0}^{n} (-1)^{l+j} \beta_j z_{2l-j-1}$, $l = 1, 2, \ldots, n$, $\beta_0 = 1$, $z_0 = 1$, $\beta_i = 0$ if $i < 0$ or $i > n$, and $z_i = 0$ if $i < 0$ or $i > n-1$.

Now, for every $k \in \{1, 2, \ldots, n-1\}$, we consider the following quadratic curve in the $\mathbb{R}^n$ space $(x_1, x_2, \ldots, x_n)$:

$$\begin{cases} c_{k+1}^2 - 4c_k c_{k+2} = 0, \\ c_l = 0, \\ l \in \{1, 2, \ldots, n+1\},\ l \neq k, k+1, k+2, \end{cases} \tag{11.18}$$

where $c_l := \sum_{j=0}^{n+1} (-1)^{l+j} a_j x_{2l-j-1}$, $l = 1, 2, \ldots, n+1$, $a_0 = 1, x_0 = 1$, $a_i = 0$ if $i < 0$ or $i > n+1$, and $x_i = 0$ if $i < 0$ or $i > n$.

For notational simplicity, introduce the following matrices

$$\mathcal{H}_a := \begin{bmatrix} a_1 & -1 & 0 & 0 & \cdots \\ -a_3 & a_2 & -a_1 & 1 & \cdots \\ a_5 & -a_4 & a_3 & -a_2 & \cdots \\ \vdots & \vdots & \vdots & \vdots & \ddots \end{bmatrix}_{(n+1)\times(n+1)},$$

$$\mathcal{H}_\alpha := \begin{bmatrix} \alpha_1 & -1 & 0 & 0 & \cdots \\ -\alpha_3 & \alpha_2 & -\alpha_1 & 1 & \cdots \\ \alpha_5 & -\alpha_4 & \alpha_3 & -\alpha_2 & \cdots \\ \vdots & \vdots & \vdots & \vdots & \ddots \end{bmatrix}_{n\times n},$$

$$\mathcal{H}_\beta := \begin{bmatrix} \beta_1 & -1 & 0 & 0 & \cdots \\ -\beta_3 & \beta_2 & -\beta_1 & 1 & \cdots \\ \beta_5 & -\beta_4 & \beta_3 & -\beta_2 & \cdots \\ \vdots & \vdots & \vdots & \vdots & \ddots \end{bmatrix}_{n\times n},$$

$$\mathcal{X}_x := \begin{bmatrix} 1 \\ x_1 \\ \vdots \\ x_n \end{bmatrix}, \quad \mathcal{X}_y := \begin{bmatrix} 1 \\ y_1 \\ \vdots \\ y_{n-1} \end{bmatrix}, \quad \mathcal{X}_z := \begin{bmatrix} 1 \\ z_1 \\ \vdots \\ z_{n-1} \end{bmatrix},$$

$$\mathcal{C}_c := \begin{bmatrix} c_1 \\ c_2 \\ \vdots \\ c_{n+1} \end{bmatrix}, \quad \mathcal{C}_{c\alpha} := \begin{bmatrix} c_{\alpha,1} \\ c_{\alpha,2} \\ \vdots \\ c_{\alpha,n} \end{bmatrix}, \quad \mathcal{C}_{c\beta} := \begin{bmatrix} c_{\beta,1} \\ c_{\beta,2} \\ \vdots \\ c_{\beta,n} \end{bmatrix},$$

where $a_i = 0$ if $i > n+1$, $\alpha_i = 0$ and $\beta_i = 0$ if $i > n$. Then, it is easy to verify that

$$\mathcal{C}_c = \mathcal{H}_a \mathcal{X}_x, \quad \mathcal{C}_{c\alpha} = \mathcal{H}_\alpha \mathcal{X}_y, \quad \mathcal{C}_{c\beta} = \mathcal{H}_\beta \mathcal{X}_z.$$

Denote by $\mathcal{H}_{a\alpha}$ the $n \times n$ matrix formed by the first n rows and the first n columns of the matrix $\mathcal{H}_a$, $\mathcal{H}_{a\beta}$ the $n \times n$ matrix formed by withdrawing the first row and the second column of the matrix $\mathcal{H}_a$, and $h_{a\beta}$ the $n \times 1$ vector formed by the last n rows and the second column of the matrix $\mathcal{H}_a$.

Denote $\mathcal{X}_{x\alpha} := [1, x_1, x_2, \ldots, x_{n-1}]^T$, $\mathcal{X}_{x\beta} := [1, x_2, x_3, \ldots, x_n]^T$, $\underline{\mathcal{C}}_c := [c_1, c_2, \ldots, c_n]^T$, and $\overline{\mathcal{C}}_c := [c_2, c_3, \ldots, c_{n+1}]^T$.

If $c_{n+1} = 0$, i.e., $x_n = 0$, let us take the following transformation

$$x_i = y_i^{\langle n+1-i \rangle_2} \left(\frac{a_{n+1}}{a_n} y_{i-1} + y_i \right)^{\langle n-i \rangle_2}, \quad i = 1, 2, \ldots, n-1, \tag{11.19}$$

where $y_0 = 1$.[8] Then it can be verified that

$$\underline{\mathcal{C}}_c = \mathcal{H}_{a\alpha} \mathcal{X}_{x\alpha} = \mathcal{H}_\alpha \mathcal{X}_y = \mathcal{C}_{c\alpha}.$$

Thus, for every $k \in \{1, 2, \ldots, n-2\}$, the quadratic curve (11.18) in the $\mathbb{R}^{n-1}$ space $(x_1, x_2, \ldots, x_{n-1}, 0)$ can be obtained by using the above transform (11.19) for the ellipse curve (11.16) in the $\mathbb{R}^{n-1}$ space $(y_1, y_2, \ldots, y_{n-1})$.

If $c_1 = 0$, i.e., $x_1 = a_1$, let us take the following transformation

$$x_i = (a_1 z_{i-1})^{\langle i \rangle_2} \left(\frac{1}{a_1} z_{i-1} + z_i \right)^{\langle i+1 \rangle_2}, \quad i = 2, 3, \ldots, n, \tag{11.20}$$

where $z_n = 0$.[9] Then, it can be verified that

$$\overline{\mathcal{C}}_c = \mathcal{H}_{a\beta} \mathcal{X}_{x\beta} + a_1 h_{a\beta} = \mathcal{H}_\beta \mathcal{X}_z = \mathcal{C}_{c\beta}.$$

---

[8] For example, when $n+1 = 4$, we have:

$$x_1 = y_1, \quad x_2 = \frac{a_4}{a_3} y_1 + y_2.$$

When $n+1 = 5$, we have:

$$x_1 = \frac{a_5}{a_4} + y_1, \quad x_2 = y_2, \quad x_3 = \frac{a_5}{a_4} y_2 + y_3.$$

When $n+1 = 6$, we have:

$$x_1 = y_1, \quad x_2 = \frac{a_6}{a_5} y_1 + y_2, \quad x_3 = y_3, \quad x_4 = \frac{a_6}{a_5} y_3 + y_4.$$

When $n+1 = 7$, we have:

$$x_1 = \frac{a_7}{a_6} + y_1, \quad x_2 = y_2, \quad x_3 = \frac{a_7}{a_6} y_2 + y_3, \quad x_4 = y_4, \quad x_5 = \frac{a_7}{a_6} y_4 + y_5.$$

[9] For example, when $n+1 = 4$, we have:

Thus, for every $k \in \{2, 3, \ldots, n-1\}$, the quadratic curve (11.18) in the $\mathbb{R}^{n-1}$ space $(a_1, x_2, \ldots, x_n)$ can be obtained by using the above transformation (11.20) for the ellipse curve (11.17) in the $\mathbb{R}^{n-1}$ space $(z_1, z_2, \ldots, z_{n-1})$.

Since we suppose that Lemma 11.17 holds for nth-order polynomials $\alpha(s)$ and $\beta(s)$, combining the discussions above, we know that Lemma 11.17 also holds for $(n+1)$st-order polynomial $a(s)$. Thus, by using mathematical induction, we complete the proof of Lemma 11.17. □

11.8 Conclusions

In this chapter, we have studied the robust synthesis problem for strictly positive real (SPR) transfer functions. The concepts of SPR regions and weak SPR regions have been introduced and their properties have been discussed. We show that the SPR region associated with a fixed polynomial is unbounded, whereas the weak monic SPR region is bounded. We then prove that the intersection of several weak SPR regions associated with different polynomials cannot be a single point. Furthermore, we show how to construct a point in the SPR region from a point in the weak SPR region. By using the complete discrimination system for polynomials, complete characterization of the (weak) strictly positive real regions for transfer functions in coefficient space is given, which answers an unsolved problem proposed by Huang, Hollot and Xu in 1990 [27]. Based on these theoretical development, we have proposed an algorithm for the synthesis of robust SPR transfer functions. This algorithm works well for both low-order and high-order polynomial families. Especially, the derived conditions are necessary and sufficient for robust SPR design of polynomial segment or low-order ($n \leq 4$) interval polynomials. Illustrative examples are provided to show the effectiveness of this algorithm. The SPR synthesis problem for high-order interval polynomials is currently under investigation.

$$x_2 = \frac{z_1}{a_1} + z_2, \quad x_3 = a_1 z_2.$$

When $n+1=5$, we have:

$$x_2 = \frac{z_1}{a_1} + z_2, \quad x_3 = a_1 z_2, \quad x_4 = \frac{z_3}{a_1}.$$

When $n+1=6$, we have:

$$x_2 = \frac{z_1}{a_1} + z_2, \quad x_3 = a_1 z_2, \quad x_4 = \frac{z_3}{a_1} + z_4, \quad x_5 = a_1 z_4.$$

When $n+1=7$, we have:

$$x_2 = \frac{z_1}{a_1} + z_2, \quad x_3 = a_1 z_2, \quad x_4 = \frac{z_3}{a_1} + z_4, \quad x_5 = a_1 z_4, x_6 = \frac{z_5}{a_1}.$$

References

1. Ackermann J, Bartlett A, Kaesbauer D, Sienel W, Steinhauser R (1993) Robust control: systems with uncertain physical parameters. Springer-Verlag, Berlin
2. Anderson BDO (1967) A system theory criterion for positive real matrices. SIAM J Control 5:171–182
3. Anderson BDO (1968) A simplified viewpoint of hyperstability. IEEE Trans Automatic Control 13:292–294
4. Anderson BDO, Bitmead RR, Johnson CR, Kokotovic PV, Kosut RL, Mareel IMY, Praly L, Riedle BD (1986) Stability of adaptive systems: passivity and averaging analysis. MIT Press, Cambridge
5. Anderson BDO, Dasgupta S, Khargonekar P, Kraus FJ, Mansou M (1990) Robust strict positive realness: characterization and construction. IEEE Trans Circuits Syst 37:869–876
6. Anderson BDO, Jury E, Manours M (1987) On robust Hurwitz polynomials. IEEE Trans Automatic Control 32:909–913
7. Anderson BDO, Moore JB (1970) Linear optimal control. Prentice Hall, New York
8. Anderson BDO, Vongpanitlerd S (1973) Network analysis and synthesis. Prentice Hall, New York
9. Barmish BR (1994) New tools for robustness of linear systems. MacMillan Publishing Company, New York
10. Barmish BR, Kang IH (1992) Extreme point results for robust stability of interval plants: beyond first order compensators. Automatica 28:1169–1180
11. Bartlett AC, Hollot CV, Huang L (1988) Root locations for an entire polytope of polynomial: it suffices to check the edges. Math Control Signals Syst 1:61–71
12. Betser A, Zeheb E (1993) Design of robust strictly positive real transfer functions. IEEE Trans Circuits Syst Part I 40:573–580
13. Bhattacharyya SP, Chapellat H, Keel LH (1995) Robust control: the parametric approach. Prentice Hall, New York
14. Bianchini G (2002) Synthesis of robust strictly positive real discrete-time systems with l_2 parametric perturbations. IEEE Trans Circuits Syst Part I 49:1221–1225
15. Bianchini G, Tesi A, Vicino A (2001) Synthesis of robust strictly positive real systems with l_2 parametric uncertainty. IEEE Trans Circuits Syst Part I 48:438–450
16. Body S, El Ghaoui L, Feron E, Balakrishnan V (1994) Linear matrix inequalities in system and control theory. Society for Industrial and Applied Mathematics, Philadelphia
17. Chapellat H, Dahleh M, Bhattacharyya SP (1991) On robust nonlinear stability of interval control systems. IEEE Trans Automatic Control 36:59–69
18. Dasgupta S, Bhagwat AS (1987) Conditions for designing strictly positive real transfer functions for adaptive output error identification. IEEE Trans Circuits Syst 34:731–737
19. Dasgupta S, Chockalingam G, Anderson BDO, Fu M (1994) Lyapunov functions for uncertain systems with applications to the stability of time varying systems. IEEE Trans Circuits Syst Part I 41:93–106
20. Dasgupta S, Parker PJ, Anderson BDO, Kraus F J, Mansour M (1991) Frequency domain conditions for the robust stability of linear and nonlinear dynamical systems. IEEE Trans Circuits Syst 38:389–397

21. Desoer C A, Vidyasagar M (1975) Feedback systems: input-output properties. Academic Press, San Diego
22. Gantmacher F (1959) Matrix theory. Chelsea, New York
23. Henrion D (2002) Linear matrix inequalities for robust strictly positive real design. IEEE Trans Circuits Syst Part I 49:1017–1020
24. Hitz L, Anderson BDO (1969) Discrete positive-real functions and their application to system stability. Proc IEE 116:153–155
25. Hollot CV, Bartlett AC (1986) Some discrete time counterparts to Kharitonov's stability criterion for uncertain systems. IEEE Trans Automatic Control 31:355–356
26. Hollot CV, Huang L, Xu ZL (1989) Designing strictly positive real transfer function families: a necessary and sufficient condition for low degree and structured families. In: Proc Mathematical Theory of Network and Systems, Amsterdam, The Netherland, 215–227
27. Huang L, Hollot C V, Xu ZL (1990) Robust analysis of strictly positive real function set. In: Proc Second Japan-China Joint Symposium on Systems Control Theory and its Applications, Osaka, Japan, 210–220
28. Ioannou P, Tao G (1987) Frequency domain conditions for strictly positive real functions. IEEE Trans Automatic Control 32:53–54
29. Kalman RE (1963) Lyapunov functions for the problem of Lur'e in automatic control. Proc Nat Acad Sci 49:201–205
30. Kharitonov VL (1978) Asymptotic stability of an equilibrium position of a family of systems of linear differential equations. Differentsial'nye Uravneniya 14:2086–2088
31. Landau ID (1979) Adaptive control: the model reference approach. Marcel Dekker, New York
32. Marquez HJ, Agathoklis P (1998) On the existence of robust strictly positive real rational functions. IEEE Trans Circuits Syst Part I 45:962–967
33. Mosquera C, Perez F (2001) Algebraic solution to the robust SPR problem for two polynomials. Automatica 37:757–762
34. Nesterov Y, Nemirovski A (1994) Interior point polynomial methods in convex programming. Society for Industrial and Applied Mathematics, Philadelphia
35. Patel VV, Datta KB (1997) Classification of units in H_∞ and an alternative proof of Kharitonov's theorem. IEEE Trans Circuits Syst Part I 44:454–458
36. Popov VM (1973) Hyperstability of control systems. Springer-Verlag, New York
37. Rantzer A (1992) Stability conditions for polytopes of polynomials. IEEE Trans Automatic Control 37:79–89
38. Taylor JH (1974) Strictly positive-real functions and the Lefschetz-Kalman-Yakubovich (LKY) lemma. IEEE Trans Circuits Syst 21:310–311
39. Wang L, Huang L (1991) Finite verification of strict positive realness of interval rational functions. Chinese Science Bulletin 36:262–264
40. Wang L, Huang L (1992) Robustness analysis of discrete systems – a geometric approach. Chinese Science Bulletin 37:1747–1752
41. Wang L, Huang L (1993) Finite verification of the characteristic specification of discrete systems. Chinese Science Bulletin 38:521–525
42. Wang L, Yu WS (1999) A new approach to robust synthesis of strictly positive real transfer functions. Stability and Control: Theory and Applications 2:13–24
43. Wang L, Yu WS (2000) Complete characterization of strictly positive real regions and robust strictly positive real synthesis method. Science in China (E) 43:97–112

44. Wang L, Yu WS (2001) On robust stability of polynomials and robust strict positive realness of transfer functions. IEEE Trans Circuits and Syst Part I 48:127–128
45. Wang L, Yu WS (2001) Robust SPR synthesis for low-order polynomial segments and interval polynomials. In: Proc American Control Conference, Arlington, VA, 3612–3617
46. Wen JT (1988) Time domain and frequency domain conditions for strict positive realness. IEEE Trans Automatic Control 33:988–992
47. Xie LJ, Wang L, Yu WS (2002) A new geometric algorithm with order reduction for robust strictly positive real synthesis. In: Proc 44th IEEE Conference on Decision and Control, Las Vegas, NV, 1844–1849
48. Xie LJ, Wang L, Yu WS, Qiu YH (2002) Robust strictly positive real (SPR) synthesis based on genetic algorithm. In: Proc 15th IFAC World Congress, Barcelona, Spain
49. Yakubovich VA (1962) The solution of certain matrix inequalities in automatic control theory. Doklady Ajkademii Nauk USSR 143:1304–1307
50. Yang L, Hou XR, Zeng ZB (1996) Complete discrimination system for polynomials. Science in China (E) 26:424–441
51. Yang L, Xia BC (1997) Explicit criterion to determine the number of positive roots of a polynomial. MM-Preprints 15:134–145
52. Yang L, Zhang JZ, Hou XR (1996) Nonlinear algebraic equations and machine proving. Shanghai Science and Education Press, Shanghai
53. Yu WS (1998) Robust strictly positive real synthesis and robust stability analysis. PhD Thesis, Peking University, Beijing, P. R. China
54. Yu WS, Huang L (1999) A necessary and sufficient conditions on robust SPR stabilization for low degree systems. Chinese Science Bulletin 44:517–520
55. Yu WS, Wang L (1999) Some remarks on the definition of strict positive realness of transfer Functions. In: Proc Chinese Conference on Decision and Control, Shenyang, P. R. China, 135–139
56. Yu WS, Wang L (2000) Design of strictly positive real transfer functions. In: Proc IFAC Symposium on Computer Aided Control Systems Design, Salford, UK
57. Yu WS, Wang L (2001) Anderson's claim on fourth-order SPR synthesis is true. IEEE Trans Circuits Syst Part I 48:506–509
58. Yu WS, Wang L (2001) Robust SPR synthesis for fourth-order convex combinations. Progress in Natural Science 11:461–467
59. Yu WS, Wang L (2001) Robust strictly positive real synthesis for convex combination of the fifth-order polynomials. In: Proc IEEE Symposium on Circuits and Systems Conference, Sydney, Australia, 739-742
60. Yu WS, Wang L (2003) Robust strictly positive real synthesis for convex combination of the sixth-order polynomials. In: Proc American Control Conference, Denver, CO, 3840–3845
61. Yu WS, Wang L, Ackermann J (2003) Solution to the general robust strictly positive real synthesis problem for polynomial segments. In: Proc 2003 European Control Conference, Cambridge, UK
62. Yu WS, Wang L, Ackermann J (2004) Robust strictly positive real synthesis problem for polynomial families of arbitrary order. Science in China (F) 47:475–489

63. Yu WS, Wang L, Tan M (1999) Complete characterization of strictly positive realness regions in coefficient space. In: Proc IEEE Hong Kong Symposium on Robotics and Control, Hong Kong, P. R. China, 259–264
64. Yu WS, Wang L, Xiang Y (2003) Robust strictly positive real synthesis of polynomial segments for discrete time systems. In: Proc 42nd IEEE Conference on Decision and Control, Maui, HI, 622–627

Index